TEXTBOOK OF IMMUNOLOGY

TEXTBOOK OF
IMMUNOLOGY

Arvind Kumar

The Energy and Resources Institute

ISBN 978-81-7993-380-0

Suggested citation

Kumar, Arvind. 2013. *Textbook of Immunology.* New Delhi: TERI

Published by
The Energy and Resources Institute (TERI)
TERI Press
Darbari Seth Block
IHC Complex, Lodhi Road
New Delhi – 110 003
India

Tel. 2468 2100 or 4150 4900
Fax 2468 2144 or 2468 2145
India +91 • Delhi (0)11
E-mail teripress@teri.res.in
Website www.teriin.org

Printed in India

Preface

The human being is constantly threatened by a variety of organisms and their molecules. The immune system, which is the human body's defence mechanism, makes a distinction between these non-self organisms and molecules and its own healthy tissues. The branch of science that analyses the structure and function of the immune system is known as immunology. It developed as a distinctive subject in the mid-20th century.

This book is an attempt to give a broad coverage on the subject. Including the latest advance information as the subject has steadily progressed during the last couple of years, the book leads to the new dimension of understanding immunology and immune system. The book provides to its reader including graduate, undergraduate and postgraduate students all the information on immunology, starting from basic to advance, in simple comprehensible language. This has been further complemented by the descriptive subtitle-based chapters that provide a concise overview of the important subject areas in immunology.

Divided into 14 chapters, the book is written in a clear and lucid manner and each topic is organized into sub-topics. For clarity the book has been illustrated with simple well-labelled diagrams.

Chapter 1, Basics of Immunity, describes the historical paradigm of immunology with the focus on sequential advancement in immunity. It gives a comprehensive précis of immunology and its components to make the reader inquisitive to know about the subject. The special attraction of this chapter is Major Histocompatibility Complex (MHC), which has been discussed in brief, and clonal selection of B cells along with cooperative interdependent function of innate and adaptive immunity. Specificity of antibodies, allergic and anaphylactic response, components of immune system, antigen presentation, and immunodeficiency have also been covered.

Chapter 2, Composition of Immune System, explains the composition of immune system, which can be described as a backbone of immunity in immunology, immunological cells, lymphocytes and lymphoid organs with their structural details. The special attraction of this chapter is the immunological cells markers.

Chapter 3, Antigens and Immunogens, details the antigenicity and immunogenicity of the antigens in the host and the various types that help in bringing out the advance descriptive knowledge without which there can be no talk on immunity. The focal points of this chapter are the nature of B-cell epitopes and T-cells epitopes, with description of pattern recognition receptors (PRRs) protein and TLR. In continuation to Chapter 3, Chapter 4, Antigen Determinants, analyses immunoglobulin proteins raised against specific antigens. It details their types and sub-types and the essential structure. Topics such as diversity of antibody which is due to shuffling of gene fragment or segment production help the reader understand how and why infinite diversity of antibodies are present in the vertebral hosts.

Chapter 5, Antigen and Antibody Interaction, examines the interaction between antigen and immunoglobulin and the various techniques developed based on it for clinical uses and application. The chapter focuses on immunological techniques and their clinical application with the inclusions of modern advance and highly research-oriented techniques, such as nanotechnology and quantum dot.

Chapter 6, Major Histocompatibility Complex, as the name suggests, describes in detail MHC structure and types including the functions. On reading this chapter, readers will get a grasp on MHC-based naming of various mice strains, linkage disequilibrium including immune responsiveness.

Chapter 7, Complement System, is a small chapter written with the aim to clarify without compromising on basic information.

Chapter 8, Regulation of Adaptive Immune Response, examines the regulation of adaptive immune system with an emphasis on B- and T-lymphocyte cells and their receptors and co-receptors.

In Chapter 9, Cell-mediated Immunity and Cytotoxicity, the role of CTLs, NK cells and other cytotoxic cells including macrophages-based immunity mechanism are discussed in brief based on the recent discoveries in this area. The experimental proof of cell mediated cytotoxicity mechanism and immunity have also been included in this chapter.

Chapters 10 and 11, Allergic Reactions or Hypersensitivity and Autoimmunity respectively, describe hypersensitivity and autoimmunity with the help of examples and causes. Allergic reactions of various allergen present in an environment based on different mechanism are written in a simplified but articulate form.

Chapter 12, Immunity to Infections and Tumours, discusses immunity to infection and the most contagious disease—cancer. Viral, bacterial and protozoan infection based diseases have also been explained along with recent techniques of cancer immunotherapy and causes of cancer.

In Chapter 13, Immunology of Graft Rejection or Transplant Rejections, transplantation of organs and graft and recent advancement in respect to immunological development have been discussed. The major focus is on immunotherapy and drug-based acceptance of graft.

Finally, Chapter 14, Cytokines, describes key immunomodulatory immunological molecules, that is, cytokines and chemokines with their receptors. The discussion on the role of cytokines in T-cell and B-cell immunity, functionality of Th 17 and Regulatory T cells (Tregs), and cytokines in cancer therapy would interest the reader. This chapter, written in detail, illustrates the recent advancement of research in this key immune regulator molecules with their sub-types Th 1 and Th 2.

I hope that the reader will find this book both enjoyable and useful, and any comments and suggestions for improvements in future editions would be gratefully received. I would like to thank all those people including reviewers, editors, associate editors, copyeditors whose endurance have helped and influenced in making this book possible. Special thanks are also due to those whose help in the production of the book has been indispensable. My parents' blessing has always been with me and empowered me when writing this book. At the end I cannot forget to bestow gratitude to all my family members especially my wife Anamika, daughter Omi Richa, and son Animesh for their patience and support.

Contents

CHAPTER ONE 1

Basics of Immunity

1.1 INTRODUCTION

A process by which the body protects against outside parasitic microorganisms or their products (antigens) is known as immunity. The study of immunity is called immunology. Normally, there are two types of immunity present in vertebrates—the first line of defence or innate immunity (Akira, Takeda, and Kaisho 2001), and the second line of defence or adaptive immunity, also known as humoral immunity. The word immunity is derived from a Latin word *immunis*, meaning "exemption" or "protection from infectious diseases".

The first available written document in history was of 430 BC. It was written by Thucydides, a historian of the Peloponnesian war, about the plague in Athens. In his writing, he mentioned that the people who recovered from the plague were able to nurse others sick from the plague itself as they were not prone to illness during the its occurrence the second time. In the 15th century, the Chinese and Turks for the first time attempted to induce immunity in normal persons using a technique known as variolation. In this technique, the dried crusts of smallpox were either introduced into the body through small cuts in the skin or by inhalation through the nose. Later, in 1718, this technique was improved by Edward Jenner, an English physician (Brostoff and Male 2001), based on the observation that milkmaids who came in contact with cowpox were immune to smallpox, a fatal viral disease. Edward Jenner observed that the fluid from cowpox might be able to protect any person from smallpox. This hypothesis was later proved accurate and was used to protect children all over the world from smallpox (Lydyard, Whelan, and Fanger 2000).

Based on this hypothesis, Louis Pasteur developed an attenuated strain (pathogenically weak cholera-causing bacterial strain) of cholera bacterium to protect chickens from cholera. He injected some old cholera bacterial strain into fresh chickens and found that they got ill but soon recovered from the disease. The cholera bacterial strains pathogenicity that weaken with age are called attenuated strain. When fresh bacterial culture was injected into the chickens earlier injected with attenuated cholera strain, it was found that the chickens were protected against the disease. Such attenuated strains were called vaccine, a term derived from the Latin word *vacca*, meaning "cow" (to honour Edward Jenner who discovered inoculation of cowpox to protect a person from smallpox) (Brostoff and Male 2001).

1.2 MECHANISM OF IMMUNITY

Emil von Behring (a German scientist) and Shibasaburo Kitasato (a Japanese scientist) experimentally proved the mechanism of immunity for the first time. They demonstrated that animals immunized with serum of diphtheria were able to develop immunity in c animals against diphtheria. The immunization is due to the presence of an active component in the serum that neutralizes and lowers the effect by precipitating the toxin, and also removes the infected yet live bacteria by agglutination. Based on the mode of action, the active component was named antitoxin, precipitin or agglutinin. Emil von Behring was awarded Nobel Prize in medicine in 1901 for his finding.

1.2.1 Humoral Immunity

In 1930, Elvin Kabat for the first time reported that gamma-globulin (γ-globulin), called immunoglobulin, of serum acts as an active component and is mainly responsible for immunological activity after infection. Immunoglobulin was later on called antibodies. Since the antibody is already present in the body fluid, or humour, this type of immunity is called humoral immunity.

1.2.2 Cell-mediated or Cellular Immunity

This concept was first introduced by Elie Metchnikoff in 1883. He demonstrated that white blood cells (WBCs), called phagocytes, are mainly responsible for the immunity of the serum. In immunized animals, these phagocytes are more active to phagocytose (ingest) microbes and other foreign particles. The other phagocytes that were identified by Metchnikoff were monocytes and neutrophils of the blood.

According to Metchnikoff, the main cause of immunity in the immunized animals is active cells rather than the serum components. This hypothesis was strengthened by the experimental proof given by Merrill Chase in 1940. He demonstrated that WBCs enable immunity against tuberculosis in guinea pigs. In 1950, lymphocytes were identified for their role in both cellular and humoral immunity.

Dr Bruce Glick, while working as a scientist in the Mississippi State University, experimentally differentiated lymphocytes into two categories—thymus-derived cells called T-lymphocytes and bursa of Fabricius (part of cloaca in birds) derived cells called B-lymphocytes. These cells have interdependent roles in both cellular and humoral immunity.

1.3 SPECIFICITY OF ANTIBODIES

The selective theory and the instructional theory are two of the earliest theories proposed to define the specificity of the antibody developed either against any organic compound or pathogenic and non-pathogenic substances.

1.3.1 Selective Theory

The selective theory was proposed by Paul Ehrlich in 1900. It states that blood cells express cell-specific receptors, called side chain receptors, to inactivate infectious agents. This theory suggests that the specificity of the receptors is determined even before exposure to the antigen. This theory was later on modified and corrected by taking into account that receptors are present both in soluble and cell-bounded forms. The binding or interactions of foreign agents to the receptors are of in "lock and key" type, as proposed earlier by Emil Fischer (in 1894) for enzyme and substrate interaction. After interaction, the infectious agent induces the cells to have more receptors with similar specificity.

1.3.2 Instructional Theory

The instructional theory was proposed by Friderich Breinl and Felix Haurowitz in 1930 and later modified by Linus Pauling in 1940. It is based on the protein folding hypothesis. According to this theory, the specificity of the antibody is determined by the antigen that provides a template to fold the antibody around itself. This gives a definite configuration to the antibody with respect to the specific antigen. Due to the availability of detailed complex structural information of protein, RNA, and DNA, it was found that it is not possible to develop such an antibody. Furthermore, this theory explains only specificity and diversity but does not explain how the antibody recognizes self and non-self. Later, in 1950, it was argued that it is not possible to develop antibodies against any structure and thus this theory was not accepted.

1.3.3 Clonal Selection Theory

Based on new experimental data in 1950 Niels Jerne, David Talmadge, and F Macfarlane Burnett redefined the selective theory as the clonal selection theory. According to this theory, the lymphocytes express membrane-associated receptors that are specific to various types of antigens. The expression of receptors on the lymphocyte membrane is not guided by an antigen, but rather appears before the exposure of the cells to the specific antigen. Later on, binding of the antigen to the receptor activates the lymphocytes to proliferate into a clone of cells having specificity similar to that of the antigen. The interaction of antigen to receptors is responsible for activation and proliferation of the cells. With a few modifications, this theory later on became the base of modern lymphocyte immunology.

1.4 COMPONENTS OF THE IMMUNE SYSTEM

The immune system is made up of two important components—the innate immune system and the adaptive immune system. In a broad sense, the innate immune system, also known as the first line of defence, provides a barrier that either prevents the entry of pathogens or eliminates the pathogens at the earliest stage. The types of barriers present include (1) anatomical barrier such as skin and mucous membrane; (2) physiological barrier such as temperature and low pH; (3) chemical barrier such as lysozyme and phagocytic/endocytic barriers; and (4) inflammatory barrier.

The innate immune system is non-specific at early stages but becomes partially specific to antigens later due to the presence of pattern recognition receptors (PRRs), leading to inflammation. On the other hand, the adaptive immune system comes into play after the entry of pathogens or its components into the body contains four main characteristics.

1. Specificity towards antigens.
2. Diversity to recognize the antigens.
3. Memory to recognize the same antigen after second encounter.
4. Differentiation of self antigens from non-self antigens.

1.4.1 Innate Immune System

The innate immune system (Kimbrell and Beutler 2001) follows a two-step activity. The first is prevention and the second is elimination. Innate immunity can be divided into two groups, as discussed in subsequent sections.

1.4.1.1 Non-specific immunity based on prevention steps

The prevention step involves two kinds of barriers to prevent the entry of pathogens into the body—the anatomic barrier and the physiological barrier. The anatomic barrier includes skin and mucous membranes and the physiological barrier includes temperature, acidic pH, and certain chemicals such as soluble lysozyme, interferon, and complement proteins.

1.4.1.1.1 *Anatomical barrier-based immunity*

Skin-based innate immunity: The skin consists of an outer epidermal layer and an inner thick dermal layer. The epidermal layer is made up of dead cells filled with keratin, which works as a waterproof layer. The dermal layer is made up of connective tissue, blood vessels, and two types of glands—the sebaceous glands and sweat glands. The sebaceous glands secrete sebum (an oily secretion) and are associated with hair follicles. Sebum lowers the pH of the outer epidermal layer to 3–5, as the layer contains lactic acid and fatty acids to prevent microbial growth. However, the layer contains some commensalism bacteria that metabolize sebum.

Mucous membrane-based innate immunity: The mucous membrane forms an external lining of the conjunctive digestive tract, alimentary canal, respiratory canal, and urogenital tract. This membrane consists of two layers—the outer epithelial layer and the inner connective tissue layer. The mucous secretion, along with saliva and tears, contain antibacterial or antiviral substances that wash away microbes that invade the outer layer. Due to its viscous nature, the mucous secreted by the outer epithelial layer entraps the invading microbes. The mucous membrane in the lower respiratory tract is normally covered with cilia (hair-like protrusions) that can propel mucous-entrapped microbes. The pathogenic microorganisms outside will have to compete with the non-pathogenic microbial flora already colonizing the mucous cilial niches, which prevent them from entering the body through the membrane.

These mucous membrane barriers are sometimes evaded by certain microbes; for example, influenza virus cells are able to pass through these barriers due to special types of surface molecules. These molecules protect the virus cells from being swept away by the cilia. Other microbes, such as gonorrhoea, develop projections (pili or fimbriae) that help them bind to the urogenital epithelial tract and pass through. The lactic acid and fatty acids present in the mucous lowers the pH, thus killing the trapped pathogens.

1.4.1.1.2 *Physiological barrier-based innate immunity*

The three physiological barriers are discussed below.

1. ***Acidic pH:*** The pH plays a significant role in innate immunity by preventing the growth and entry of microbes and by killing them after invasion into tissues.
2. ***Temperature:*** The temperature surrounding the invaded area plays a crucial role in removing the invaded microbes (rise in body temperature recorded is normally more than 37°C).
3. ***Chemicals involved in innate immunity:*** Damaged tissues due to infection from microbes release serum proteins in large amounts to control the infection. These immediately released proteins are known as acute phase proteins.

The liver releases proteins, known as C-reactive protein (CRP), which bind to the C-polysaccharide of the cell wall of bacteria and other prokaryotic pathogens (Medzhitov and Janeway 2000). This leads to complement-mediated killing of microbes.

Histamine is another chemical released in huge amounts on damaged tissue. Histamine causes increased blood permeability and vasodilation of capillaries.

Damaged tissues release molecules that can activate the inactive peptides, called kinins, of blood plasma and also activate the cascade of enzymes present in the plasma to form insoluble strands of fibrin to form blood clot. Activated kinins have histamine-like permeability and vasodilation effects, while blood clot prevents spread of the infection. Bradykinin is a type of kinin responsible for the activation of skin pain receptors. The following chemicals and chemical-based mechanisms are involved in innate immunity.

- Histamine: role in vasodilation.
- Collectin: causes damage to bacterial cell wall.
- Interferon: kills viral infection.
- Kinin or bradykinin: pain receptor activator.
- Proteases: split the complement system; the split complement subunits bind to the bacterial wall to help in opsonization.
- CRP: binds to the C-polysaccharide of bacteria.
- Fibrine: helps in blood clotting.

As mentioned, the physiological-based innate immunity components are temperature, pH, soluble lysozyme, interferon, and complement proteins. These components are important for innate immunity to control pathogenic invasion; for example, chickens prevent anthrax infection due to high body temperature by innate immunity. The low pH of a chicken's stomach due to gastric acidity controls microbial growth of the anthrax bacteria after infection and prevents their colonization. Newborns are normally susceptible to infection due to less acidity of the stomach.

Lysozyme (hydrolytic enzyme) present in eye tear and mucous secretion cleaves the peptidoglycan layer of bacterial cell wall to control the growth of invading bacteria. Similarly, interferons (group of proteins) control viral growth, while its inactive complements that circulate in the body gets activated and function as the effector system, enabling the binding of antibodies to the microorganisms and subsequently killing the microbes. In the same way, collectin (a surfactant) kills bacteria by destroying lipid membranes (Lawson and Reid 2000).

1.4.1.1.3 *Phagocytosis-based innate immunity*

The phagocytosis-based innate immunity is mediated by the macrophage to form a phagosome that fuses with the lysosome to kill bacteria. This is an important barrier or type of innate immunity to control infected microbes. Phagocytic cells that are involved in the ingestion of extracellular microbes and toxic materials or other foreign particles are blood monocytes, tissues macrophages, neutrophils, and so on. The process of absorption of solid particles by phagocytic cells is called endocytosis, while of absorption of liquids is called pinocytosis. The microbes or solid particles engulfed are enclosed by a plasma membrane to form large vesicles called phagosomes. Endocytosis might be receptor dependent or receptor independent. Receptor-independent phagosomes fuse with lysozomes so that the lysosomal enzymes can digest captured materials.

1.4.1.1.4 *Inflammation-based innate immunity*

Redness, swelling, heat, and pain are signs of inflammation. It is due to leakage of vascular fluid or serum proteins, antibacterial substances, and movement of phagocytic cells to the site of infection. As described by Roman physician Celsus in the 1st century AD, inflammation has four signs/symptoms: (1) redness (*rubor*), (2) swelling (*tumour*), (3) heat (*calor*), and (4) pain (*dolor*). Greek physician Galen, in the 2nd century, described the fifth sign/symptom as loss of function (*functio laesa*). These five signs/symptoms incite three major events in the inflammation upon infection by microbes. Infection causes damage of cells and tissues at the infected sites, releasing vasoactive and chemotactic factors. Increased blood flow, capillary permeability, and influx of phagocytes, lymphocytes, and WBCs at the infected sites are initiated by these released factors. The higher serum protein contains complement and antibodies of exudates (damaged tissues fluid) released after damage and swelling (oedema) of tissues to destroy the infected microbes by antibacterial activity. The phagocytic cell movement is a three-step process—margination, diapedesis/extravasation, and chemotaxis. Margination is the attachment of cells to the wall of blood; diapedesis/extravasation is the movement of cells from the capillary walls to the endothelial cells; and chemotaxis is a migration of phagocytes to the damaged tissues sites.

1.4.1.2 Specific innate immunity

Specificity is based on a receptor called germline encoded PRR. It is partially specific because it lacks memory cells (Akira, Takeda, and Kaisho 2001). Its characteristic features are recognition of bacteria, activation of downstream signals, and transcription factor (TF) activation, which leads to the activation of genes to synthesize immunomodulator proteins (such as cytokines and chemokines).

1.4.1.2.1 *Conserved motifs pattern recognition receptor-based innate immunity*

Microbes have specified unique types of molecules that are not present in eukaryotes. PRR-based recognitions are established on these conservative ideas of microbes and are hence highly specific to recognize the immune system in eukaryotes and help destroy invading microbes only. This property of innate immunity is known as pattern recognition property. There are many small molecules involved in the pattern recognition activity, along with a class of receptors called Toll-like receptors (TLRs). Toll receptors (TRs) were originally identified in Drosophila and hence similar types of receptors are also present in eukaryotes, called as toll-like receptors, or TLRs. Every component of bacterial strain has specific types of receptors and, therefore, that many types of TLRs. At present, up to 19 TLRs are present. Apart from TLRs, other PRRs found are mannose binding lectin (MBL), CRP, scavenger receptor (SR), and lipopolysaccharide binding receptor (LBR). PRRs are present in the blood stream, tissue fluids, and on cell surfaces of macrophages, neutrophils, and dendritic cells. MBL triggers complement-mediated lysis of infected cells and tissues.

1.4.2 Adaptive Immune System

The innate immune system is generally similar in most organisms, while the adaptive immune system differs from species to species and even on the basis of foreign antigen specificity. The main characteristic features of adaptive immunity are (1) nature of antigens; (2) recognition of antigens; (3) development of memory against antigens; and (4) recognition of self/non-self antigens.

The nature of any antigen defines the specificity of the antibody. This specificity enables the antibody (Landsteiner 1947) to distinguish different protein molecules. This in turn enables the immune system to recognize various foreign antigens, thus empowering it to deal with antigen diversity. The recognized antibody develops memory cells to generate lifelong immunity against the particular antigen. The immune system adopts this normal principle to recognize the antigens as self and non-self to prevent any complication in the body. Failure in recognition of self and non-self antigens by the immune system leads to various diseases called autoimmune diseases.

The two major groups of cells taking part in adaptive immunity are lymphocytes and antigen-presenting cells (APCs), both of which are associated to each other and MHC (major histocompatibility complex) specific associated peptide complex (that is, MHC-I and MHC-II associated peptide).

1.4.2.1 Lymphocytes

Lymphocytes are WBCs generated in the bone marrow by a process called haematopoiesis. Lymphocytes move, along with blood, and usually reside in the lymphoid organs of the body. On the basis of the maturation site, lymphocytes are divided into B-lymphocytes and T-lymphocytes.

B-lymphocytes develop and mature in the bone marrow. Maturation of B-lymphocytes is actually rearrangement of B-cell receptor (BCR) genes. The fully matured B-lymphocytes develop a special type of antigen binding receptor on the membrane surface (called antibody molecules) or BCR before leaving the bone marrow. In the bone marrow, B cells, or B-lymphocytes, develop BCR without interacting with the antigen. When mature but naïve B cells come in contact with the antigen, interaction takes place between the similar antigen and membrane-bound antibody. This interaction leads to the differentiation of B-lymphocytes into memory B cells and effector B cells. Memory B cells are structurally similar to the parent B-lymphocytes but has longer lifespan. Effector B cells, also known as plasma cells, are specialized B-lymphocyte cells that can produce antibodies (effector molecules of humoral types of immunity). In its short lifespan (few days), these cells secrete large amounts of antibodies, with a synthesis rate of normally 2000 molecules/s.

On the other hand, T cells develop in the bone marrow but mature within the thymus gland. In course of maturation, these cells develop specific antigen binding molecules called T-cell receptors (TCRs) on the surface of the membrane by the rearrangement of TCR genes. These receptors are able to recognize only those antigens associated with MHC as a peptide molecule. MHCs are genetically diverse glycoproteins associated with the cell membrane.

Based on expression specificity, MHC molecules are divided into two categories—class I MHC molecules and class II MHC molecules. Class I MHC molecules are expressed on the surface of the nucleated cells of almost all vertebrates. Class I MHC molecules are structurally made up of an α-heavy chain linked to a β2-microglobulin invariant protein. An α-chain is composed of three domains—α_1, α_2, and α_3. On the other hand, class II MHC molecules are composed of one α-chain and one β-chain of glycoprotein. These molecules are expressed only by APCs. Upon interaction of antigen-associated MHCs of APCs to T-lymphocytes, the latter differentiate and proliferate into memory T cells and effector T cells. T-lymphocytes can be divided into two subpopulations based on their membrane glycoprotein types and functionality—T-helper (Th) cells and T-cytotoxic (Tc) cells. There is a third type of T-lymphocytes known as T-suppressor (Ts) cells, which are normally not well distinct from the other two T-lymphocytes. CD4 membrane-bound glycoprotein is present on the Th cells, while CD8 membrane-bound glycoprotein is present on the Tc cells. Functionally, the Th cells interact with the antigens-class II MHC complex molecules and are activated as effector cells. These effector cells secrete small proteinous molecules, called cytokines, and other growth factors. In the presence of secreted cytokines, the Tc cells interact with antigen–class I MHC complex molecules to differentiate and proliferate into cytotoxic T-lymphocytes (CTLs), which act as effector cells. The function of the CTLs is to kill and eliminate infected cells such as virus-infected cells, tumour cells, and cells of a graft. CTLs do not secrete cytokines but target altered-self cells that exhibit foreign antigen–class I MHC complexes.

1.4.2.2 Antigen-presenting cells

Antigen-presenting cells express membrane-bound class II MHC molecules associated with processed external proteins and deliver signal to the activate Th cells. Examples of APCs are macrophages, B-lymphocytes, and dendritic cells. The mechanism of activation is very simple. In the first step, the APCs internalize protein antigen. In the second step, the antigen is processed into smaller peptide fragments and are then present on the surface of the membrane, in association with class II MHC molecules. In the third step, the Th cells interact with the antigen–class II MHC complexes, leading to their activation.

1.4.2.3 Recognition of antigens

The main immunological cells, B cells and T cells, are unable to recognize whole bigger-sized protein antigens. They can, however, recognize discrete sites that constitute a separate entity; these sites are called epitopes, or antigenic determinants. An epitope is the most active region on the complex antigen because it is specifically recognized by BCRs and/or TCRs. The recognition pattern of both BCRs and TCRs are not similar. BCRs recognize only epitopes, while TCRs recognize epitopes even if they are in complex with MHC molecules associated with the membrane of the APCs. There are different types of epitopes. Bacterial or viral surface displayed soluble proteins, glycoproteins, polysaccharides, and lipopolysaccharides that are recognized by B cells or, when processed and presented with class II MHC-associated peptides, are recognized by Th cells. Class I MHC-associated peptides alter self cells (either due to virus infection or cancer proteins) are recognized by Tc cells.

1.4.3.4 Generating specificity of B cells and T cells

Membrane-bound antigen-binding receptors (antibodies) determine B-cells antigenic specificity against particular types of antigens. Specificity against any antigen starts due to the random rearrangement of a series of gene segments that encode antibody molecules expressed on the surface of maturing B cells in bone marrow. Moreover, specificity of B cells is determined by the rearrangement of gene segments of specific mature B cells to produce membrane-associated antibody called BCR. Heavy and light chains of a surface antibody is encoded by separate single functional gene sequences for a heavy chain and for a light chain to provide identical specificity of clone of mature B cells. Thus, fully mature B-lymphocytes are antigenically committed to a single epitope, and each B-lymphocyte has distinct separate specificity for a single epitope of an antigen. The B-cell population matures in the bone marrow by expressing unique antibodies on the membrane surface. Thus, about 10^{10} or more different antigenic specific B-cell population are present collectively. There is also an elimination of B-cells–antibody complex population that recognize self-components. This is to avoid the production of self-reactive antibodies called autoantibodies.

The maturation of T cells is similar to that of B cells, starting with the rearrangements of a series of gene segments. In T cells, there is rearrangement of TCR gene segments that encode antigen-binding receptors on the membrane. A total number of 10^5 TCRs express on T-lymphocytes that have identical specificity towards a particular antigen. Such rearrangement capabilities of TCR genes generate approximately 10^9 different T-lymphocytes with special types of antigenic specificities. This unique feature provides enormous diversity to T-lymphocytes for various antigens. At the end, there is a selection process in the thymus to eliminate T-lymphocytes carrying self-reactive receptors.

1.4.3.5 Major histocompatibility complex

Major histocompatibility complex is a multiple loci genetic complex encoding class I MHC and class II MHC molecules. Although MHC molecules can recognize antigens, they do not have fine specificity similar to that of BCRs or TCRs of lymphocytes. Instead, MHCs can bind only to antigenic peptides that are the resultant of intracellular degradation of antigenic protein molecules. An antigenic cleft is created on class I MHC and class II MHC molecules at the distal region, that is, away from the cell membrane. The antigenic cleft is due to variation in the amino acid sequences of the end-residing alleles. Such properties enable the antigenic cleft with different specificity for the different types of antigens.

1.5 PROCESSING AND PRESENTATION OF ANTIGENS

Processing of antigens is a procedure by which complex antigens are broken or degraded into small pieces or fragments. Presentation is a display of the fragments on the surface of the APCs.

1.5.1 External antigens

Processing of foreign antigens is essential so that it can be displayed in the form of a complex with class II MHC molecules and can thus be recognized by T cells. The whole procedure is known as processing and presentation of complex antigens on the surface of APCs. APCs are macrophages, dendritic cells, and B cells that degrade external or exogenous protein antigens into smaller peptide fragments through phagocytosis or endocytosis. These fragments of peptides form a complex with class II MHC molecules inside the endolysosome (endosome + lysosome) of APCs and the complex is then exported to the surface (Plate 1). These complexes are recognized by CD4 displaying T cells called Th cells. Hence, CD4 displaying Th cells are class II MHC restricted cells.

1.5.2 Internal Antigens

There are some cells that produce internal or endogenous antigens, either upon viral infection (viral nucleic acids) or due to alteration in cell properties, for example, cancerous (cancer marker proteins) cells. These endogenous antigens are processed and moved to the endoplasmic reticulum (ER) to form a complex with class I MHC molecules. This complex first moves to Golgi complex and then finally to the cell surface membrane (Figure 1.1). Class I MHC molecules are expressed by almost all the nucleated cells. The CD8 expressing T cells, called Tc cells, recognize only class I MHC-antigen displayed complex. Thus, CD8 expressing Tc cells are called class I MHC molecules restricted.

1.6 CLONAL SELECTION OF B CELLS

Immunocompetent animals hold large numbers of T-lymphocyte and B-lymphocyte clones reacting with antigens. The membrane-bound receptors of these clones determine the specificity against antigens. However, lymphocyte specificity is determined before contact with the antigen by random rearrangements of TCR genes and BCR genes of T-lymphocytes and B-lymphocytes, respectively. The importance of antigen is obvious when antigenically committed T cells and B cells come in contact and get activated, and proliferate and expand the populations of specific lymphocytes by a process called clonal selection. Thus, clonal selection is defined as the population of proliferated T-lymphocytes or B-lymphocytes, as a result of antigenic stimulation of the parental T-lymphocytes or B-lymphocytes with same antigenic specificity. In this process, the

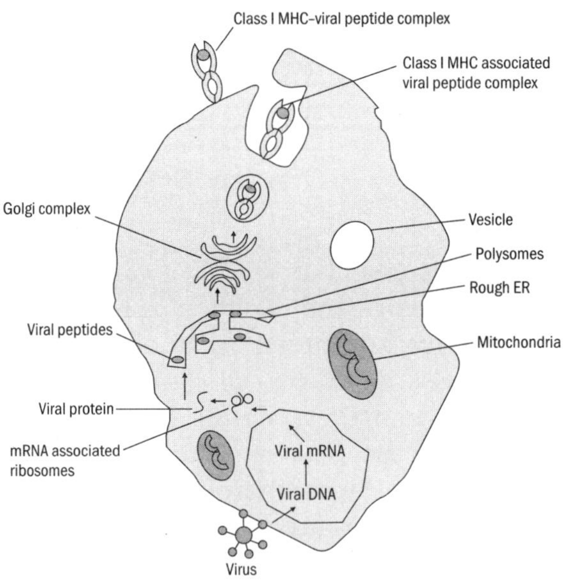

Figure 1.1 Processing and presentation of endogenous proteins by antigen-presenting cells

Note Class I MHC peptide association occurs in endoplasmic reticulum and then passes through the Golgi complex. It is present on the cell surface.

single cell upon stimulation by a specific antigen's epitope divides repeatedly into clone cells (that is, cells make a clone selecting a specific antigen's epitope, which then activates them to divide by interacting with the receptor) (Figure 1.2).

This hypothesis gives an idea and understanding of the specific self and non-self antigens. During the clonal selection process, self and non-self cells are distinguished and the selection process is accomplished by the elimination or functional suppression of lymphocytes having self-reactive receptors on its membrane surface. Two types of lymphocytes are developed during the clonal selection process—plasma cells and memory cells. Plasma cells have shorter lifespans and can produce antibodies with the ability to confer primary responses to the host against infection. On the other hand, memory cells have longer life spans and play a crucial role in future protection against similar antigens/infections (that is, memory cells have the ability to confer secondary response to the host).

Primary response is of low magnitude, short duration, and takes longer to response against antigens (normally 10–17 days). Secondary response is of higher magnitude, more rapid and with shorter response time (2–7 days) against antigens that attack the host a second time around. This difference is due to the time taken by naïve cells to grow into plasma cells during primary response for the production of antibodies. Thus, plasma cells have a longer lag phase and take more time to identify and respond to antigens while memory cells respond quicker since they do not take time to identify the antigens striking a second time (Plate 2). Moreover, the memory cells in secondary response generate plasma cells more rapidly and in larger numbers, thus producing a larger amount of antibodies. On the other hand, the naïve cells in primary response generate a lesser number of plasma cells, thus taking more time and producing less amount of antibody (Plate 2).

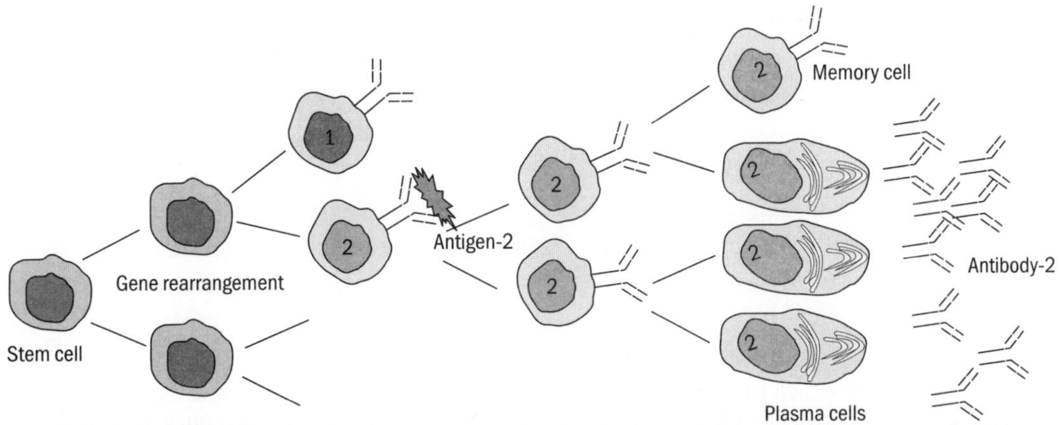

Figure 1.2 Maturation of antigen specific B cells by gene rearrangement and proliferation and differentiation into antigen-2 specific memory cells and plasma cells

Note Plasma cells produce antigen-2 specific antibody-2

Similar to B cells, clonal selection of T-lymphocytes is subjected to the proliferation and development of two types of T cells—effector T-lymphocytes and memory T-lymphocytes. Effector T cells are first to encounter the foreign antigen and are involved in primary response, which is slower and has a longer lag phase. On the other hand, memory T cells take part in secondary response, which is faster and has a shorter lag phase.

1.7 COOPERATION BETWEEN INNATE IMMUNITY AND ADAPTIVE IMMUNITY

Innate and adaptive immune systems work in cooperation with each other to produce combined immune responses, which are more effective than either immune system acting independently. Interaction and cooperation between the systems is done by the immune components (small protein molecules) that play a significant role in both types. Such cooperation is normally visible during the interaction between macrophages and microbes. Interaction between the two systems produces soluble protein molecules that facilitate participation of the adaptive immune system in eliminating microbes. The activated macrophage secretes cytokines, with which the adaptive immune system takes part against cellular microbe infection (and infection from other pathogens). In this process, certain T cells secrete cytokines upon encounter with the antigen, which enable macrophage to kill the engulfed microbes more effectively in oxygen-dependent and oxygen-independent mechanism. Opsonization is a process in which the antibody binds with the antigen to enable the macrophage to phagocytosis more readily. This is another example of cooperation between the innate and adaptive immune systems.

1.8 DIFFERENCE BETWEEN INNATE AND ADAPTIVE IMMUNE SYSTEMS

It is essential that the innate and adaptive immune systems are interdependent and work in cooperation in order to protect the body from infectious diseases. Although the systems are interdependent, they also differ from each other. The major differences are in response timing, specificity, and ability to respond to secondary infection by the same microorganisms or pathogens.

Innate immunity is normally rapid, with a respond time of less than 1 h. This is possible because this immunity system utilizes pre-existing responding components against a limited amount of foreign substances. Innate immunity system remains constant and does not have an ability to enhance response during the immunity processes. Specificity against foreign antigen or microbes is limited and fixed during secondary infection or exposure to antigens; the secondary response remains identical to the primary response. The cells normally involved are macrophages, monocytes, dendritic cells, neutrophils, eosinophils, epithelial cells, and mucosal cells. The cells secretion products and hair also contribute in innate immunity.

Adaptive immunity is slower in response, which normally takes days, because it acts according to the nature of the foreign antigens. It has to develop measures to recognize a wider range of foreign antigens. This immune system can improve the response against antigen or microbes during the immunity processes. Specificity against the foreign antigens and microbes is highly diverse, and secondary response against an identical antigen is much stronger and rapid than primary response. The cells normally involved are APCs, B-lymphocytes, T-lymphocytes, and their products.

Although in vertebrates, both innate and adaptive immune systems cooperate with each other to provide an effective barrier against infection, it cannot be concluded that the immune molecules involved are essentially strong for protection. Most essential immune system molecules in vertebrates are antibodies, T-lymphocytes, and B-lymphocytes, which are absent in invertebrate phyla. However, invertebrates are able to protect themselves from microbial populations. Insects and plants have a strong innate immune system that is not dependent on lymphocytes to provide immunity (Fritig, Takeda, and Kaisho 1998; Otvos 2000). The interior spaces of many insects (for example, fruit flies and cockroaches) and plants have well-established innate immune system. In fruit flies (*Drosophila melanogaster*), the interior body spaces synthesize small peptides after bacterial and mould infection. These peptides have antibacterial and antifungal properties, which eliminate the infected microbes and fungi in the body cavity. Fruit flies also have TRs on their cell surfaces, which recognize the microbes and trigger antimicrobial activity either by synthesizing small peptides with antimicrobial and antifungal (drosomycin) activities or by activating the cells to remove foreign molecules (Ganz and Lehrer 1997). These mechanisms are confirmed by the fact that mutant strains of fruit flies are unable to synthesize drosomycin and are thus prone to fungal infection.

In plants, immunity to infection is conferred by the production of a number of antimicrobial peptides and non-peptides. These organic compounds (phytoalexins) protect the plants from microbial pathogens (Fritig, Takeda, and Kaisho 1998). The peptides synthesized digest the microbial membrane and cell walls of the pathogens to prevent bacterial growth. Even in infected sites, some of the plants respond against the pathogens by isolating healthy cells and strengthening their cell walls for protection.

1.9 IMMUNITY MALFUNCTION AND IMMUNODEFICIENCY

Sometimes, normal immune systems are unable to protect the host either due to malfunction of the immune system or due to immunodeficiency. Inappropriate and misdirected activities of the immune system lead to discomfort, life-threatening diseases, and even death. The obvious consequences of inappropriate immunity include allergic or anaphylactic response, autoimmune diseases, grafts non-acceptance and graft-versus-host disease, and immunodeficiency.

1.9.1 Allergic or Anaphylactic Response

Common antigens of allergy are plant pollen grains, food substances such as peanuts, tree nuts (for example, almonds, cashews, and walnuts), and hair and dandruffs of animals. In 1902, Charles Richet tried to immunize a dog against jellyfish (*Physalia physalis*) toxins and found that rather than protecting the animal, the toxins increased its sensitivity. Sometimes, such antigen vaccinations were fatal. Based on this observation, Richet concluded that immunization or vaccination normally causes phylaxis (protection). But in the case of physalia toxin, the opposite of phylaxis (anaphylaxis) occurs. Anaphylaxis means lethal sensitivity to the antigen on repeat exposure to the same. In 1913, Richet was awarded the Nobel Prize in Physiology or Medicine for his research on anaphylaxis.

The symptoms of allergy or anaphylaxis are sneezing, wheezing, asthma, hives or dermatitis, and in extreme cases due to inflammation, strangulation. The mechanism involved in allergic or anaphylactic response is dependent on IgE (immunoglobulin E) production. The specific allergen activates B cells to produce a specific antibody (IgE) that can attach to the mast cells to release or secrete molecules, which causes allergic symptoms and in worst cases leads to inflammation of lungs and asthmatic attack. Anaphylactic response mostly occurs within an hour of exposure to the allergen. The most common and effective treatment is the use of injectable epinephrine.

B-cells receptor upon interaction with allergen converts B cells into plasma cells, which produces IgE antibody. The IgE then binds with the receptor present on mast cells to release allergen molecules. This process is called degranulation.

B-cells receptor + Allergen → Plasma cells → IgE production → Mast cells + IgE → Release of allergen molecules

1.9.2 Autoimmune Diseases

Sometimes due to malfunction in immunity, the immune system is unable to differentiate self and non-self cells. This causes damage to cells, tissues, and organs, leading to a number of chronic diseases. Examples of these diseases are (1) multiple sclerosis caused due to attack of immune cells on tissues of the brain and central nervous system; (2) Crohn's disease caused due to attack of immune cells on gut tissues; and (3) rheumatoid arthritis caused due to attack of immune cells on the joints of legs and arms. The causes of autoimmune diseases are mostly triggered by environmental factors.

1.9.3 Grafts Non-acceptance and Graft-versus-host Disease

Transplantation of organs from one person to another is treated by the body as invasion of foreign cells or organs. Hence, the immune system triggers to eliminate the transplant. Similarly, the transplanted cells or tissues consider the new host as foreign and react to cause graft-versus-host fatal disease. It is possible to treat this disease with immune suppressor drugs, but this could lead to other consequences and make the host more infection prone.

The discovery of human blood groups allowed for successful transfusion of blood; this discovery solved the major problems of transplantation. In 1930, Karl Landsteiner was awarded the Nobel Prize in Physiology or Medicine for his discovery of different human blood groups. In 1980, G Snell, J Dausset, and B Benacerraf were awarded the Nobel Prize in Physiology or Medicine for the discovery of MHC, which plays a key role in understanding transplantations. Finally, in 1990, E D Thomas and J Murray were awarded the Nobel Prize in Physiology or Medicine for their contribution in the advancement of research to decrease the immunity-based rejection in transplantation (that is, minimizing the risk of transplant rejection).

1.9.4 Immunodeficiency

Immunodeficiency occurs in a host either due to genetic abnormality or due to loss of immune function. The latter is due to damage of biological or chemical agents. Examples of immunodeficiency diseases are severe combined immunodeficiency (SCID), acquired immune deficiency syndrome (AIDS), and disease due to non-synthesis of IgA. SCID affects both B cells and T cells. AIDS occurs due to infection of retrovirus called human immunodeficiency virus (HIV) into Th cells to damage immunity. So far, treatment has not been found to cure AIDS.

AT A GLANCE

- Protection against infecting microbes, organisms, foreign substances, protein biomolecules, and other antigens is known as immunity. It is of two types in all the vertebrates—innate and adaptive.
- Innate immunity is initially non-specific but later on becomes partially specific to infection. It constitutes a first line of defence by various barriers such as anatomic, physiological, phagocytosis, and inflammations.
- Innate and adaptive immunity work in cooperation with each other, independently. The innate immune system produces and passes signals to activate the adaptive immune system to control infections.
- The adaptive immune system follows four measures: (1) specificity, (2) diversity, (3) memory, and (4) differentiation of self with non-self.
- The above four features of adaptive immune system is contributed by antibodies produced by mature B cells, or plasma cells; memory B cells; and T-cell receptors (TCRs).
- TCR recognizes only those antigens that are processed and presented by class I and class II major histocompatibility complex (MHC).
- The cells that process and present the antigens are known as antigen-presenting cells, or APCs. Macrophages, B cells, and dendritic cells are APCs to process and present exogenous

or extracellular antigens in association with class II MHC molecules. Class I MHC molecules present endogenous or intracellular antigens, for example, viral or tumour proteins produced inside the cells.

- T-lymphocytes are of two types—$CD4^+$ T-helper cells, or Th cells, and $CD8^+$ T-cytotoxic cells, or Tc cells. Th cells interact with class II MHC associated antigens to secrete cytokines to regulate immunity, while Tc cells interact with class I MHC associated antigens to generate cytotoxic T-lymphocytes (CTLs) to kill the target cells.
- The immune system eliminates extracellular antigens by humoral responses (that is, mediated by antibodies), while endogenous antigens are eliminated by cell-mediated immunity.
- Innate immunity is mostly prevalent in insects, earthworms, and higher plants where adaptive immunity is completely lacking. On the other hand, vertebrates have both immunities—innate immunity as the first line of defence and adaptive immunity as the second line of defence.
- Malfunction or disruption of the immune system (that is, incomplete failure line of defence) leads to diseases such as asthma and allergy, while its complete failure leads to autoimmune diseases due to treatment of self cells as non-self to kill or eliminate them.

QUESTIONS

1. What do you understand by innate and adaptive immunity? How does both cooperate with each other to boost up the immune system after infection?
2. What is innate immunity? Discuss the various measures adopted in the innate immune system to protect a body from infection. Why is innate immunity considered partially specific in nature?
3. What is adaptive immunity? Discuss the mechanism adopted in the adaptive immune system to protect the body.
4. What are antigen-presenting cells? Describe the importance of antigen-presenting cells in immunity.
5. What are external and internal antigens? Discuss the processing of both types of antigens with the help of a suitable diagram.

OBJECTIVE QUESTIONS

A. Fill in the blanks.

a. Blood vessels and sebaceous glands are present in _______ layer of skin.
b. Outer most layer of skin is called_______.
c. Acute phase response secreted protein is known as _______.
d. Mucous secretions contain _______ as hydrolytic enzyme.

e. Antibacterial secretion of skin is _______ in nature and called _______.
f. Molecule having antiviral activities is called _______.
g. Molecules having vasodilation properties are called _______.
h. Fluid accumulate in intercellular space to cause swelling is known as _______.
i. Vesicle that contains ingested materials is called _______.
j. Attachment of phagocytic cells the wall of endothelial cells is known as _______.
k. Dead cells, fluid and digested materials accumulate to form _______.
l. Acute phase proteins are called _______.
m. Phagocytes migrate into the tissues through endothelial cells wall is called _______.

B. Match the following.

i. Class I MHC	a. Interferon
ii. Class II MHC	b. BCR
iii. T-helper cells	c. Intracellular antigens
iv. Cytotoxic T-lymphocytes	d. Extracellular antigens
v. B cells	e. $CD8^+$
vi. Secreted antibody	f. Memory cells
vii. T cells	g. Allergy or asthma
viii. Immune dysfunctions	h. Non-specific
ix. Adaptive immunity	i. TCR
x. Innate immunity	j. $CD4^+$

CHAPTER TWO 2

Composition of Immune System

2.1 INTRODUCTION

The immune system is composed of cells and organs that provide immunity to the host body against microbial infection. Functionally, there are two types of organs in the human or vertebrate body that constitute the immune system: primary lymphoid organs and secondary lymphoid organs. Primary lymphoid organs provide the micro-environment for the development and maturation of lymphocytes. Secondary lymphoid organs provide the site where captured antigens interact with mature lymphocytes. Both primary and secondary lymphoid organs are connected to each other by blood vessels and lymphatic systems. Blood and lymph contain white blood cells (WBCs) (or leukocytes) and leukocytes respectively, which take part in providing immunity. Except lymphocytes, all other cells present in the blood or lymph are accessory cells that serve to activate lymphocytes. Only lymphocytes possess and contribute diversity, retain memory of the antigen, and have the ability to recognize self-antigens and non-self-antigens—key features of the adaptive immune system (Banchereau, Briere, Caux, *et al.* 2000; Liu 2001). The accessory cells help in adaptive immunity by activating lymphocytes and enhancing the phagocytosis ability to remove antigens. These cells produce effector molecules to boost immunity. For example, T-lymphocytes secrete cytokines upon activation, which have immunomodulatory roles in the regulation of immune responses (Ward, Loeb, Soede-Bobok, *et al.* 2000).

2.2 CELL DEATH: NECROSIS AND PROGRAMMED CELL DEATH

Necrosis occurs due to abnormal physiological conditions. It causes the influx of water and extracellular ions into the cell, followed by swelling of the mitochondria and the entire cells. It finally leads to the rupture and lysis of the cell membrane. Necrosis is associated with the release of lysosomal enzymes, which cause extensive tissue damage and inflammatory response.

Apoptosis is a process of cell damage caused by the cell's own demise under normal physiological conditions. The morphological and biochemical changes that occur in apoptosis are chromatin aggregation, nuclear and cytoplasmic condensation creating convolution in cells, cytoskeleton modifications leading to membrane blebbing (vesicular outgrowth of membrane), chromatin condensation and segregation, fragmentation of DNA into smaller pieces, and partition of cytoplasm and nucleus apoptotic bodies (membrane-bound vesicles) without inflammatory response. Apoptotic bodies first separate from the cell as cell fragments, then swell and get lysed. This phase is known as secondary necrosis.

Shedded apoptotic bodies are immediately phagocytosed by macrophages. This ensures the non-spreading of proteolytic enzymes, including other lytic enzymes, oxidizing molecules, and cationic proteins, thus protecting the surrounding tissues and preventing inflammations mediated by apoptosis at the apoptotic site.

Apoptosis has a very significant importance in maintaining haematopoietic stem cells (HSCs). It sustains and keeps the number of haematopoietic progenitor cells constant. Apoptosis is also important in bringing about essential immunological processes such as tolerance and destruction of specific target cells by cytotoxic T cells as well as natural killer (NK) cells. This kind of cell death is common in those cells that have completed their lifespan.

Apoptosis is characterized by the initiation of some specific genes such as p53, Fas, and many proteases genes. It leads to the expression of p53 protein, Fas protein, and proteases proteins, also known as caspases. These proteins are a part of a cascade of reactions leading to apoptosis. There are some anti-apoptotic genes that get activated to stop apoptotic cell death. The genes that inhibit apoptosis include Bcl-2 (B-cell lymphoma 2) family genes and Bcl-xL. One of the adverse consequences of the overproduction or uncontrolled production of Bcl-2 protein is its ability to transform lymphoid cells into cancerous lymphoma cells by inhibiting programmed cell death (PCD) cell signals. This happens due to the translocation of Bcl-2 gene into the immunoglobulin heavy chain causing transcriptional activation of the gene, which then leads to the overproduction of Bcl-2 protein by lymphoma cells.

In contrast to apoptosis, necrosis is a process in which there is death of injured cells. The distinct morphological changes that occur in necrosis are clumping of chromatin, swelling of cell organelles, and flocculation of mitochondria. The necrotic cells finally disintegrate, releasing intracellular contents that trigger inflammation of local tissues.

2.2.1 Regulation of B-cells and T-cells Numbers

Apoptosis controls the numbers of activated B cells. Activated B cells have increased expression of cytokine receptors and decreased expression of Bcl-2 protein. Bcl-2 is known to regulate and control apoptosis. An increased amount of Bcl-2 prevents apoptosis, while a decreased amount makes the naïve cells or memory cells prone to apoptotic cell death. Thus the reduced expression of Bcl-2 leads to the removal of excess activated B cells by the process of PCD. The numbers of T cells are also controlled by the same procedure. The process is explained in Chart 2.1.

2.3 CELLS OF THE IMMUNE SYSTEM

There are two major groups of cells in the immune system: lymphocytes and phagocytes. These two major groups of cells and other specialized cells are derived from HSCs. HSCs differentiate into the epithelial lining of the respiratory tract, gastrointestinal and genitourinary systems to build up a strong line of defence system against the foreign invasion of pathogens and their products (Weissman 2000). These cells take part in immunity by cell-to-cell interaction or by cytokines interaction, or by both procedures. Cytokines are present either in soluble form or in the membrane-bound condition with receptors. Immune cells or cells taking part in the immune systems are distinguished from other non-immune cells by cluster of designation (CD) markers other than their shapes and appearances. HSCs are totipotent in nature and are found in yolk sac, spleen, liver, and bone marrow. Some HSCs are also present in the mesenchymal areas of embryonic and fetal cells of animals. The HSCs of bone marrow give rise to four major groups of cells lineages: (1) erythrocytes or erythroid, (2) platelets or megakaryocytic cells, (3) granulocytes (including mononucleate phagocytes) or myeloid, and (4) lymphocytes. Antigen-presenting cells (APCs) are mostly derived from myeloid

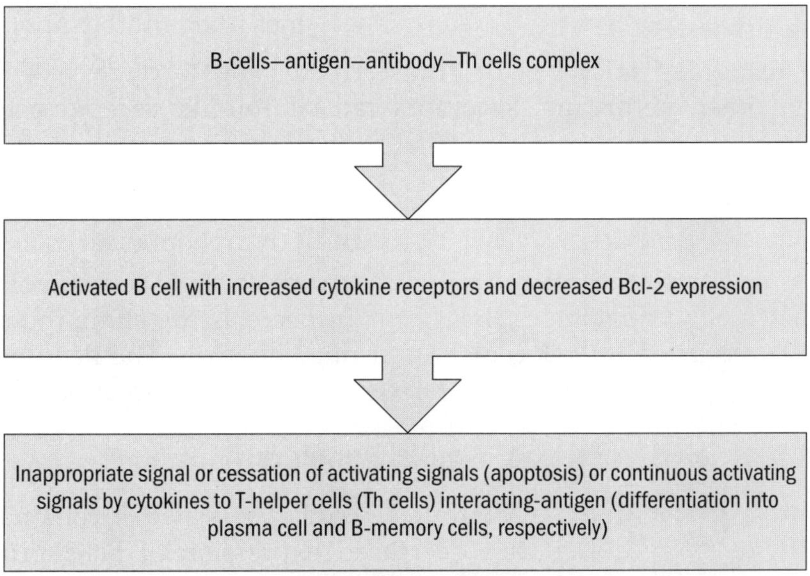

Chart 2.1 Process explaining how apoptosis leads to differentiation

precursor cells. Protection from external pathogens is normally done by granulocytes and lymphocytes. Both lineages are very important for the function of the immune system. Overall there are two grouping of cells based on the immune system: innate immune system cells and adaptive immune system cells.

2.3.1 Innate Immune System Cells

Innate immune system cells are monocytes/macrophages and polymorphonuclear granulocytes. A lobed and irregularly shaped nucleus is referred to as a polymorphic nucleus. Polymorphonuclear granulocytes are neutrophils or polymorphonuclear neutrophils (PMNs), basophils, and eosinophils categorized on the basis of their staining capabilities with acidic and basic dyes. WBCs constitute roughly 60%–70% of the blood stream. The structure of cells is given in Figure 2.1.

2.3.1.1 Monocytes/macrophages

Blood-circulating monocytes can migrate through blood vessel walls into various organs to differentiate into macrophages. In humans, monocytes are 10–18 μm in diameter, with a horseshoe-shaped nucleus, mitochondria, lysosomal granules, and pinocytotic vesicles. Lysosomal enzymes present include peroxidase and many acid hydrolases, which digest and kill phagocytosed microbes, cells containing viruses, and cancerous cells. These cells last for months to years. They can squeeze between the endothelial cells of blood vessels by a process called diapedesis. The migration from blood vessels to surrounding tissues occurs in two steps—adhesion (called margination) to the endothelial cells of blood vessels followed by extravasation, that is, effusion or squeezing from blood vessels to surrounding tissues.

Monocytes recognize microbes or transformed cells through receptors that are present in their outer membrane. Receptors can be Toll-like receptors, IgG receptors,

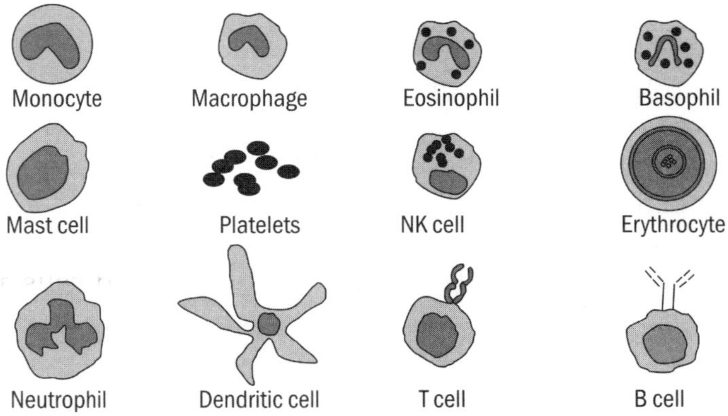

Figure 2.1 Innate and adaptive immune system cells

complement receptors, and mannose receptors. These receptors recognize the sugars or lipids of the microbial membrane, which monocytes then adhere to, followed by ingestion, killing, and digestion.

2.3.1.1.1 *Differentiation of monocytes into macrophages*

Monocytes circulate in blood, while macrophages reside in tissues. Promonocytes are precursors of monocytes. They leave the bone marrow and enter into the blood stream where they mature into monocytes. After a certain period, monocytes differentiate into tissue macrophages or dendritic cells after activation. Differentiated monocytes that become macrophages adopt the following changes.

- Size increases 5–10 folds.
- Number and complexity of intracellular organelles increase.
- Phagocytic activity increases many folds.
- Level of hydrolytic enzymes goes up.
- Secrete a number of soluble factors.
- Move to specific tissues to become fixed macrophages or remain free to move from one place to another by amoeboid movement.

Activated macrophages express class II major histocompatibility complex (MHC) molecules to function as APCs, secrete inflammatory mediators/cytotoxic proteins, and activate T cells with more potential. The most potent activator of macrophages is interferon gamma (IFNγ) secreted by Th cells.

Macrophages are named depending on the tissues in which they are present, for example, alveolar macrophages in the lungs, histiocytes macrophages in connective tissues, Kupffer cells macrophages in the liver, mesangial cells macrophages in the kidney, microglial cells macrophages in the brain, and osteoclasts macrophages in the bone.

2.3.1.1.2 *Macrophages killing mechanisms of ingested pathogens*

a. Oxygen-independent killing mechanism of pathogens

This is mediated by the synthesis of many enzymes, antimicrobial peptides, and cytokines. Lysozymes and hydrolytic enzymes that function without oxygen to degrade microbial cell wall are synthesized. The antimicrobial peptides and cytotoxic peptides produced by macrophages are collectively known as defensins. Defensins are approximately 30-amino-acid-long cationic peptides rich in cysteine. Each peptide, which contains six or more cysteine, is stabilized by intramolecular disulphide bonds and forms a circular molecular structure. The cytokine that kills pathogen-infected cells and tumour cells is called the tumour necrosis factor α (TNFα) (Nathan and Shiloh 2000).

b. Oxygen-dependent killing mechanism of pathogens

Oxygen-dependent killing is mediated by reactive oxygen intermediates (ROIs), reactive nitrogen intermediates (RNIs), and monochloramine (NH_2Cl). After phagocytosis caused by the activation of macrophages in a process called respiratory burst, the membrane-bound oxidase acts on oxygen to form ROIs such as superoxide anion (O_2^-) and hydroxyl radicals (OH^+), which are highly toxic to ingested microbes/pathogens. After the fusion of lysosome to phagosome, hypochlorite anion (ClO^-) is produced from hydrogen peroxide (H_2O_2) and chloride ions by the activity of myeloperoxidase. The ClO^- ion is highly toxic to microbes (Nathan and Shiloh 2000).

When the macrophage is activated by lipopolysaccharide (LPS), muramyl dipeptide (MDP) and IFNγ, nitric oxide synthase (NOS) is expressed. The NOS oxidizes L-arginine to form nitric oxide (NO) and L-citrulline. NO alone, or along with superoxide ions, has very strong antimicrobial and anti-pathogenic (against bacteria, fungi, parasitic worms, and protozoa) activity.

2.3.1.2 Polymorphonuclear granulocytes

Normally called granulocytes, these cells consist of bone-marrow-released neutrophils (production rate, 7 million/min). Neutrophils are 10–20 μm in diameter with multi-lobed nuclei. Phagocytic cells comprise 95% of blood-circulating granulocytes and live only for two to three days in blood vessels. From blood vessels, these cells can move to the surrounding tissues by diapedesis, similar to the monocytes/macrophages. Migration of granulocytes occurs in two steps—receptor-mediated adhesion of granulocytes with ligands on endothelial cells followed by squeezing to the surrounding tissues. Adhesion is helped by the chemokine (chemoattractants) interleukin-8 (IL-8), intercellular adhesion molecules (ICAMs), and vascular cell adhesion molecules (VCAMs) expressed on endothelium and fibronectin. Granulocytes play an important role in inflammation and cannot inherent specificity for antigens. Their importance is more prevalent in a person with reduced WBCs or genetic defects, which leads to the loss of ability to extravasate in response to chemoattractants (Liu 2001).

Neutrophil chemoattractants are C5a subunits of activated complements, fibrinolytic/kinin-derived factors, leucocytes/platelets products, and bacterial products. Neutrophils contain two types of granules: (1) azurophilic granules called primary granules composed mainly of lysosomes (acid hydrolases, myeloperoxidase and muramidase or lysozyme) and (2) specific granules called secondary granules containing lysozyme and lactoferrin. The antibiotic proteins present in both types of granules are defensins, seprocidins, cathelicidins, and bacterial permeability inducing (BPI) agents. Phagosomes (vacuoles that contain microbes) fuse with lysosomes to form an independent body called phagolysosome, where the killing and digestion of microbes take place. Fcγ-receptor-mediated activation of neutrophils releases cytotoxic granules in type III hypersensitivity reactions to kill and eliminate microbes.

2.3.1.3 Eosinophils

Eosinophils comprise 2%–5% of leucocytes in a healthy person and contain bilobed nuclei. Its cytoplasmic granules are membrane-bound organelles that have different electron density than the matrix and stain with acidic dyes (eosin). When activated by stimuli, eosinophils degranulate to kill large unphagocytosed pathogens, thus providing specialized immunity against parasitic worms. Enzymes histaminase and aryl sulphatase, released from the eosinophils along with degranulates, inactivate products of mast cells, such as, histamine and leukotrienes. The released eosinophil products down-regulate inflammatory response and granulocyte migration at invasion sites.

2.3.1.4 Basophiles and mast cells

Basophiles account less than 0.2% of leucocytes. Morphologically, basophiles differ from mast cells, but both are similar in many properties. Both contain randomly distributed granules, which can be stained with basic dyes. The granules in both cases contain heparin, leukotrienes, histamine, and eosinophil chemotactic factor of anaphylaxis (ECF-A). Both cells are stimulated by allergens leading to degranulation. Allergens first form a complex with IgE antibody and then bind to mast cells or basophil cells through the IgE receptor FcεRI. FcεRI is a high-affinity Fc receptor specific to IgE. In response, cytoplasmic granules fuse with the cytoplasmic membrane of basophiles and mast cells. They then release the content outside the cells that contain histamine. Histamine increases allergy but, at the same time, facilitates in developing immunity against parasites by increasing inflammation.

2.3.1.5 Platelets

Platelets are derived from megakaryocytes and contain granules. The production rate of platelets in humans is 1011 platelets/day. Most of the produced platelets are sequestered in the spleen. Platelets express low-affinity IgE receptors (FcεRII or CD23), IgG receptors (CD32 or FcγII), and class I MHC. Apart from the above receptors, platelets also contain factor VIII receptors, GpIIb/IIIa complex (CD41) cytoadhesion molecules, and GpIb/Ix complex (CD42) receptors. CD41 and CD42 receptors are responsible for the binding of fibrinogen/fibronectin/vitronectin and Von Willebrand factor (vWF), respectively. The vWF also binds to CD51 receptors. Platelets adhere to the damaged sites of vascular tissues. They aggregate there and release serotonin and fibrinogen to increase capillary permeability, complement activation, and leukocytes attraction.

2.3.1.6 Natural killer cells

Natural killer cells comprise 15% of lymphocytes and are identified by CD16 (FcγRIII) and CD56 monoclonal antibodies (mAb) (Liu 2001). The presence of CD16/CD56 molecules and the absence of CD3 molecules confirm the presence of NK cells in the blood lymphocytes. CD16 molecules are also expressed by neutrophils, macrophages, and γδT cells. Although some T cells have CD16/CD56 molecules as well, they also contain CD3 molecules, which is absent in NK cells. Most T cells are positive to CD3/CD8. T-cell and are never expressed by NK cells. IL-2 activates NK cells because most NK cells contain IL-2R β-chain receptors (70 kDa). Some resting and activated T cells having large granular lymphocytes (LGLs) also express IL-2R β-chain. All cells that respond to IL-2 are known as lymphokine-activated killer (LAK) cells. LAK cells are able to kill and remove fresh tumour cells.

Natural killer cells normally recognize and kill virus-infected cells, including certain tumour cells. They mostly target MHC molecules, non-expressing cells or modified MHC molecule cells. NK cells are also involved in antibody-dependent cellular cytotoxicity (ADCC). The most important cytokine released by activated NK cells is IFNγ.

2.3.1.7 Origin of innate immune system cells

All innate immune system cells originate from precursor cells initially. They then differentiate and mature into respective cells according to the need. The initial precursor cells are pluripotent haemopoietic stem cells (PHSCs). PHSCs give rise to two kinds of precursor cells: (1) lymphoid stem cells from which lymphatic cells develop and (2) colony-forming units (CFUs) from which CFU-granulocytes, erythrocytes, monocytes, megakaryocytes (CFU-GEMM) develop with the help of specific cytokine interaction. CFU-GEMM gives rise to precursor cells for granulocytes, erythrocytes, monocytes, and megakaryocytic cells, which mature in the presence of a specific ratio of CSFs and various cytokines. Further differentiation into other cells is also guided by the presence of specific cytokines in a particular ratio. As all immune cells differentiate from the same precursor cells, they acquire specific receptors for further growth (Weissman 2000).

2.3.1.8 Markers of innate immune system cells

Differentiating cells may acquire or lose different differentiation factors or markers at successive stages during development (Table 2.1). For example, CD34 is an early maturation marker that disappears when neutrophils and monocytes/macrophages mature. CD38 and MHC-II molecules express on CFU-GM but disappear from neutrophils. MHC-II molecules continue expressing on monocytes. CD14, CD33, and CD35 (CR1) are exclusive markers for monocytes, while CD15, CD41, and CD42 are exclusive markers for platelets. CD23 (FcεRII) and CD32 (FcγRII), both cell markers, are present on monocytes, macrophages, granulocytes, mast cells, and platelets. Macrophages exclusively acquire CD68 along with CD16 (FcγRIII) molecules on the cell surface during differentiation; they are absent on its precursor monocytes.

Table 2.1 Markers denoting the cell types

Cell types	*Markers on cell surfaces*
B cell	CD5, CD21 (CR2), CD32 (FcγRII), CD35 (CRI), CD40, CD45
Th cells	CD2, CD3, CD4, CD5, CD28, CD45
Tc cells	CD2, CD3, CD5, CD8, CD28, CD45
NK cells	CD2, CD8, CD16 (FcγRIII), CD45, CD56

2.4 ADAPTIVE IMMUNE SYSTEM CELLS

2.4.1 Antigen-presenting Cells

Antigen-presenting cells, which are immunostimulatory leucocytes, are mainly found in skin, lymph nodes, spleen, underneath mucosal epithelial cells, and thymus. Those present in the skin epidermis and in mucosa are called Langerhans cells and are characterized by special granules called Birbeck granules. Langerhans cells are rich in MHC class II molecules and migrate as veiled cells via afferent lymphatic vessels into the paracortex of the draining lymph nodes. In lymph nodes, they come into contact with T cells, loss their Birbeck granules characteristic, and are called interdigitating dendritic cells (IDCs).

Antigen-presenting cells of thymus are also known as IDCs and are abundant in the medulla. IDCs play a crucial role in removing those T cells that are active against self-antigens. The process of removing self-antigen-reacting T cells by IDCs is known as "negative selection". Follicular dendritic cells (FDCs) are found in the germinal centres of primary and secondary B-cell-rich follicles of lymph nodes, spleen, and mucosa-associated lymphoid tissues (MALT). FDCs are derived from mesenchymal cells and present antigens to B cells. They are non-migratory and establish stable intracellular connections through desmosomes. These dendritic cells interact with antigens via CD21 and CD35 complement receptors because they lack class II MHC molecules on the surface. They are able to express Fc receptors. Class II MHC molecules-associated APCs are also present within the germinal centre called the germinal centre dendritic cell (GCDC). GCDCs are migrating cells, which interact with T cells in the germinal centre.

Most APCs derived from bone marrow are still not well defined. In vitro, monocytes lose their phagocytic capacity in the presence of GM-CSF and IL-4 to differentiate into class II MHC positive APCs with morphology similar to dendritic cells. B cells and macrophages that are class II MHC positive and present antigens to T cells upon activation are also placed in the APC category.

CD2 and CD1a markers are exclusively present on GCDC and Langerhans cells, respectively. Out of these APCs, only macrophages have phagocytic and non-specific esterase properties.

2.4.2 Lymphocytes

The primary lymphoid organ (Plate 1a) is the thymus, while the central lymphoid organ is the postnatal bone marrow. The secondary lymphoid organs are spleen, lymph nodes, and MALT. Lymphocytes are produced by the primary lymphoid organ. They then migrate to the secondary lymphoid organs. Lymphoid tissues comprise only 2% of the total body weight. There are normally 2 × 1012 lymphoid cells present in an adult human. The lifespan of these cells is very long, and they can remain in the body as memory cells for many years. The blood contains two types of lymphocytes: (1) a small-sized (6 μm) lymphocyte with no granules and a round nucleus with a high nucleus–cytoplasm (N:C) ratio and (2) a large-sized (10 μm) cell with granules, indented nucleus, and a low N:C ratio. The larger lymphocyte contains azurophilic granules in the cytoplasm and, hence, is called an LGL.

2.4.3 Lymphocyte Markers

Cell markers are represented in the form of CD, which refer to clusters or groups of mAbs having the ability to bind to specific molecules. Research is in progress to find the mAb-binding pattern to different leucocytes. The mAbs having similar characteristics are grouped together and assigned a CD number. Based on their specific features, molecular markers are further categorized; for example, lineage markers denote specific cell lineage. T-cell lineages are designated by CD3 markers. Maturation markers are acquired during the differentiation stages with the loss of preliminary markers; for example, CD1 is a marker of developing T cells but not of mature T cells. Activation markers appear after the cells are activated and are not present in non-activated cells; for example, CD25 is present only in activated cells. But in some cases, non-maturation markers lost during maturation reappear after activation of the cells, for example, CD10 of B cells. In other cases, activation markers may be present even in non-activated cells, but in low density they become more pronounced upon activation, for example, MHC class II molecules of monocytes. The density of class II MHC gets elevated after activation with IFNγ in monocytes (Rothenberg 2000).

Two types of resting lymphocytes have been identified in blood on the basis of morphological structure and markers on the cell surface, T cells and B cells. Markers for T cells and B cells are given in Table 2.1.

2.4.4 T cells

T cells may be with granular or non-granular cytoplasm and may contain gall bodies. Granular T cells have well-developed Golgi apparatus and mitochondria but lack gall body structure. The granular pattern is shown under a light microscope after staining for acid phosphatase. Non-granular T cells contain primary lysosomes and lipid droplets. Upon non-specific esterase staining, these cells look like a single spot in a light microscope. A gall body is a cluster of primary lysosomes attached with lipid droplets. T cells are either Th cells or T-cytotoxic cells (Tc cells). Normally, 95% population of Th cells and 50% population of Tc cells are small, non-granular with a high N:C ratio. LGL morphology is not shown by Tc cells.

2.4.4.1 T-cells antigen receptors

There are two types of T-cell receptors (TCRs): α- and β-chain polypeptides containing TCRs, and γ- and δ-chain containing TCRs. In both types of TCRs, two chains are linked by disulphide bonds to provide stability. Both receptors are linked with a CD3 complex to form a CD3–TCR complex. CD3 is composed of a set of five polypeptides; approximately 95% populations of blood T cells are αβT cells and the rest 5% populations are γδT cells.

2.4.4.2 Classification of αβT cells

The αβT cells (cells having TCR with α-, β-chains) are further subdivided into two distinct non-overlapping groups—one containing CD4 markers and the other containing CD8 markers. CD4-marker-containing T cells are mostly Th cells, which induce or help in the induction of immune responses (hence called Th cells), while CD8-marker-containing T cells are called Tc cells. CD4 T cells recognize antigens associated with class II MHC molecules, while CD8 T cells recognize antigens associated with class I MHC molecules. There is another group of αβT cells that does not contain both CD4 and CD8, and can be placed in either Th-cells group or Tc-cells group.

The αβT cells are further classified on the basis of the cytokine secreted upon activation; Th cells are subdivided into Th-1 and Th-2 cells. Th-1 cells secrete IL-2 and IFNγ, while Th-2 cells secrete IL-4, IL-5, IL-6, and IL-10. Cytokines secreted by these two cells have different functional importance in the generation of immune response. Cytokines secreted by Th-1 cells take part in the removal and combating of viruses, bacteria, and parasitic infections. Th-2 cells cytokines target B-lymphocyte cells and enable them in proliferation and antibody production to protect the body from free-living microbes after infection.

The γδT cells are of two types—one that circulates in the blood and the other that are present in intraepithelial lymphocytes (IELs). IELs are $CD8^+$ and have TCRs that are prejudiced towards bacterial or viral antigens called superantigens.

There are also T cells that either share some markers with other cell lineages or have markers that are present on all T cells. These common markers expressed on all T cells are called Pan T-cell markers. They are also found on other cell lineages. For example, CD2 molecules in combination with TCR–CD3 complex and membrane-associated glycoproteins are able to activate T cells after binding with appropriate ligands. CD2 molecules are also associated with CD3 NK cells. T cells also get activated by CD5, which are able to bind with CD72. CD72 are present on most T cells and NK cells.

2.4.4.3 Suppressor T cells

Suppressor T cells suppress immune responses by killing APCs through the secretion of TGFβ (called suppressive cytokines) or through affecting signal transduction pathway or through idiotype cross talk. Both $CD4^+$ and $CD8^+$ cells can take part or act in the suppression of immune responses.

2.4.5 B cells

Resting B cells lack gall bodies, granular bodies, and LGL morphology. The characteristic features of these cells are scattered monoribosomes in the cytoplasm along with rough endoplasmic reticulum strands. Since activated B cells synthesize immunoglobulin, they have rough endoplasmic reticulum, free polyribosomes, and Golgi apparatus. The free Golgi apparatus is responsible for the glycosylation of immunoglobulin.

2.4.5.1 B-lymphocytes markers

The characteristic feature of B-lymphocytes is the presence of CD19 surface antigen. This antigen is normally present in the early stage of maturation on bone marrow progenitor cells and shows its presence during all stages of B-cell maturation. B cells are constitutively produced and comprise 5%–15% of the total circulating lymphoid pool in humans. The remaining B cells are described as B1 cells and can be characterized as $CD5^+$. B1 cells are further divided into two subgroups—B1a cells ($CD5^+$) and B1b cells ($CD5^-$). Phenotypically, B1a cells are characterized as B220lowCD19highIgMhighIgDlow $CD43^+$Mac-1^+CD5int, while B1b cells are characterized as B220lowCD19highIgMhighIgDl ow$CD43^+$Mac-$1^+$$CD5^-$. These cells are generated in the bone marrow during the neonatal period. Recently, it has been characterized in the adult bone marrow of mice. B1b cells also provide long-lasting T-cell-independent humoral immunity against pathogen apart from providing earliest protection against pathogen by secreting antibodies. A majority (approximately 90%) of B cells express IgM and IgD immunoglobulin isotypes on their membrane surface. The antigen-binding sites of these immunoglobulin isotypes are identical. The rest of the B cells (approximately 10%) express other types of immunoglobulins such as IgG, IgA or IgE. The location of B cells containing specific immunoglobulin will differ in the body; for example, IgA-positive B cells are located in the intestinal mucosa. B cells contain specific receptors on their surface called B-cell antigen receptor (BCR) complex.

B-cell antigen receptors are heterodimer and consist of Igα (CD79a) and Igβ (CD79b). TCR/CD3 complexes are involved in the cellular activation of B cells. Most B cells are associated with class II MHC receptors, which interact with T cells. Class II MHC molecules normally consist of I-A and I-E in B cells of mouse and HLA-DP, DQ, and DR antigens in humans. B cell membranes also contain complement receptors C3b and C3d. C3b and C3d are also known as CD35 and CD21, respectively. Complement receptors are functionally involved in B-cell activation and their homing.

Other important markers used in the identification of B cells are CD19, CD20, and CD22. CD19/CD21 interacts with complement–antigen complex to activate B cells by antigen-associated antibody receptors. CD40, CD45, CD72, and CD78 molecules are other B-cell markers, which generate specificity in B cells. The CD23 marker is a low-affinity receptor for IgE on B cells and binds antigen–IgE complex. Thus the presence of CD23 on B cells enhances antibody response, increases the production of soluble IgE, and inhibits membrane-bound IgE production. CD23 induces B-cell proliferation and apoptosis.

During the course of development (ontogeny), a majority of B cells express the CD5 marker and can be placed in a separate group called B1 cells or $CD5^{+}B$ cells. The differentiation pathway for these cells probably differs from the differentiation pathway of conventional B cells (B2 cells). Both B1 cells and B2 cells also differ from each other in their phenotype, localization, and self-renewal properties. B1 cells ($CD5^{+}B$ cells) are found in pleural and peritoneal cavities, produce natural antibodies IgM in maximum amount, and also produce some amount of IgA/IgG and the Pan B-cell marker B220 on their cell surfaces. The produced antibodies are polyreactive and show specificities for self-antigens, such as phosphatidylcholine, ssDNA, ribonucleoproteins, and cell surface Thy-1 antigen. B1 cells also differ from B2 cells in their functional responses to outer stimuli. Although the BCRs present in both B1 cells and B2 cells are similar in their composition, they differ in their function. The BCRs are engaged in the proliferation of B2 cells but not B1 cells. B1 cells and B2 cells also differ in induction of various signaling pathways; for example, STAT-3 activation is constitutive in B1 cells but induced in B2 cells. The locations of B1 cells differ from that of B2 cells. B1 cells, as mentioned before, are predominantly present in pleural and peritoneal cavities. On the other hand, B2 cells are predominantly present in the spleen and lymph nodes. They express intermediate levels of IgM and IgD and a high level of B220 but lack CD5. Both cells also differ in their survival capabilities. While B1 cells survive for extended periods in ex vivo culture, B2 cells rapidly die in the absence of external stimuli. The numbers of B1 cells and B2 cells vary with the age of a human being. At an early age, that is, in the first year of life, there is a sharp increase in the number followed by progressive decrease in absolute number of B-lymphocytes in elderly people. This might be due to the depletion of precursor cells in the bone marrow and alteration in T-cell activity.

2.4.5.2 Functions of natural antibodies

Natural antibodies provide the first line of defence against pathogens, help in clearing damaged self-components, form an idiotype network in the immune system, and react against autoantigens, such as ssDNA, Fc of IgG, components of cytoskeleton, and phospholipids.

2.5 CLASSIFICATION OF CELL MARKERS BASED ON STRUCTURE

Based on their distinct molecular structures, cell markers are divided into different families. Markers of different molecules and the receptors of cells are given in Table 2.2.

2.5.1 Immunoglobulin Superfamily

The immunoglobulin superfamily consists of CD2, CD3, CD4, CD8, CD28, class I MHC, class II MHC, and so on. The structure of these markers is similar to the structure of immunoglobulins.

2.5.2 Integrin Family

The integrin family consists of α- and β-chains; hence, the members of the family are called heterodimeric molecules. It is divided into several subfamilies based on β-chains. Each subfamily shares a common β-chain but differs in a specific α-chain that is unique for that particular family. The β2-integrins subfamily has CD18 as its β-chain but different α-chains to make the functional lymphocyte antigens. For example, CD18 β-chain gets associated with CD11a, CD11b or CD11c or αd to form the functional lymphocytes antigens LFA-1, Mac-1, p150,95 and αdβ2 cell surface molecules respectively, on leukocytes. The other subfamily β1-integrins comprises common CD29 β-chain and various peptides as α-chain to develop very late activation markers.

Table 2.2 Markers denoting cell molecules and receptors

Function	Marker
Adhesion molecule	CD2, CD4, CD8, CD56
Signal transduction	CD2, CD3, CD4, CD8, CD40, CD45
Receptor for Fc region of IgG	CD16
Receptor for complement (C3d)	CD21
Receptor for B7 molecule of APCs	CD28
Receptor for complement (C3b)	CD35

2.5.3 Selectins Family

The selectins family consists of CD62 expressed on leukocytes (L), activated endothelial cells (E), and platelets (P). It has a specificity similar to that of lectin for a variety of sugars expressed on glycosylated membrane. An example of this family is CD43.

2.5.4 Proteoglycans Family

CD44 has a number of glycosaminoglycans (GAG)-binding sites (chondroitin) that bind to extracellular matrix components (hyaluronic acid).

Apart from the above-mentioned superfamily, family, and subfamily, there are other superfamilies and families, which are important makers of cells, directly and indirectly involved in immunity. They include TNF and NGF receptor superfamily, the seven transmembrane spanning segment (tm7), the four membrane spanning segment (tm4), and CD20.

2.6 PRIMARY LYMPHOID ORGANS AND TISSUES

The lymphoid system is made up of lymphocytes, accessory cells (macrophages and APCs), and epithelial tissues. These are either arranged into encapsulated organs or accumulated into diffused lymphoid tissues. Lymphoid organs are classified into either primary (central) organs, where the lymphocytes are produced and where they mature, or secondary (peripheral) organs, the functional sites of lymphocytes.

Primary lymphoid organs are thymus and bone marrow in most mammals (Plate 1a). In rabbits, it includes Peyer's patches and the appendix. In birds, the primary lymphoid organ is bursa of Fabricius.

2.6.1 Thymus

Thymus is a bilobed organ located in the thoracic cavity, above the heart and below the thyroid gland. Each lobe is capsulated and divided into lobules. The lobules are separated from each other by connective tissue called trabeculae. Each lobule consists of an outer cortex and an inner medulla. The cortex contains relatively immature proliferating thymocytes. The more mature thymocytes are found in the medulla. Differentiation gradient of immature to mature thymocytes starts from the cortex to the medulla. Mature thymocytes express CD44 receptors in the medulla; these receptors are normally absent in cortical thymocytes. CD44 receptors are found on all trafficking cells and are used to bind extracellular matrix components.

Thymic lobules contain three types of epithelial cells, namely, epithelial nurse cells (present in the outer cortex), cortical epithelial cells (which form a network), and medullary epithelial cells (organized into clusters). Thymic lobules also contain macrophages (derived from bone marrow) and IDCs. All epithelial cells, IDCs, and macrophages express MHC molecules essential for selection and development of T cells. Thymic involution begins with age within the cortex leading to its complete disappearance leaving medullary remnants. The thymus is at its largest actual size at puberty in humans and begins to shrink throughout life in a process known as atrophy. This suggests that there is a correlation between aging and decline in immune responses. Stress and pregnancy increase thymic atrophy or shrinkage of thymus in women.

2.6.2 Development of Thymus

Thymus develops from the third pharyngeal pouch as a rudiment of epithelial tissues containing blood-borne stem cells. In a mouse, it develops from two embryonic layers—the ectoderm of the third branchial cleft (which develops into the thymic cortex epithelium) and the endoderm of the third pharyngeal pouch (which develops into the thymic medulla epithelium).

2.6.3 T-cells Maturation in Thymus

T cells change their phenotypes, and receptors of T cells develop a wide diversity during maturation in thymus. To recognize a variety of antigens, the αβ and γδ TCR genes undergo somatic recombination to produce functional genes for different TCRs. Less than 1% of T cells express γδ TCR cells while approximately 95% of T cells differentiate into αβ TCR cells. Phenotypically, three types of T cells are present in three developmental stages.

1. *Early stage* The phenotype is $CD44^{+}/^{-}CD25^{+}CD3^{+}CD2^{+}CD5^{+}CD71^{+}CD38^{+}CD4^{-}CD8^{-}$. At this stage, TCR genes remain in germ line configuration. The proliferation markers CD71 and CD38 are not T-lineage specific. $CD44^{+}/^{-}$ means that CD44 markers are positive in the first phase but become negative in the second phase. After appearance, CD3 cells are committed to become T cells.
2. *Intermediate stage* The phenotype is $CD44^{-}CD25^{-}CD3^{-}CD1^{+}CD4^{+}CD8^{+}$. Both α- and β-chains are expressed at this stage.
3. *Mature stage* The phenotype is $CD44^{+}CD38^{-}CD3^{+}CD1^{-}CD62L^{+}CD4^{+}CD8^{-}$ or $CD44^{+}CD38^{-}CD3^{+}CD1^{-}CD62L^{+}CD4^{-}CD8^{+}$. There are two types of set of cells—either $CD4^{+}$ or $CD8^{+}$. CD44 receptors that help in migration and homing to peripheral lymphoid tissues are expressed in almost all T cells at this stage.

2.6.4 Other Forms of TCRs during Development

In the absence of TCRα, TCRβ can associate with CD3 or 33 kDa glycoprotein in pre-T cells as a surrogate chain (pre-Tα-chain) to generate the signal that drives development, that is, proliferation, maturation, and selection of T cells in thymus.

2.7 POSITIVE SELECTION AND NEGATIVE SELECTION OF T CELLS

Based on the presence of the type of TCRs, T cells go through a selection process during the developmental stage in thymus. Such a selection process is known as "thymic education". It occurs in two steps—positive selection step and negative selection step.

2.7.1 Positive Selection

In the positive selection process, thymic cortical epithelial cells function as effector cells, and TCR-bearing T cells, which can bind self-MHC with low affinity, are selected to survive and proliferate. The unselected cells, which do not bind to specific TCR-bearing T cells, die via apoptosis. The selection occurs through CD4 or CD8. TCRs specific for class II MHC retain $CD4^+$ T cells and loses $CD8^+$ T cells. Similarly, TCRs specific for class I MHC retain $CD8^+$ and lose $CD4^+$ T cells. The unselected CD4 or CD8 T cells die via apoptosis. T cells bind weakly to self-MHC/self-antigen, so they do not get activated and cause autoimmune diseases.

2.7.2 Negative Selection

In negative selection, those T cells are eliminated that react strongly with self-antigen/components rather than self-MHC in the thymus. Such cells are eliminated in the deeper cortex by the negative selection process. Only those T cells that fail to recognize self-antigen are allowed to proceed for further development. The remaining T cells die via apoptosis. The cells that escape negative selection may react with self-antigens and cause autoimmune diseases.

2.8 SECONDARY LYMPHOID ORGANS

Secondary lymphoid organs (Plate 1a) are classified on the basis of the body regions they defend. Spleen responds predominantly to blood-borne antigens. Lymph nodes (Plate 1b) mount immune responses to antigens circulating in the lymph, entering through the skin (subcutaneous lymph nodes) or through mucosal surfaces (visceral lymph nodes). Tonsils, Payer's patches, and other MALT react to antigens that enter via the surface mucosal barriers. Bone marrow is both a primary and a secondary lymphoid organ.

Secondary lymphoid organs are either encapsulated systemic organs, such as spleen and lymph nodes, or non-encapsulated, being only an accumulation of lymphoid tissues. Lymph nodes protect the body from those antigens that enter the lymphatic vessels through skin or internal surfaces. They protect by producing antibodies in the lymphatic vessels or by cells present at that site.

2.8.1 Naming of Lymphoid Tissues

Lymphoid tissues associated with mucosal surfaces are called MALT, those associated with gut are called gut-associated lymphoid tissue (GALT), and those associated with the respiratory tract are called bronchus-associated lymphoid tissue (BALT). Lymphoid tissues are also associated with genital organs or urinary tract. The mechanisms for controlling antigens are by secretary IgA antibodies (sIgA), which are the most abundant antibodies in the body and reach the lumen of tracts via mucosal epithelial cells.

2.8.1.1 Spleen

The size of the adult spleen is 13 × 8 cm, and it weighs around 180–250 g. This organ is situated at the upper left side of the abdomen, behind the stomach, and near the diaphragm.

The periarteriolar lymphoid sheath (PALS), a lymphoid tissue surrounding the central arteriole of the spleen, is composed mainly of T cells. The germinal centre of the secondary lymphoid follicle contains memory cells, dendritic cells, and phagocytic macrophages (Banchereau, Briere, Caux, *et al.* 2000; Liu 2001). On the other hand, the primary follicle contains virgin B cells. The germinal centre is surrounded by a marginal zone and red pulp. The white pulp area of the spleen contains splenic tissues.

The other splenic tissues, known as red pulp, consist mainly of erythrocytes, platelets, granulocytes, lymphocytes, and plasma cells. Apart from immunological functions, the spleen carries out the process of haemocatheresis (destruction of aged platelets and erythrocytes). The red pulp tissue also contains venous sinuses and cellular cords. The cellular cords trap circulating cells, which are finally phagocytosed by macrophages.

2.8.1.2 Lymph nodes and lymphatic system

The human lymph node is kidney shaped or round, normally 2–10 mm in diameter. The lymph is a colourless fluid full of lymphocytes (WBCs), salts, proteins, glucose, and urea, which moves through valved vessels called lymphatic vessels. The presence of valves prevents the backward flow of lymph in the vessels. Lymph is pumped into the nodes or glands by the contraction and relaxation of surrounding muscles. The typical lymph node (Plate 2) is encapsulated by connective tissues and organized into three main areas—cortex (B-cell area), paracortex (T-cell and APCs area), and medulla full of lymph cords (T cell, B cell, plasma cells, and macrophages area). The cortex zone B cells express surface IgM, IgD, and CD44 providing evidence that these cells are virgin and active.

2.8.1.3 Tonsils

Tonsils are of three kinds—palatine, pharyngeal (known as adenoid in diseased condition), and lingual (contains Waldeyer's ring). Tonsils are rich in lymphoid tissue, T-cells zone, and large secondary follicles.

2.8.1.4 Peyer's patches

Peyer's patches, which appear round or oval, are nodules of 30–40 lymphoid follicles. These nodules are found in the lowest portion of the small intestine called ileum. A specialized intestinal epithelium cell called an M cell, which is present above Peyer's patches, transports pathogens or antigens to lymphoid follicles. M cells are flattened cells lacking microvilli projection. They have deep invaginations or pockets filled with B cells, T cells, and macrophages. These macrophages, dendritic cells, B cells, and T cells present in Peyer's patches follicles generate immune responses within the mucosa of intestine against pathogenic microorganisms or antigens. Sometimes, pathogens or antigens striking the M cells get attached, internalized, and transported to the pocket. These pathogens and antigens activate the B cells present in the invagination or pocket of M cells. The activated B cells differentiate into plasma cells. Plasma cells then leave the follicles, move to the sub-mucosa, and secrete antibodies called secretary IgA. The secreted IgA is transported across the epithelial cells into lumen to interact with antigens. Thus mucosa is rich in IgA. The whole process is summarized in Chart 2.2.

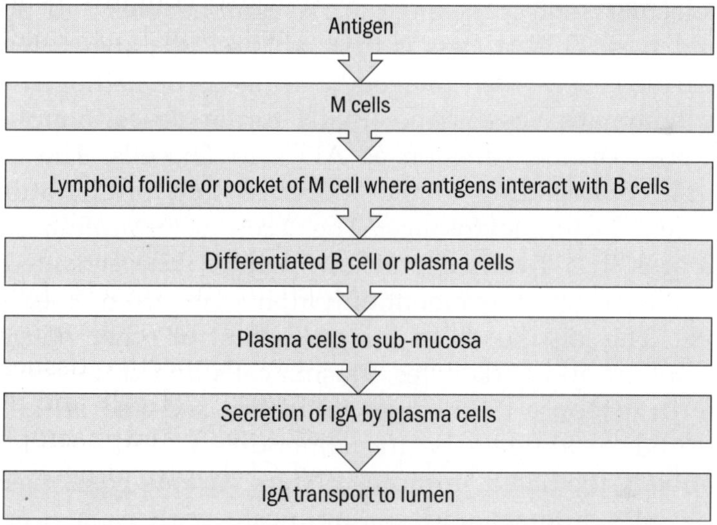

Chart 2.2 Summary of activities within Payer's patches

2.8.2 Lymphoid Tissues Associated with Skin

Lymphocytes present in the epidermal layer of skin are called intraepidermal lymphocytes. Intraepidermal lymphocytes are rich in $CD8^+$ T cells and sometimes also express $\gamma\delta$ TCRs. These T cells combat antigens, which enter through the skin. The internal dermal layer contains $CD4^+$ and $CD8^+$ T cells and macrophages.

The keratinocytes, which are specialized epidermal cells, also take part in immunity by secreting cytokines or by inducing the expression of class II MHC molecules to function as APCs. A type of dendritic cells called Langerhans cells, present in the matrix of epidermis, internalize antigen by phagocytosis or endocytosis. After internalization of antigens, Langerhans cells migrate to lymph nodes to differentiate into IDCs. The IDCs express class II MHC molecules to activate naïve Th cells.

2.9 NETWORKING OF IMMUNE SYSTEM

The immune system is composed of many organs, tissues, and cells distributed throughout the body at different locations. Even though all the components are far away from each other, in the case of infection, they communicate and collaborate with each other, working in cooperation to boost the immunity and counter the pathogens. Once tissues or cells get damaged after infection or injury, the resultant is inflammation. Inflammation leads to vasodilation, increase in blood flow, and capillary permeability. Many chemokine and chemotactic-signalling molecules are secreted at the site of infection. Such molecules send a message to the lymphocytes and phagocytic cells to move to the affected sites. This process is first initiated by Langerhans cells, epidermal layer residing dendritic cells, which phagocyte antigens or pathogens. These dendritic cells then move to lymphatic vessels and finally to the closest lymph nodes. The cells are class II MHC positive and function as APCs to Th cells. The antigen presented is recognized by Th cells leading to the activation and proliferation of Th cells to secrete activated T-cells factors (cytokines). The T-cell factors initiate the production of antibodies from B cells in a T-cells-dependent manner. The activated Th cells release chemotactic factors that trigger movement of lymphocytes from blood circulation to the nearest lymph node. This also leads to the movement of other phagocytic cells, such as neutrophils, basophils, and leukocytes. Plasma cells and differentiated B cells move to the infected site in response to chemotactic factors secreted at the site of infection. The plasma cells produce antigen-specific antibodies to kill pathogens or eliminate antigens in an antibody-mediated immunity. Thus various cells, tissues, organs, and their secreted molecules interact and communicate with each other to control the infection of pathogens and foreign particles.

AT A GLANCE

- Cells participating in the immune system are white blood cells (or leukocytes) and lymphocytes. The lymphocyte cells have immunological specificity, diversity, memory, self-recognition or non-self-recognition.
- Different cells, tissues, and organs arise from different stem cell population. Stem cells divide to generate other stem cells and differentiate to produce various types of cells, tissues or organ-specific stem cells.
- Haematopoiesis is the process through which leukocytes develop from multi-potent haematopoietic stem cells. Proliferation and differentiation into different blood cells based on different lineage of genes are helped by various growth factors, cytokines, chemokines, and transcription factors.
- Programmed cell death keeps the balance between each of the lineages of cells.
- Lymphocytes are of three different types: B cells, T cells, and natural killer (NK) cells. These cells differ from each other on the basis of their functions. Except NK1 T cells that have T-cell receptors (TCRs) and marker proteins on the membrane, other NK cells are devoid of receptors.
- Naïve B- and T-lymphocytes not exposed to antigens rest in the Go phase of the cell cycle and are smaller in size. After interaction with antigens, lymphocytes become larger, differentiating first into lymphoblasts and then into memory cells and effector cells.
- Macrophages and neutrophils are phagocytic cells that engulf and kill or degrade microbes and foreign antigens. Microbial phagocytosis is helped by opsonization, that is, attachment of antibody to the cell surface.
- Macrophages when activated secrete adaptive immune system regulating molecules and inflammation-mediating molecules. They process and present class II major histocompatibility complex (MHC)-associated peptide antigen to be recognized by the TCRs of T-helper cells (Th cells).
- Non-phagocytic cells are mast cells and basophils. Both cells play a crucial role in allergic reactions.
- Except follicular dendritic cells (FDCs), other types of dendritic cells phagocyte antigens to process and present them on its cell surface in association with class II MHC molecules. Class II MHC molecule-peptide complex activates interacting Th cells to produce costimulatory molecules on its surface. FDCs activate B cells but have no role in Th-cell activation.
- Lymphocytes reacting with self-antigens get eliminated in primary lymphoid organs such as thymus, which eliminates those thymocytes that produce mature T cells recognizing self-MHC.
- The lymphatic system accumulates fluid that circulates via the left subclavian vein in tissue spaces to deliver antigens to lymph nodes.
- Secondary lymphoid organs provide sites where lymphocytes can get activated upon interaction with antigens to differentiate into effector and memory cells. Examples of secondary lymphoid tissues are lymph nodes, spleen, clusters of follicles, Peyer's patches present in the intestine and lymphoid tissues. The respective lymphoid tissues trap antigens from nearby sites such as spleen, which is a trapping centre for blood-borne antigens.
- After infection, antigens reach the distant areas of lymph nodes via lymph and dendritic cells to involve far away cells, organs, and tissues. The activated B cells and T cells in the infected area produce immunomodulatory molecules to attract lymphocytes, leukocytes, and other components from far away to the site of infection. This is to produce active response against infection.

- Primitive vertebrates, such as jawless fishes, possess only gut-associated lymphoid tissues. These primitive vertebrates lack B cells and T cells. Thus they are devoid of an adaptive immune system. Jawed vertebrates contain different variety of lymphoid tissues, B cells, and T cells to create a network of active adaptive immunity.

QUESTIONS

1. What do you mean by lymphoid organs? How many types of lymphoid organs are present in the human body?
2. Write the name, location, and immunological functions of primary and secondary lymphoid organs.
3. What are the primary roles of spleen and thymus to boost the immune system? Discuss their interdependence role in immunity.
4. What is phagocytosis? Discuss the oxygen-dependent and independent killing mechanisms of microbes with suitable diagrams.
5. What is positive and negative lymphocyte selection? Discuss in terms of T-cells selection.
6. How many types of cells are involved in the immune system? How do they differ from each other?

OBJECTIVE QUESTIONS

A. Fill in the blanks.

1. The blood-circulating cells that participate in immune response are called ________ cells.
2. Body cells, tissues, and organs arise from ________ cells.
3. Lymphocytes are ________ cells, ________ cells, and ________ cells.
4. Naïve B- and T-lymphocytes, before antigen interaction, reside in ________ phase.
5. An antibody helping in phagocytosis is called ________ .
6. Phagocytic macrophage cells present in the liver are called ________ .
7. Monocyte-derived non-phagocytic cells are known as ________
8. T cells mature in ________, while B cells mature in ________
9. ________ cells are the first cells that arrive at the inflammation site.
10. Thymocytes originate from ________

11. Blood-circulating cells that differentiate into macrophages in tissues are called ________

B. Match the following.

i. Primary lymphoid organs	a. Cytokines
ii. Secondary lymphoid organs	b. Deliver antigens to lymph nodes
iii. Hematopoiesis inducer	c. Intestine
iv. Lymphatic system	d. Lymphocytes mature
v. Peyer's patches	e. Deliver antigens to spleen
vi. Blood circulating system	f. Lymphocytes interact with antigens
vii. Neutrophils	g. Gut associated lymphoid tissues
viii. Primitive vertebrates	h. Non-phagocytic cells
ix. Mast cells	i. Lymphoid cells
x. Jawed vertebrates	j. Phagocytic cells

CHAPTER THREE 3

Antigens and Immunogens

3.1 INTRODUCTION

Antigens are substances with a molecular weight of approximately 100,000 Da. They elicit the synthesis of antibodies by mature effector B cells called plasma cells. They also activate T cells to become effector T cells (CTLs, ThS) for cell-mediated immune responses. Antigens convert B cells and T cells into memory B cells and T cells respectively, for secondary immune response. The molecular properties of these antigens decide the immune response. The antibody produced by plasma cells reacts only with that particular antigen against which it has been synthesized. To behave like antigens, the substance must be recognized as non-self or foreign because, in general, living organisms (vertebrates) do not produce antibodies against their own (self) proteins or other self-components. Antigens are mostly proteins, but they might also be polysaccharides, polypeptides, nucleoproteins or lipoproteins. Antigens must contain, at least, the following two properties: immunogenicity and antigenicity.

Immunogenicity is the ability to activate or stimulate either B cells or T cells to produce a specific antibody or effector T cells. For example, CTLs and ThS, respectively.

Antigenicity is the ability to react or bind specifically with the produced antibody or cell surface receptors. "Specific" means the antigen will be able to react only with its corresponding antibody in a highly selective fashion and not with other antibodies produced against another antigen (Tainer, Getzoff, Paterson, *et al.* 1985). All immunogens are antigens, but not all antigens are immunogens; for example, haptens are unable to induce immune response on its own but can induce antibody production in combination with carrier proteins.

3.2 IMMUNOGEN DETERMINING FACTORS

The immune system does not recognize all macromolecules equally as infectious agents or antigens. The most potent antigens and humoral immunogens in this context are either proteins, ranking first, or polysaccharides, which come second. Other molecules include haptens, which are immunogenic only in combination with either carrier proteins or carrier polysaccharides. Cell-mediated immunogens are mainly lipids and glycolipids.

These macromolecules are not directly recognized as immunogens in most cases. They have to be processed into small peptides and then presented along with major histocompatibility complex (MHC) molecules on the cell membrane surface before being recognized by the specific cells, that is, either B cells or T cells. The immunogenic properties of any antigen are determined by the biological system. It is not an intrinsic property of any antigen.

3.3 PROPERTIES OF ANTIGENS

The main characteristic features of antigens that make them immunogens are as follows: (1) non-self nature of macromolecules, (2) size of macromolecules, (3) chemical composition and complexity of non-self macromolecules, and (4) ability to be processed and presented.

3.3.1 Non-self Nature of Macromolecules

Only non-self macromolecules or substances that are foreign to the immune system can elicit either humoral or cell-mediated immune responses. This is accompanied by *tolerance* or *unresponsiveness* towards self-components. These characters are acquired during lymphocyte development when the body components are exposed and recognized as self-components. The non-exposed components during the developmental stage of lymphocytes are regarded as non-self or foreign components by lymphocytes. The degree of non-self nature or foreignness depends on the phylogenetic distance between the two species. The more the phylogenetic distance, that is, the more the non-self character of antigens, the greater or stronger the immunogenic behaviour. For example, bovine serum albumin (BSA) is non-immunogenic in cows when injected due to the self nature, but it is immunogenic in rabbits, which is phylogenetically distant to BSA, and even more immunogenic in chicken, which is phylogenetically more distant.

During evolution, the macromolecules that remain conserved are not very immunogenic in most of the species. Some self-components that remain isolated from the immune system, when injected even into the same organism, behave like immunogens, such as sperm and corneal tissues.

3.3.2 Size of Macromolecules

The average size of macromolecules to become immunogens must be around 100 kDa; it normally ranges from 15 kDa to 150 kDa. Bovine gamma globulin (BGG) and tetanus toxoid (TT), which are used as immunogens in most experiments, have a molecular size of 150 kDa. BSA and ovalbumin (OVA) are 69 kDa and 44 kDa in size, respectively. The smallest size of common experimental antigens of sperm whale myoglobin (SWM) is 17 kDa and that of hen egg white lysozyme (HEL) is 15 kDa (Demotz, Grey, Appella, *et al.* 1989). A macromolecule of size less than 5 kDa to 10 kDa is non-immunogenic or a very poor immunogen since it cannot be phagocytosed and presented by cells.

3.3.3 Chemical Composition and Complexity of Non-self Macromolecules

Copolymers (polymers of different amino acids and sugars) are more immunogenic then homopolymers (polymers of a single amino acid or sugar). This is because copolymers are more complex than homopolymers. The immunogenicity of primary, secondary, and tertiary proteins is determined by their structural complexity; hence, it will not be same. Tertiary proteins are more immunogenic than primary and secondary proteins, while secondary structures will be more immunogenic than primary structures.

3.3.4 Ability of Macromolecules to be Processed and Presented

Molecules that are easily phagocytosed, processed (degraded), and presented along with MHC molecules are more immunogenic than those that are not phagocytosed and processed and, hence, not presented. This is why large and insoluble macromolecules that are phagocytosed, processed, and presented easily are more immunogenic than smaller soluble ones, which cannot be phagocytosed and presented. An example is the polymers of D-amino acids, which cannot be degraded by enzymes and, hence, are less immunogenic than the polymers of L-amino acids.

3.4 VARIOUS FORMS OF ANTIGENS

3.4.1 Lipid Antigens

Lipids are antigens but non-immunogenic like haptens since they cannot trigger immune response alone. They need to attach themselves with suitable carrier protein molecules, such as keyhole limpet hemocyanin (KLH) or BSA. Lipid–protein complexes can elicit effector B cells to synthesize highly specific antibodies against various lipid molecules in the complex, such as steroids, fatty acid derivatives, and fat soluble vitamins (vitamin E). The antibody synthesized against leukotrienes lipid is used to evaluate asthma patients. The level of immunosuppressive steroid prednisone is evaluated routinely in blood to determine the success of transplantation. Similarly, T cells are able to recognize those lipid compounds (such as glycolipids and some phospholipids) that are presented as CD1 complexes (Grey, Sette, and Buus 19892). CD1 is structurally similar to class I MHC molecules and is recognized by T-cell receptors (TCRs) when presented as a lipid–CD1 complex.

3.4.2 Bacterial Antigens

Bacterial antigens are of two types: (1) macromolecules associated with cell components like bacterial plasma membrane, periplasmic space, wall or envelope, and (2) by-products or secondary metabolites excreted into the environment (extracellular antigen) by microbes or pathogens.

3.4.2.1 Cellular antigens

3.4.2.1.1 Surface-associated antigens

1. *H-antigens* H-antigens are associated with flagella and, hence, found only in motile bacteria. "H" is derived from the German word *Hauch* meaning *breath,* which signifies the antigen's similarity in appearance to that of motile bacterial species on solid agar-agar media. H-antigens consist of flagellin, a protein subunit polymer. Flagellins from different bacterial species are antigenically distinct and are often characteristic of a particular bacterial species.
2. *O-antigens* O-antigens are bacterial cell wall associated antigens, normally referred to as somatic antigens. "O" is derived from the German word *Ohne hauch* meaning *without breath* referring to non-flagellated bacterial species. The O-antigen of Gram-negative bacteria is an endotoxic lipopolysaccharide (LPS) located at the outer surface of the outer membrane. The bacterial LPS consists of three regions—lipid-A, core polysaccharide, and O-specific side chain. Lipid-A (composed of glucosamine disaccharide, β-hydroxy fatty acid, and phosphate) is responsible for toxic and pyrogenic properties and is constant in nature.

   The core polysaccharide region is variable, and its structure varies from one bacterial species to other. For example, the *Salmonellae* group of bacteria has the same core structure, but it differs from other bacterial groups, such as *Pseudomonadaceae* and *Escherichia coli,* which have different core types designated as R_1 to R_4 and K_{12}. Two unusual sugars are 2-keto-3-dexoy-D-manno-octulosonic acid (KDO) and heptose (such as L-glycero-D-manno-heptose).

   O-specific side chain consists of oligosaccharide repeating units with or without single hexose branch substituents and normally accompanied by non-reducing termini. It has very high numbers with a great potential for antigenic variability. Currently, almost 169 recognized O-serogroups of *E. coli* are available. The presence and absence of O-specific side chain determine the nature of strains; in its presence, the strain will be smooth Gram negative, while in its absence or when lost, the strain will be rough Gram negative with a change in specificity of antigen–antibody reaction.

3.4.2.1.2 *Surface-capsular antigens*

Capsular polysaccharides are important antigen features of many Gram-negative and Gram-positive bacteria. They represent the dominant surface antigen masking other antigens such as LPS or teichoic acids. Their variability determines the serological types in bacterial strains; for example, *Streptococcus pneumoniae* has 83 serological types on this basis. Capsular polysaccharides are virulent determinants for several bacteria and, hence, are known as Vi antigens. In the case of *E. coli,* the capsular antigen is known as K-antigen. The antigenic formula of various bacterial strains is determined on the basis of the presence of different antigens. For example, *E. coli* has the following antigenic formula: O55:K5:H21. There are at least 83 antigenically distinct serotypes for *S. pneumoniae,* 12 for *Neisseria meningitidis,* 11 for *Staphylococcus aureus,* and 6 for *Haemophilus influenzae.* In some bacteria, a single antigenic capsular type takes the place of multiple antigenic types; for example, Vi antigen of *Salmonellae* and hyaluronic acid capsules of *Streptococci* group A and C.

3.4.2.1.3 *Other surface antigens*

- *Fimbrial antigens* Fimbriae are known as adhesive filamentous appendages of polypeptide subunits. Type I fimbriae are characterized by mannose-sensitive haemagglutinating, adhesive, and pellicle-forming properties in many bacteria. Type II fimbriae show similarity with type I fimbriae in number and dimensions but do not confer haemagglutinating, adhesive, and pellicle-forming properties. The fimbriae protein is an important antigenic determinant in type I and type II bacterial strains. Their molecular weights are approximately 20–30 K.
- *Streptococcal M, R, and T proteins* M, R, and T proteins of *Streptococcal* sp. are major components of cell wall. Protein M determines the virulence factor because it impedes bacterial phagocytosis. R- and T-antigens occur near the cell surface and are attached to the cell wall.
- *Staphylococcal* protein *Staphylococcal* cell wall protein A (40–60 kDa) and protein M are important virulence determinants in humans causing many diseases related to lungs. Protein A interferes with innate immunity by binding with TNFα receptors and also with the IgG Fc region. As protein A binds with the Fc region of immunoglobulins, the bacteria attach to IgG molecules in the wrong orientation on their surface, thus disrupting opsonization and phagocytosis.
- *Teichoic acid* Teichoic acid is an important component of the bacteria cell, which acts as an antigenic determinant. Currently, lipoteichoic acid (LTA) is used often as an

important antigen because it is possible to inactivate it. The platelet-activating factor acetylhydrolase (PAF-AH), a phospholipase, selectively inactivates staphylococcal and pneumococcal LTAs by removing their acyl-2 chain but not affecting LPS and phosphatidylcholine.

- *Outer membrane proteins* There are other bacterial membrane components that have immunogenicity, but they are yet to be determined experimentally.

3.4.3 Extracellular Antigens

Extracellular antigens are secondary metabolites or degraded products, such as toxins, coagulase, membrane-damaging cytolysins, hyaluronidase, haemolysin, neuraminidase, RNase, DNase, proteases, and so on.

3.5 ROLE OF BIOLOGICAL SYSTEM IN IMMUNOGENICITY

Immunogenicity is not solely guided by the nature of antigens or immunogens but also by the genotype of the biological system, dose of antigens, route of administration to the host, and adjuvants (helping agents). These characteristic features also affect the ability and intensity of immunogenicity.

3.5.1 Genotype of Recipient Biological System

Two different genotypes of inbred strains of animals have different immune responses to the similar type of polypeptide or other immunogens because of their different genetic constitution. The immune response in different animals is controlled by the gene located in a region of MHC. This gene in two different strains of animals will produce different products, causing a difference in immune response, that is, one will show either more immune response than the other and vice versa. Since MHC is related to the processing and presentation of antigens to T cells (Grey, Sette, and Buus 1989) or B cells, it plays a key role in determining the immune response towards particular antigens. The F1 generation of two strains having high and low immune responses will show intermediate type of immune response following Mendel's law. The genes responsible for controlling immune responses were mapped in the MHC region using the backcross analysis technique. The genes responsible for B-cell receptor (BCR) and TCR and other regulatory proteins also affect immune response.

3.5.2 Dosage and Administration Route

Only appropriate amounts (doses) of immunogens elicit immune response. Low dose or insufficient amount either fails to activate B-lymphocytes or T-lymphocytes or triggers the development of immunological tolerance. High dose of immunogen will also trigger immunological tolerance in the animal. Apart from appropriate dose, the frequency of administration of antigens also affects the immune response. A single dose will induce a weak immune response, while repeated injections over a certain period of time (weeks or days), followed by a last dose or booster dose, increase clonal proliferation of B cells or T cells, enabling strong immune response in the recipient animal. In experimental animals, immunogens are injected by routes other than the digestive system tract, that is, parenterally (*para*: around; *enteric*: gut). Following are the routes commonly used for the administration of immunogens to experimental animals.

- Into a vein: intravenous (IV)
- Into the skin or dermis: intradermal (ID)
- Into a muscle: intramuscular (IM)
- Into the peritoneal cavity: intraperitoneal (IP)
- Beneath or under the skin: subcutaneous (SC)

The selection of administration route is based on which organs or cell populations need to be involved in the immune response. The IV route affects the spleen, while the subcutaneous route affects the closely situated lymph nodes.

3.5.3 Helping Substances

Some substances are mixed with immunogens just before injection to enhance immunogenicity. These substances are known as *adjuvants*, a word derived from the Latin word *adjuvare*, meaning help. Adjuvants are essential when either the amount of antigen is very low or the antigen has low immunogenicity. Although the detailed mechanism of how adjuvants function to enhance the immune response is still not fully deciphered, we do know that they exert the following one or more effects after administration.

3.5.3.1 Increased exposure time of phagocytose

Antigens stand firm or continue to exist for a longer period than the expected time, that is, persistence of antigen is prolonged. There is also an increase in the exposure time of antigen due to slower release at the injection site. The adjuvant causes precipitation of antigen, thus increasing the chances of phagocytosis by phagocytic cells, for example, alum (aluminum potassium sulphate) and water-oil as adjuvant. Water-oil adjuvant is known as *Freund's incomplete adjuvant* (antigen in water + mineral oil + emulsifying agent, such as mannide monooleate). It breaks oil into small droplets surrounding antigens, thus allowing antigens to be released more slowly into the injection site.

3.5.3.2 Enhanced costimulatory signals

Freund's complete adjuvant (FCA) contains an additional ingredient called heat-killed *Mycobacterium tuberculosis* (muramyl dipeptide), which activates macrophages to enhance phagocytic activity and express higher level of class II MHC molecules and B7 membrane molecules. Class II MHC molecules present antigens to T-helper cells (Th cells). B7 molecules of antigen-presenting cell (APC) membranes bind to CD28 cell surface protein of Th cells to enhance costimulatory signals and T-cell-mediated immune response. Costimulatory signals are also enhanced by Freund's incomplete adjuvant but with lower effect than FCA.

3.5.3.3 Local inflammation is increased (granuloma formation)

Both adjuvants—alum and Freund's adjuvant—trigger local inflammation and infiltration of macrophages. Infiltered-activated macrophages form a mass of cells called granuloma and increase activation of Th cells.

3.5.3.4 Non-specific proliferation of lymphocytes is stimulated

Non-specific proliferation of lymphocytes is stimulated by adjuvants such as synthetic polyribonucleotides (poly IC/poly AU), *Bordetella pertussis,* and bacterial LPS. It leads to antigen-induced lymphocytes clonal selection.

3.6 ANTIGENIC DETERMINANTS (EPITOPES) AND PARATOPES

All antigens or immunogens macromolecules are made up of one or more than one discrete sites, which are to be recognized by lymphocytes. These discrete active sites of immunogens macromolecules are known as *epitopes* or *antigenic determinants* (Plate 1) (Laver, Air, Webster, *et al.* 1990). Antibodies are developed or synthesized against these discrete sites by B-lymphocyte plasma cells. The antigen-specific membrane receptors on T-lymphocytes or synthesized antibodies bind to the specific epitope of antigenic macromolecules. B cells and T cells (Grey, Sette, and Buus 1989) recognize different epitopes on the same immunogens due to different specific-binding receptors. B cells bind soluble or free antigens in solution, whereas T cells bind only those antigens that are presented with MHC molecules on the surface of APCs or altered self-cells (Grey, Sette, and Buus 1989). It has been experimentally proved that BCRs recognize epitopes of the amino-terminal portion of immunogen (Laver, Air, Webster, *et al.* 1990) (for example, glucagon, 29-amion-acid long human hormone) to develop antibodies, while T cells recognize epitopes residing in the carboxyl-terminal part of immunogen (Grey, Sette, and Buus 1989). The epitope sites in proteins are formed due to primary, secondary, tertiary, and quaternary structures, while in polysaccharides, branched chains or multiple branches form conformations of active sites or epitopes (Laver, Air, Webster, *et al.* 1990).

Paratopes are those parts of an antibody that interact with the respective antigenic determinant or epitope against which the antibody has been raised. It is present in the Fv region, that is, the variable region of antibody consisting of heavy chain variable region and light chain variable region (Hv/Lv).

3.7 PROPERTIES OF B-CELL ACTIVE SITES

The properties of B-cell active sites are decided by the nature of antigen epitopes. B-cell epitopes play an important role in vaccine design, immunodiagnostic tests, and antibody production. Hence, it is important to know the various features and types of B-cell epitopes (Laver, Air, Webster, *et al.* 1990) for a better understanding of their immunological and diagnostic uses.

- The size of antigen epitopes recognized by B cells must not be larger than the size of the antibody-binding site. The bonds between antigen epitopes and antibody paratopes are of the weak non-covalent type.
- The sequence of amino acids in epitopes, along with the chemicals of the micro-environment, determines the shapes of epitopes that can be recognized by the antibody. The complementary nature of epitopes and paratopes brings the interacting sites close to each other providing strong bond strength.
- Haptens and other smaller ligands such as carbohydrates and small oligonucleotides bind within a groove of the antibody. The smaller peptides normally fold into a compact structure within the pocket of the antibody to bring its amino terminal close to the carboxyl terminal. For example, angiotensin II, a small octapeptide, contacts 14 residues of the antibody's groove by van der Waals (non-covalent) contact.
- Sometimes antibodies make contact with epitopes across a large flat face. The protrusions on epitopes or antibodies in flat-face interaction are matched by corresponding depressions on the interacting antibodies or epitopes (Stanfield and Wilson 1995). The total number of amino acids of peptide of epitopes matches with a similar number of residues in the binding sites of antibodies.
- The binding patterns of smaller peptides are not similar to that of larger peptides or that of globular proteins. In globular proteins, larger areas of epitopes interact with antibodies. In smaller peptides, they first fold to fit into the pocket of an antibody and then interact. Larger peptides do not fold to fit into pockets; rather, they follow large flat-face interaction patterns with the antibody (Stanfield and Wilson 1995).

3.8 NATURE OF B-CELL EPITOPES PROTEIN

The native proteins on B-cell epitopes are composed of *hydrophilic* amino acids, and they protrude making them normally accessible to membrane associated or free antibodies. The hidden amino acid sequences of interior proteins are also often hydrophilic amino acids, but they do not act as epitopes until they first get denatured. The interaction interface between antibody and antigen shows many complementary protrusions and depressions associated with each other through hydrogen or ionic bonds (Stanfield and Wilson 1995).

3.8.1 Types of B-cell Epitopes

There are two main types of B-cell epitopes: linear or sequential or continuous epitopes and conformational or non-sequential or discontinuous epitopes. The primary structure of the first type of epitope is linear and is composed of a continuous sequence of antigens. The second type of epitope has a tertiary structure composed of discontinuous amino acid sequences of antigens. There are also other types of B-cell epitopes: cryptotopes, neotopes, and mimotopes.

In general, protein antigens contain both linear and conformational B-cell epitopes. Linear or sequential epitopes are shorter peptides (6–8 amino acids) corresponding to a contiguous amino acid sequence of a protein, for example, SWM. Conformational epitopes are composed of amino acids that are non-contiguous in primary protein sequence but are brought into close proximity within the folded tertiary protein structure, for example, HEL (Demotz, Grey, Appella, *et al.* 1989) (Plate 1). The native behaviours of both linear and conformational B-cell epitopes change when they get denatured or fragmented. The SWM has five sequential B-cell epitopes (15–21 amino acids, 56–62 amino acids, 80–86 amino acids, 113–119 amino acids, and 146–151 amino acids). The fragmentation of SWM will yield five fragments each, retaining one sequential B-cell epitope. This was proved by experimental observation that antibody binds with each of the five fragments. In non-sequential HEL protein (Demotz, Grey, Appella, *et al.* 1989), the single-stranded polypeptide chain forms a tertiary structure or final conformation by four intrachain disulphide bonds. This makes almost eight different epitopes. The destruction of disulphide bonds causes a loss of all non-sequential epitopes; hence, the antibody developed against the native protein will not bind to the denatured protein. Corresponding to the specific HEL loop (for example, 64–80 amino acids) (Demotz, Grey, Appella, *et al.* 1989), synthetic open- and closed-loop epitopes can be constructed (Plate 1). By inhibition experiment, it has been proved that only closed synthetic loops having disulphide bonds maintain the structure. The corresponding native peptide loop structure inhibits binding of anti-loop antiserum.

3.8.2 Cryptotopes, Neotopes, and Mimotopes

Cryptotopes, neotopes, and mimotopes are viral protein epitopes, which have the following unique properties.

- *Cryptotopes* Cryptotopes are also known as cryptic epitopes and defined as hidden epitopes or antigenic determinants of protein or virion present on subunits of surface. Cryptotopes become antigenically active only when protein aggregates or virion gets dissociated or broken to expose their surface, for example, HIV gp120.
- *Neotopes* Neotopes are specific for the quaternary structure and absent in dissociated viral subunits, thus called neotopes. In other words, these epitopes are absent from the constituent monomeric subunits.

- *Mimotopes* Mimotopes are proteinous structures that mimic the epitope structure. They are present in both dissociated and intact viral particles and subunits. Mimotopes are present at only one extreme end of the particle containing 5′-end of the RNA. They are used in phage display libraries and vaccine development.

3.9 LOCATION OF B-CELL EPITOPES

B-cell epitopes are normally localized in flexible or most mobile regions of antigens. These epitopes are mobile to maximize the complementarities and effective binding of the antibody. Effective binding of antibody is also supported by the flexibility of epitopes, but such binding affinities are lower than the binding of epitope to rigid non-mobile regions.

3.10 IMMUNODOMINANT AND OVERLAPPING EPITOPES

The old dogma of immunology that globular proteins contain small numbers of epitopes has now been replaced by the fact that the maximum area of the surface of globular proteins is immunogenic. Monoclonal antibodies developed against BSA have the ability to bind to different overlapping antigen-binding profiles. For example, there are 64 developed monoclonal antibodies against BSA or myoglobin or HEL that can bind to 25 different epitopes on BSA or myoglobin or HEL, respectively (Demotz, Grey, Appella, *et al.* 1989). In fact, in a single animal, certain epitopes of an antigen are immunogenic, but other epitopes may not be. Certain epitopes induce more vigorous or pronounced immune response than other epitopes in a specific animal. Such epitopes are known as immunodominant epitopes.

3.11 T-CELL EPITOPES DERIVED FROM ANTIGENS

Only native proteins can elicit a secondary cell-mediated response after primary immunization. The denatured protein cannot do so. In contrast, both native and non-denatured proteins can elicit secondary cell-mediated response. T cells do not recognize soluble native antigens but rather recognize antigen that has been processed into antigenic peptides and are presented in combination with MHC molecules. Hence, the conformational change due to denaturation does not affect T-cell epitopes.

3.11.1 Formation of Trimolecular Complex by Antigenic Peptides

Antigenic peptides combine with TCR and class II MHC molecules to form a trimolecular complex of antigen-TCR-MHC class II. MHC molecule interaction and binding with antigen is non-specific, that is, one MHC molecule can bind with different peptides of different varieties (Demotz, Grey, Appella, *et al.* 1989; Stanfield and Wilson 1995). For example, one MHC can bind to OVA, haemagglutinin, and lambda repressors.

3.11.2 Processing of Antigens

Two different types of antigens—endogenous and exogenous—are processed in two different pathways by APCs. Dendritic cells, B cells, and macrophages, known as professional APCs, constitutively express class II MHC molecules. Skin fibroblast cells, thymic epithelial cells, thyroid epithelial cells, vascular endothelial cells, pancreatic β cells, and brain glial cells are called non-professional APCs. Endogenous antigens are processed into cytoplasm, while exogenous antigens are processed by the endocytic pathway.

3.11.2.1 Processing and presentation of endogenous antigens

Endogenous antigens are generated inside cells by infected virus or cancerous cells. These antigens are processed by a cytosolic-mediated pathway. After processing, the smaller peptides are presented along with class I MHC on the cell surface of APCs as follows.

Viral or tumour cell-generated antigens → Proteasome + adenosine triphosphate (ATP) + Ubiquitin complex acts on antigens → Generation of peptides by processing antigens → ATP-activated TAP1 and 2 rough endoplasmic reticulum (RER) receptor-mediated transfer of peptide to RER → Association and presentation of peptide with class I MHC to cell surface

3.11.2.1.1 Role of ubiquitin in the processing of endogenous protein antigen

First, ubiquitin covalently binds to the amino ($-NH_2$) terminal of protein at the lysine-rich site to form $-NH_2$-C=O-ubiquitin complex, the energy for which is provided by ATP. The protein–ubiquitin complex is degraded by proteasome. Proteasome is a protease complex of 26S with four rings and a diameter of 10–50 Å, which degrades protein into peptides.

3.11.2.1.2 Assembly of class I MHC before binding to peptide

The assembly of class I MHC is mediated by the participation of a chaperone protein *calnexin*. Calnexin binds to the α-chain of class I MHC to enable it to fold. It is released after the binding of β2-microglobulin to the α-chain of class I MHC. Then β2-microglobulin-class I α-chain complex binds to another chaperone protein *calreticulin* and *tapasin*. Tapasin is a TAP-associated protein that facilitates binding of RER-transported peptide with the help of β2-microglobulin to the α-chain of class I MHC. Later, the associated protein exits from RER through the Golgi body with the dissociation of calreticulin and tapasin leading to the presentation of class I MHC-associated peptide on the surface of APCs. Another chaperon ERp57 helps in forming disulphide bonds between two class I chains in a manner yet to be confirmed.

3.11.2.2 Processing and presentation of exogenous antigens

Phagocytosed or internalized antigens first pass through acidic compartments having pH between 4.5 and 6.0. They pass from a pH of 6.0–6.5 in early endosome to a pH of 5.0–6.0 in late endosome to finally a lysosome having a pH of 4.5–5.0, leading to the presentation of class II MHC–peptide complex to the surface of APCs, for example, B cells and macrophages.

3.11.2.2.1 Assembly of class II MHC molecules

Inside the RER, freshly synthesized class II MHC having α-class II MHC chains and β-class II MHC chains gets associated with an invariant chain to form a class II MHC-invariant complex. This is to stop the premature binding of processed peptides to class II MHC molecules. Later, the invariant chain gets digested to leave class II-associated invariant chain peptide (CLIP) associated with class II MHC molecule occupying the groove. The association takes place with the help of non-classical class II MHC molecule known as HLA-DM, which acts as a catalyst to replace CLIP with processed peptide. The exchange of peptide and CLIP is prevented or inhibited by HLA-DO, which binds to HLA-DM. It has been experimentally proven that high acidity decreases the chance of DM-DO association. This subsequently increases the chance of association of peptide to class II MHC molecules with the replacement of CLIP in the lysosome of B cells in preference with other APCs.

3.11.2.2.2 Non-peptide antigen presentation

Protein antigens are presented by CD1 molecules, which come in non-classical class I MHC molecules. The structure of a CD1 molecule is very similar to a class I MHC molecule. It presents in association with β2-microglobulin. CD1 genes, which are five in number, are found located on chromosome-1, rather than being present within the MHC. These five CD1 genes are present in two groups: (1) CD1A, B, C, and E in group I and (2) CD1D in group II. Both humans and rabbits have all these five genes, that is, both group I and group II genes, while rodents have only group II genes. CD1 presents *M. tuberculosis* cell wall lipids, such as mycolic acid, and glycolipids, such as lipoarabinomannan. The processing pathway of non-protein antigens followed by a CD1 molecule is different from the class I and class II pathways. It has the ability to process non-protein antigens in TAP-deficient cells. These non-protein-presenting CD1 cells, especially CD1d, are recognized by natural killer (NK) T cells. Other CD1 cells are CD1a and Cd1b.

3.11.2.2.3 Important facts about antigen processing and presentation

Antigen processing and presentation is inhibited by chloroquine and leupeptin. Ementine is known to inhibit viral protein synthesis. Chloroquine drug is also known to differentiate between the processing of exogenous antigens and endogenous antigens in APCs.

3.11.2.2.4 *Recognition of epitopes by T cells*

T cells recognize only internal peptides. Before recognition, these peptides, which are the processed products of macromolecules, must be presented at the surface of APCs or altered self-cells. J. Rothbard experimentally proved that T-cell epitopes of macromolecules reside inside macromolecules (Laver, Air, Webster, *et al.* 1990).

3.12 HAPTEN–CARRIER CONJUGATES ANTIGENICITY

As mentioned earlier, hapten itself is not immunogenic, but when it is conjugated chemically with a large protein molecule, called a carrier, it forms an immunogenic hapten–carrier complex. This complex contains multiple copies of the hapten in a single carrier protein molecule. It can develop three types of antibodies (Figure 3.1) in animals immunized with it: (1) antibodies against the hapten epitope, (2) antibodies against the unaltered epitope of the carrier protein, and (3) antibodies against the new epitope developed due to conjugation of hapten with the carrier protein, that is, the hapten–protein complex. Hapten-based experiments were performed by Karl Landsteiner in the 1920s and 1930s to test the cross-reactivity of various haptens. Hapten-based experiments have the following advantages.

- Chemically defined antigenic determinant is available.
- Antigenic determinant's structure can be modified, and its immune specificity can be studied.
- Anti-hapten antibody reactivity can be tested with other haptens, which are slightly different structurally, to check cross-reactivity. Such a reaction of antibodies is called cross-reaction.

Examples of some important haptens are dinitrophenol, aminobenzene, o-aminobenzoic acid, m-aminobenzoic acid and p-aminobenzoic acid, peptide hormones, and steroid hormones. Anti-hapten antibody against human chorionic gonadotropin

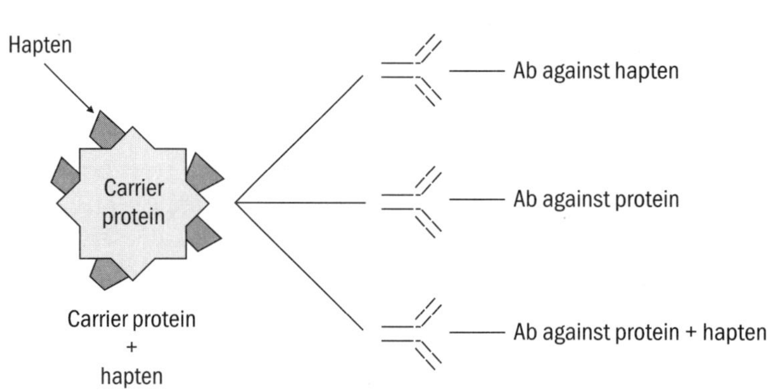

Figure 3.1 Development of antibody against carrier protein, hapten, and conjugate

(HCG) is used to test pregnancy. Landsteiner found that antibodies developed against aminobenzene or its derivatives react only with the original hapten and did not cross-react with any other derivates.

Note Hapten without carrier protein cannot trigger antibody formation.

3.13 EPITOPE MAPPING

Epitope mapping is done by protein microarrays, that is, by identifying the binding sites of the epitope of antibodies on antigens. It can be helpful in the discovery of new therapeutics, vaccines, and diagnostics. It is done by the following methods.

- *X-ray co-crystallography* is time consuming and requires high amount of purified proteins.
- *Peptide scanning* is faster but expensive and is used only for linear epitopes. It has provided the first library of overlapping sequences of short peptide sequences.
- *Site-directed mutagenesis* is slow and requires more labour. Amino acids sequences in peptides are recognized by loss of reactivity. It is used to map complex proteins, for example, integral membrane protein, G protein coupled receptors (GPCRs), and envelope protein of virus.

3.14 PATTERN RECOGNITION RECEPTOR PROTEIN

Pattern recognition receptors (PRR) recognize particular types of molecular patterns present in the microbe's motifs and are involved in innate immunity (Peiser, Mukhopadhyay, and Gordon 2002). These receptors recognize motifs that are highly conserved and found only within the microbial species; they are absent in the host and thus are highly specific. The patterns of macromolecules recognized by these receptors are sugars, proteins, lipoproteins, and nucleic acids. Receptors also have a strong ability to distinguish between self- and non-self-molecules; they cannot develop autoimmune diseases like adaptive immunity. These receptors are present in both soluble (in blood and tissue fluid circulation) and non-soluble forms (attached to cell membranes of macrophage, neutrophils, and dendritic cells). Types of PRRs are mannose-binding lectin receptors (MBLR), C-reactive protein receptors (CRPR), LPS-binding protein receptors (LBPR), and complements involved in innate immunity (Peiser, Mukhopadhyay, and Gordon 2002). They are present either in blood stream or in tissue fluids. MBLR, CRPR, and complements promote *opsonization* of pathogens or antigens. These receptors can also activate the complement system to target pathogens and alter cells to trigger *complement-mediated lysis*. Apart from soluble receptors, there are cell-membrane-associated receptors such as *scavenger receptors* (SRs) on macrophages and dendritic cells. SRs help cells in binding and phagocytosis of pathogens. Two important types of PRRs include the Toll receptor present in *Drosophila* and its mammalian homolog Toll-like receptor (TLRs), which recognizes only different microbial products patterns (Underhill and Ozinsky 2002). Both the receptors are type I transmembrane proteins with an ectodomain consisting of leucine-rich repeats (LRRs) and one or two cysteine-rich regions.

3.14.1 Toll-like Receptors

Toll-like receptor proteins are similar to the protein present in fruit flies (*Drosophila melanogaster*), encoded by the Toll-gene and discovered in 1985 by Christiane Nüsslein-Volhard. The receptor's name comes from its homology with the Toll receptors of fruit flies. Microbes and their products crossing physical barriers, such as skin and mucosa, are finally recognized by TLRs. In response, TLRs activate inflammatory-based immune responses. Jules A. Hoffmann and his colleagues discovered the role of Toll in developing immunity against fungal infection in *Drosophila* for the first time. In 1994, human TLRs were discovered by Nomura and his colleagues. Bruce A. Beutler and his colleagues discovered the function of TLRs using mice as a model. They found that mice unable to respond to LPS had non-functional TLR4 due to mutation. TLRs are highly specific to pathogen-associated molecules, which are also evolutionarily conserved.

Toll-like receptors are themselves conserved during evolution and do not easily change. Examples of some conserved pathogenic molecules that are specifically recognized by certain specific TLRs are LPS, lipoprotein, lipopeptides, lipoarabinomannan, flagellin, dsRNA of virus, unmethylated CpG islands of bacteria, and viral DNA. After binding with the above molecules, TLRs get activated and transduce signals to cause transcriptional activation, leading to the synthesis and expression of cytokines and chemokines that finally result in inflammation and the drawing in of phagocytic cells. Thus Toll receptors or TLRs play an important role in sensing invading pathogens in insects, mammals, and plants (Underhill and Ozinsky 2002).

As of now, 14 TLRs—TLR1 to TLR 15 (TLR14 has not been discovered)—have been discovered located either on the cell surface or in cell compartments. The TLRs located on the cell surface include TLRs 1, 2, 4, 5, 6, 10, and 11, while the TLRs located in the cell compartment are TLRs 3, 7, 8, and 9. The locations of TLRs 12, 13, and 15 are unknown. These TLRs have different adaptors to pass the signal to the target site. They bind and are activated by different specific ligands, which are different components of bacteria (Table 3.1).

3.14.2 Signalling Pathway

There are two signalling pathways in TLRs—MyD88 dependent and MyD88 independent but type I IFN-dependent pathways. Signalling pathways via TLRs originate from conserved cytoplasmic Toll/IL-1 receptor (TIR) domains. TIR-domain-associated

Table 3.1 Toll-like receptors and their binding specifications

Receptor/location/adaptor protein	*Ligand/location*	*Ligand recognition effect*	*Cell type*
TLR1/cell surface/MyD88/MAL	Triacyl lipopeptides/ bacteria	Delivery to receptor on cell membrane	B-lymphocytes, macrophages, monocytes, dendritic cells
TLR2/cell surface/MyD88/MAL	Glycolipids/bacteria Lipoproteins/bacteria Lipoteichoic acid/bacteria HSP70/host cells Zymosan/fungi, Yeast Oxidative stress Synthetic Pam_3CSK_4 $MALP_2$, PGN	Attracts and activate phagocytes; secretes several cytokines	Macrophage, monocytes, myeloid dendritic cells, mast cells
TLR3/cell compartment/ TRIF	dsRNA, poly I:C/viruses	Triggers production of interferon cytokines	Dendritic cells, B-lymphocytes
TLR4/cell surface/MyD88/MAL/ TRIF/TRAM	LPS/bacteria HSP/bacteria and host Fibrinogen/host Heparin sulfate/host Hyaluronic acid/host Taxol	Attracts and activate phagocytes; secretes several cytokines	Macrophage, monocytes, myeloid dendritic cells, intestinal epithelium
TLR5/cell surface/ MyD88	Flagellin/bacteria	Attracts and activates phagocytes; secretes several cytokines	Macrophage, monocytes, dendritic cells, intestinal epithelium
TLR6/cell surface/ MyD88/MAL(TIRAP)	Diacyl lipopeptides/ mycoplasma	Attracts and activates phagocytes; secretes several cytokines	Macrophage, monocytes, mast cells, B-lymphocytes
TLR7/cell compartment/ MyD88	ssRNA/virus, imidazoquinoline/ synthetic compound, loxoribine (guanosine analogue)/synthetic compound	Attracts and activates phagocytes; secretes several cytokines, including interferons	Macrophage, monocytes, plasmacytoid dendritic cells, B-lymphocytes
TLR8/cell compartment/ MyD88	ssRNA/virus, synthetic compounds	Triggers secretion of interferon	Macrophage, monocytes, dendritic cells, mast cells
TLR9/cell compartment/ MyD88	Unmethylated CpG DNA/bacteria	Attracts and activates phagocytes; secretes several cytokines	Macrophage, monocytes, plasmacytoid dendritic cells, B-lymphocytes
TLR10/cell surface	Not discovered yet		Macrophage, monocytes, B-lymphocytes,
TLR11/cell surface/ MyD88	Profilin/toxoplasma		Macrophage, monocytes, liver cells, kidney, bladder epithelium

Contd...

Table 3.1 Contd...

Receptor/location/adaptor protein	Ligand/location	Ligand recognition effect	Cell type
TLR12/NA	Not discovered yet		
TLR13/NA	Not discovered yet		
TLR15/NA	Not discovered yet		

adaptor proteins such as MyD88, TIRAP/Mal, TRIF, and TRAM regulate the signalling pathways. TRIF mediates induction of IFNβ in place of MyD88 in TLR3 and TLR4 signalling pathways; thus these pathways are not dependent on MyD88. TIRAP/Mal is an adaptor protein in TLR2 and TLR4 mediated by MyD88-dependent signalling. TRAM is specifically associated with the TLR4-mediated TRIF-dependent pathway.

3.15 TLR-RELATED PROTEINS

There are three other protein molecules—*RP105, Nod1,* and *Nod2*—structurally and functionally similar to the TLR protein molecules and play an important role in responses to LPS. RP105 is a type I transmembrane protein with an ectodomain consisting of LRRs and one cysteine-rich region, but it does not contain a TIR domain like TLRs. It is expressed primarily on mature B cells. Nod1 and Nod2, on the other hand, are cytosolic proteins functionally equivalent to TLRs. They are composed of one and two N-terminal *caspase recruitment domains* (CARD), respectively. Both contain a central nucleotide-binding domain and multiple LRRs at the C-terminus. LRR has a regulatory function on the activity of CARD. CARD, an effector domain upon dimerization, interacts with RIP-like interacting CLARP kinase (RICK), which in turn activates NFκB.

3.16 IMPORTANT FINDINGS

- John A. Tainer, *et al.* discovered the mobility of epitopes in the same protein by comparing them with the positions of known B-cell epitopes.
- P. G. H. Gell and Baruj Benacerraf (1959) suggested the qualitative difference between the T-cell and B-cell responses to antigens.
- J. Rothbard experimentally proved that T-cell epitopes of macromolecules reside inside the macromolecules.
- Karl Landsteiner (1920s, 1930s) discovered antigenicity of haptens in the hapten–carrier protein complex.

AT A GLANCE

- All antigens are not immunogenic and hence all antigens are not immunogens; but all immunogens are antigens.
- The immunogenic properties of any antigen are determined by foreignness to host, molecular size, chemical composition, complex nature, doses used, antigens being processed and presented, genetic nature of the host, administration route, and adjuvant type.
- Generation of T-cell epitopes occurs against the processing and presentation of antigens in small peptides in the form of class I or class II major histocompatibility complex (MHC) peptide complex displayed on the surface of antigen-presenting cells (APCs). T-cell activation is done by the interaction between the T cell receptors (TCRs) of T cells and the peptide-forming complex, which is displayed with MHC on APCs or altered cells.
- Haptens are very small molecules unable to induce immunogenicity alone. Hence, they cannot bind to any antibody alone. But in combination with carrier proteins, haptens are immunogenic and produce three types of antibodies—against haptens, against carrier proteins, and against hapten-carrier complexes.
- The hapten-carrier complex is used to test allergic responses of drugs, for example, penicillin.
- Dendritic cells, B cells, and macrophages are called professional APCs.
- Skin fibroblast cells, thymic epithelial cells, thyroid epithelial cells, vascular endothelial cells, pancreatic β cells, and brain glial cells are called non-professional APCs.
- Endogenous antigens are processed into cytosols and proteasomes and assembled with class I MHC molecule in which transfer is mediated by rough endoplasmic reticulum (RER).
- Acidic endocytic-mediated processing of exogenous antigens is done to present the peptide in association with class II MHC molecules by replacing the invariant part of class II-associated invariant chain peptide (CLIP).
- Bacterial cell walls are sources of non-peptide antigens. These antigens are presented by CD1 molecules in class I and class II independent pathways.
- Antigen processing and presentation is inhibited by chloroquine and leupeptin. Chloroquine is also used to differentiate the processing of exogenous and endogenous antigens.
- Synthesis of viral protein is inhibited by emetine.

QUESTIONS

1. What do you understand by antigens and immunogens? Describe the factors responsible for determining the antigenicity of any antigen.
2. What are haptens? Discuss their importance with suitable examples.
3. What are pattern recognition receptors? Discuss their importance with suitable examples of their specificity towards antigens.
4. How do biological systems and the route of administration on recipient animals affect immunogenicity of any antigen? Discuss with suitable examples.

5. What are B-cell and T-cell epitopes? Discuss cryptotopes, neotopes, and mimotopes with examples.
6. What are antigens? Discuss various forms of antigens. Describe bacterial types of antigens in details.
7. Why it is difficult to develop a vaccine against HIV, which causes AIDS?

OBJECTIVE QUESTIONS

A. Fill in the blanks.

a. Pattern recognition receptors recognize _________ present only in microbes but absent in host.

b _________ are small molecules that cannot induce immune response alone but are immunogenic in combination with _________ .

c. B-cell epitopes are _________ or _________ .

d. B-cell epitopes bind to _________ .

e. T-cells epitopes are generated by _________ .

f. Class I and Class II MHC recognize only _________ epitopes.

g. Soluble types of epitopes are _________ .

h. Endogenous antigens are processed into peptides within _________ by _________ and assemble with _________ molecules in the _________

i. Exogenous antigens are degraded into peptide within _________ compartments and get associated with _________ molecules.

j. Peptide gets associated to clefts of class II MHC molecules after replacement of _________ chain.

k. Major source of non-peptide antigens are _________ .

l. Non-peptide antigens are _________ .

m. Non-peptide antigens are presented by _________ molecules.

n. Differentiation between the processing of exogenous antigens and endogenous antigens is done by using drug _________

o. Viral protein synthesis is inhibited by _________

p. Antigen processing is inhibited by _________ and _________ .

B. Match the following.

i. Adjuvants	a. MHC of recipient
ii. Genotype affects immunity	b. Increases immunogenicity

iii. Intra peritoneal	c. T-cell epitopes
iv. Peptide–MHC complex	d. Route of administration of antigen
v. Penicillin	e. Non-immunogenic
vi. Non-foreign molecules	f. Hapten
vii. PRR	g. Adaptive immunity
viii. Antibody	h. Innate immunity
ix. i.v.	i. Polyclonal antibody
x. Protein molecules	j. Intra venous
xi. Non-peptide antigen	a. Non-classic molecule HLA-DM
xii. Peptide binding to cleft of class II	b. β2-microglobulin
xiii. TAP1 and TAP2	c. Multifunctional protease complex
xiv. Calnexin replaced by	d. Binds to amino terminal end of protein
xv. CLIP	e. ATP dependent peptide transport into RER
xvi. Ubiquitin	f. Inhibitor of MHC assembly
xvii. Endocytic pathway	g. CD1 molecules
xviii. Proteasome	h. Leupeptin
xix. Protease inhibitor	i. Acidic condition dependent
xx. Chloroquine	j. Class II MHC

CHAPTER FOUR 4

Antigen Determinants

4.1 ANTIBODIES

Antibodies are serum proteins called *immunoglobulins* or *glycoproteins* that bind to specific antigens or immunogens against which they have been produced. They are known as glycoproteins because carbohydrate modulates are added in the Fc region of the antibody. Addition of carbohydrates is known as *glycosylation* of antibodies. Membrane-associated antibodies are found on B cells to confer antigenic specificity and also to help them in clonal proliferation. Blood is composed of cellular fractions, that is, red blood cells (RBCs), leukocytes, platelets, and plasma (fluid fraction, macromolecules, fibrinogen, and other blood-clotting components). The fluid fraction of plasma contains approximately 20% antibodies of the total serum protein.

4.1.1 Serum Composition

Serum protein is composed of a mixture of proteins such as *albumin, globulin,* and *fibrinogen*. Globulin is made up of mainly α1-, α2-, β-, and γ-globulins. Using the electrophoresis technique, A. Tiselius and E. A. Kabat, for the first time, experimentally proved that serum proteins contain antibodies that are normally present in the α-, β-, and γ-globulin regions. Tiselius and Kabat, using ovalbumin protein, immunized rabbit and divided the collected serum into two aliquots for two experiments. In the first experiment, electrophoresis of the first aliquot was done and four peaks were found. One peak was of albumin, and the other three peaks were of α-, β-, and γ-globulins. The second aliquot was used for precipitation reaction with ovalbumin. The precipitate was removed, and the remaining serum protein, on electrophoresis, revealed a drop in the γ-globulin peak. Drop in the γ-globulin peak was an evidence of the presence of serum antibodies. Gamma-globulin was called an immunoglobulin. It contains maximum IgG class antibodies.

4.2 SITES FOR THE FORMATION OF ANTIBODIES

Antibodies are formed in the lymphoid tissue, that is, in the spleen, gut-associated lymph tissues, lymph nodes, and bone marrow, in response to antigens. Large pools of lymphocytes are present in the body in response to different antigens leading to the synthesis of antibodies. The best-fitting antigens to the receptors present on the surface of lymphocytes cause activation by the interaction of T-lymphocytes to produce antigen specific antibodies (Wedemayer, Patten, Wang, *et al.* 1997; Wilson and Stanfield 1994).

4.2.1 Structure of Antibody

Antibodies are heterodimer molecules consisting of four peptide chains—two identical light (L) chains and two identical heavy (H) chains (Plate 1). The light chains are smaller peptides of 25 kDa and get folded into two domains, while the heavy chains of larger polypeptides (50 kDa) get folded into four or five domains. The N-terminal domains of each chain constitute the variable (V) regions, and the rest of the domains constitute the constant (C) regions. Immunoglobulins are Y-shaped structures made up of two identical Fab arms composed of a light chain and a heavy chain variable part, including the first constant region of a heavy chain and the Fc stem region along with other constant domains. The Fab regions have antigen-binding sites (Schroeder and Cavacini 2010), while the Fc region is responsible for the elimination of antigens (Plate 2, paths a and b). Each light chain is attached to a heavy chain by disulphide bonds and non-covalent bonds (salt linkages, hydrogen bonds, and hydrophobic bonds) to form an H–L chain. Two H–L chains are attached to each other by disulphide bonds and non-covalent interactions to form a four-chain heterotetramer $(H\text{–}L)_2$ antibody structure (Wedemayer, Patten, Wang, *et al.* 1997). Antibodies of different classes and subclasses differ in numbers and positions of disulphide bonds.

Approximately 110 amino acids of the amino terminal of antibodies chains are highly variable and differ among antibodies; this region is called the variable region. Variable regions of light chains are denoted by V_L and that of heavy chains by V_H; they are together known as the Fv region. Variable regions are also known as complementarity-determining regions (CDRs) or hyper-variable regions. They are the antigen-binding sites of any antibody. The specificities of two antibodies differ from each other due to differences in the sequence at the CDRs (Schroeder and Cavacini 2010). Beyond variable regions, the structures of light and heavy chains are called constant regions and denoted by C_L and C_H, respectively.

4.3 GLYCOSYLATION

Attachment of carbohydrates to the constant region of an antibody at a particular site is called *glycosylation* of the antibody. All antibodies contain carbohydrate at the conserved positions in the constant regions of the heavy chains. Each isotype of the heavy chain has a distinct array of N-linked carbohydrate structures, which affects assembly of proteins, secretion, and functional activity. Up to 20%–30% of rabbit and human IgGs contain carbohydrates associated with Fab regions. The N-linked carbohydrates vary in structure due to processing and can contain high mannose, multiple-branched or biantennary complex oligosaccharides (Schroeder and Cavacini 2010; Wedemayer, Patten, Wang, *et al.* 1997; Wilson and Stanfield 1994).

4.3.1 Importance of Glycosylation

Glycosylation occurs in the Fc region (with a few exceptions where it also occurs in the Fab region) and is essential for effector functions. There should be different effector functions for glycosylated or non-glycosylated antibodies. Glycosylation affects the structure of the antibody and has an important role in its (antibodies) clearance from the serum and complement activation. The intrinsic solubility of non-glycosylated antibody is lesser than the glycosylated antibody. Non-glycosylated antibodies are less stable than the glycosylated antibodies. Glycosylation of antibodies affects their activity and stability; hence, this process is used in therapy to cure many diseases (Schroeder and Cavacini 2010).

4.4 ENZYMATIC AND CHEMICAL-BASED STUDY OF ANTIBODY STRUCTURE

Gamma-globulin produces high and low molecular weight fractions. A low molecular weight fraction contains IgG immunoglobulin of approximately 150 kDa. Enzymes or chemicals that digest IgG to generate fragments have been used to study its basic structure. *Papain* enzyme generates three fragments—two identical fragments of 45 kDa and one dissimilar fragment of 50 kDa. The two identical fragments are able to bind to antigens; hence, they are called the Fab fragment (Fab denotes "fragment, antigen, binding"). The dissimilar fragment forming a crystal in cold storage is called the Fc fragment (Fc denotes "fragment, crystallizable"). It cleaved just above the interchain disulphide bonds joining the heavy chains (Plate 2, path a). The second proteolytic enzyme *"pepsin"* generates two 100 kDa Fab-like fragments, which bind antigen and are denoted as $F(ab')_2$ fragments. They cleave just below the interchain disulphide bonds joining the heavy chains. The Fc fragment is digested into multiple fragments and hence not recovered (Plate 2, path b). *Mercaptoethanol* cleaves disulphide bonds, that is, reduction and alkylation generate two polypeptide chains of size 50 kDa, which are heavy chains, and two polypeptide chains of size 25 kDa, which are light chains (Plate 2, path c). Whether Fab, $F(ab')_2$, and Fc are made of heavy chains or light chains or both has been proved by using antisera from goats immunized with either Fab or Fc fragments of IgG of rabbit. The antisera of Fab react with both heavy and light chains, while the antisera of Fc react only with heavy chains. Based on the results obtained, it has been concluded that Fab contains both heavy and light chains, while Fc contains only heavy chains (Schroeder and Cavacini 2010; Wedemayer, Patten, Wang, *et al.* 1997; Wilson and Stanfield 1994).

4.5 SEQUENCING OF IMMUNOGLOBULIN

The serum antibody is a highly heterogeneous protein and, therefore, was not suitable for sequencing in the earlier days. In fact, sequencing of antibody was not possible as there were insufficient amounts of homogeneous serum proteins. Later sequencing became possible due to the discovery of *multiple myeloma.* Multiple myeloma cells are *cancerous plasma cells* that produce antibodies of a single molecular species for an unlimited period of time without any activation by antigens. The antibodies produced are almost similar to normal antibodies, but based on its source (myeloma cells), they are called myeloma proteins. Myeloma proteins constitute 95% of the serum antibody in myeloma patients. Those containing excess of light chains in the urine of patients are called *Bence–Jones proteins*, named after its discoverer. Clones of malignant plasma cells develop when mineral oil is injected into the peritoneal cavity of mice. They are designated as *plasmacytomas.* Since the production of *plasmacytomas* is induced by mineral oil, these cells are also called mineral-oil-induced plasmacytomas cells (MOPCs) (Hesslein and Schatz 2001).

Bence–Jones proteins are used for sequencing of light chains and heavy chains of antibodies. Half of the light chain, starting from the amino-terminal side, consisting of 100–110 amino acids, is called the variable region since the sequence varies among different Bence–Jones proteins. The rest of the chain containing the carboxyl-terminal end is called the constant region. It contains two basic amino acid sequences. This suggested that there are two types of light chains (LC)—*kappa* (κ) and *lambda* (λ). A single antibody contains either κ-LC or λ-LC but never both types of light chains. B-lymphocyte expresses only one class of light chain during its whole lifespan, either κ or λ. The κ-LC to λ-LC ratio is 60 to 40 or 64 to 36 in a healthy human, while the κ-LC to λ-LC ratio is 95 to 5 in mice, meaning that κ-LCs predominate in mice. Chickens contain only lambda chains. The myeloma protein has been used to study heavy chain sequencing, and its pattern is found almost similar to light chains. The amino-terminal end contains 100–110 amino acids with a highly variable sequence; hence, it is called the variable region. The remaining carboxyl-terminal part contains five different types based on different basic sequence patterns—μ, δ, γ, ε, and α. Each type determines the class of immunoglobulin, that is, μ for IgM, δ for IgD, γ for IgG, ε for IgE, and α for IgA. Each of these heavy chains is called an *isotype.* On the basis of the length of amino acid sequences, isotypes are divided into two groups: (1) 330-amino-acid-long constant region for δ, γ, and α; and 440-amino-acid-long constant region for μ and ε. Each type of antibodies has either κ-LC or λ-LC. Alpha and gamma are further divided into sub-isotypes due to minor differences in amino acid sequences; α has sub-isotypes α1 and α2, that is, subclasses IgA1 and IgA2 (Hesslein and Schatz 2001), while γ has sub-isotypes γ1, γ2, γ3, and γ4, that is, subclasses IgG1, IgG2, IgG3, and IgG4 in human and γ1, γ2a, γ2b, and γ3 in mice (Schroeder and Cavacini 2010).

4.6 FINE STRUCTURE OF IMMUNOGLOBULIN

Immunoglobulins are proteins, and the final functional structure is determined by a quaternary structure. They contain primary, secondary, tertiary, and quaternary structures (Schroeder and Cavacini 2010).

- *Primary structure* The primary structure is formed by amino acid sequences present in the variable and constant regions of the heavy and light chains.
- *Secondary structure* The secondary structure is formed by the back-and-forth folding of polypeptide chains as antiparallel β-pleated sheets (Figure 4.1).
- *Tertiary structure* Polypeptide chains of secondary structure connect to outside chains of β-pleated sheets to form a globular domain structure.
- *Quaternary structure* Adjacent heavy and light chains in the globular tertiary structure interact to create a functional domain, which is an effector molecule and performs as a binding antigen specifically.

4.7 DOMAINS OF IMMUNOGLOBULIN

Both light and heavy chains of immunoglobulin contain several 110-amino-acid-long homologous units. Each homologous unit is called a domain, which is a loop-like structure due to the formation of intrachain disulphide bonds. Light chains contain two domains—one variable domain (V_L) and one constant domain (C_L). Heavy chains contain four to five domains—one variable domain (V_H) and three to four constant domains (C_H1, C_H2, C_H3, and C_H4) based on the class of antibodies.

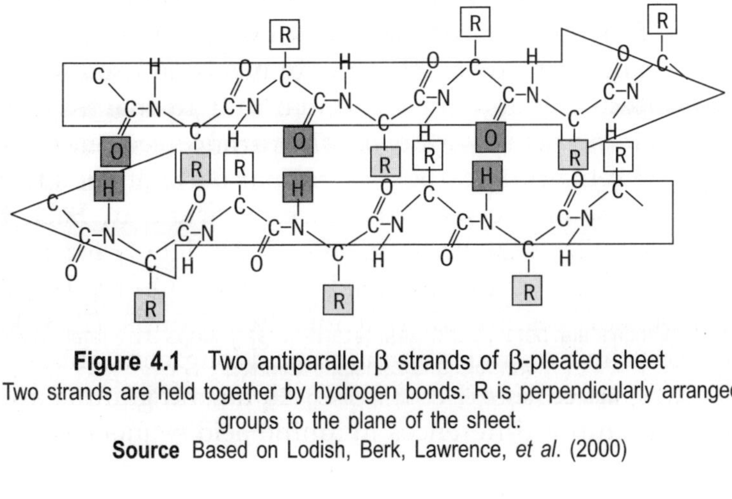

Figure 4.1 Two antiparallel β strands of β-pleated sheet

Note Two strands are held together by hydrogen bonds. R is perpendicularly arranged amino groups to the plane of the sheet.

Source Based on Lodish, Berk, Lawrence, *et al.* (2000)

4.8 X-RAY CRYSTALLOGRAPHY OF DOMAINS

The structure of a domain structure revealed by X-ray crystallography consists of a sandwich of two β-pleated sheets containing two antiparallel β-strands (Plate 3a). The antiparallel β-strands are held together by hydrogen bonds that form between the $-NH_2$ groups of one β-strand with the –COOH groups of the adjacent β-strand to form an *immunoglobulin fold* (Figure 4.2). The β-strand consists of alternative hydrophobic (facing inside orientation) and hydrophilic (facing outward orientation) amino acids. The two β-sheets are held together by hydrophobic bonds and conserved disulphide bonds. The variable domain of a β-sheet is slightly longer due to the presence of an extra pair of β-strands and an extra loop sequence than the constant domain of the β-sheet.

A β-sheet is analogous to a sandwich made up of two pieces of breads (β-strands) containing *butter* (hydrophobic bonds) in between and a *tooth pick* (disulphide bonds) holding them together (Figure 4.1). Immunoglobulin folds facilitate non-covalent interactions between domains of two β-sheets facing each other. Interactions are between identical domains, that is, C_H2/C_H2, C_H3/C_H3, and C_H4/C_H4, and between two non-identical domains, that is, V_H/V_L and C_H1/C_L (Wedemayer, Patten, Wang, *et al.* 1997).

4.9 VARIABILITY OF CDRS

Variability is defined as the total number of different amino acids at a given position divided by the frequency of the most common amino acid at the given position. Suppose sequences of 100 heavy chains are compared, and the data collected showed that *serine* is present at position 7 in 51 sequences of heavy chains. Hence, the frequency of the most common amino acid is 51 ÷ 100 = 0.51. In the rest of the sequences (100 – 51 = 49),

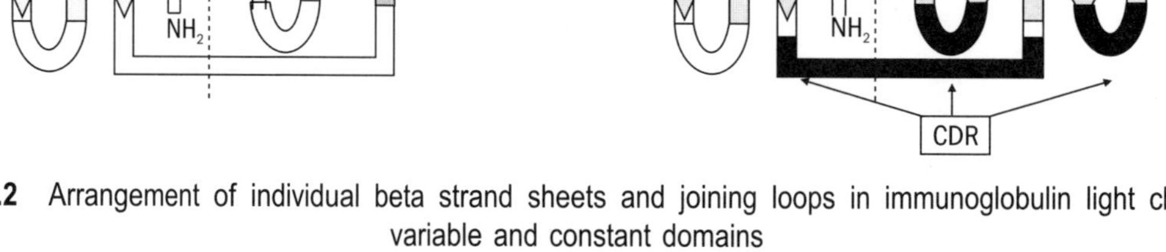

Figure 4.2 Arrangement of individual beta strand sheets and joining loops in immunoglobulin light chain with variable and constant domains
Source Based on Schifter, Girling, Ely, *et al.* (1973)

the amino acid present at position 7 is tryptophan, tyrosine or histidine. The variability at position 7 would be 7.8 (that is, 4/0.51). The maximum variation is present in the sequence of V_L and V_H domains, which corresponds to loops involved in the joining of β-strands. Such regions are called *hyper-variable regions* and are sites for the binding of antigens. The binding sites are actually complementary to the epitope structure. These areas are known as CDRs (Plate 3). The remaining less variable regions of V_L and V_H domains are known as *framework regions*. There are three heavy chains and three light chains, a total of six CDR regions present in a loop. The variations in the length and the amino acid sequence of the six CDRs in each Fab fragment determine the specificity of the antibodies. Framework regions act as scaffolds and support the six loops. They are non-variable, while CDRs differ in orientations in different antibodies. By X-ray crystallography, it has been confirmed that CDRs of antibodies have antigen-binding sites (Wedemayer, Patten, Wang, *et al.* 1997). More regions of the V_H domain take part in the binding of antigen than the binding regions of the V_L domain, that is, the V_H domain has a dominant role in the binding of antigens. Again by using X-ray crystallography, it has been proved that antigen binding induces change in the conformation of either antibody or antigen or both for a closer fit (Stanfield and Wilson 1995). For example, neuraminidase and anti-neuraminidase interaction causes change in both the epitope and its binding site. Haemagglutinin binding does the same thing.

4.10 CONSTANT REGIONS DOMAIN AND THEIR FUNCTIONS

Heavy chains in IgA, IgD, and IgG contain three homologous domains (Plate 4a) and a non-homologous hinge region, while IgE and IgM contain four domains and no hinge region (Plate 4b). IgA, IgD, and IgG have C_H1/C_H1, hinge regions have C_H2/C_H2 and C_H3/C_H3, while IgE and IgM have C_H1/C_H1, C_H2/C_H2, C_H3/C_H3, and C_H4/C_H4. The C_H1/C_H1 domain is at the amino-terminal end, while the C_H3/C_H3 and C_H4/C_H4 domains (in IgE and IgM) are in the carboxyl-terminal end. The C_H2/C_H2 domain of IgE and IgM corresponds to the hinge region of other antibody subtypes, but its functionality is not similar to the hinge region. In fact the function of C_H2/C_H2 in IgE and IgM is still unknown. The C_H2/C_H2 domain of IgA, IgD, and IgG, and the C_H3/C_H3 domain of IgE and IgM is separated by oligosaccharide side chains and, hence, is more accessible to an aqueous environment. These domains are involved in the activation of complements by IgG and IgM.

4.10.1 Hinge Region

It is an extended peptide sequence present between C_H1 and C_H2 domains in α, δ, and γ heavy chains. The hinge region is rich in *proline* residues and provides flexibility to the two Fab arms of IgA, IgD, and IgG so that they can bind with antigens. Another amino acid prominent in this region is *cysteine*. Due to it being a proline-rich region, it is more prone to cleavage by proteolytic enzymes such as papain and pepsin.

4.10.2 Domains of Fab Regions

The important domains of the Fab region are C_H1 and C_L. They provide antibody diversity by facilitating random associations between V_H and V_L domains. Both domains give flexibility to the antibody to interact with antigen. They make the antibody flexible by increasing the rotation of the Fab arms and also by holding the V_H and V_L domains together with interchain disulphide bonds. Both domains are common in all isotypes of antibodies.

There are two types of antibodies—membrane-bound immunoglobulin (mlg) and secreted immunoglobulin (slg) (Kerr 1990). Both domains differ from each other in their structure and functions.

4.10.3 Membrane Bound Immunoglobulin

Different classes of mlgs are expressed in B cells at their different developmental stages. Immature B cells express mlgM, while mature B cells express both mlgM and mlgD. A memory B cell expresses mlgM, mlgG, mlgA or mlgE (Reth 1995). The carboxyl-terminal end of an mlg domain contains the following regions.

- A variable length spacer sequence of 26 amino acids in all different isotypes.
- A constant length hydrophobic transmembrane sequence in all different isotypes.
- A variable length short cytoplasmic tail in all different isotypes.

4.10.4 Secretary Immunoglobulin

Secretary immunoglobulin has a hydrophobic amino acid sequence at the carboxyl end. Examples of secretary antibodies are IgA and IgM. As evidence suggests, IgM is the antibody secreted at the early stage of development and is gradually replaced by the secretion of IgA in adults (Kraehenbuhl, Kerneis, Berdoz, *et al.* 1995). The intestine is the largest lymphoepithelial organ that contains cells which are able to absorb, process, and present various types of antigen B cells, T cells, and antibody-producing plasma cells. Intestinal plasma cells produce more antibodies against environmental antigens than do all the other lymphoid organs (bone marrow, spleen, and lymph nodes) together. Epithelial receptors selectively transport secreted polymeric IgA into the lumen. IgA is the most heterogeneous immunoglobulin because it is present in multiple molecular forms, such as monomeric, polymeric, and secretory. The slgAs are multivalent (have four to eight antigen-binding sites); they are thus more effective than mlgAs. The slgA is also resistant to proteolytic enzymes due to the unique α-chain structure in its primary structure (Kraehenbuhl, Kerneis, Berdoz, *et al.* 1995).

IgA also shows anti-inflammatory properties. It prevents adherence of toxin and viruses and inhibits antigen absorption by forming intraluminal immune complexes. An abundance of glycans on the Fc region of the antibody IgA helps to aggregate enterobacteria-carrying type I fimbriae by the interaction of mannose-binding fimbrial lectin and high mannose-type glycan side chains. This prevents enterobacteria from adhering to epithelial cells. The sIgA is produced by two separate B-cell populations (Kraehenbuhl, Kerneis, Berdoz, *et al.* 1995; Schroeder and Cavacini 2010). One of them is the B1-lymphocyte, earlier known as $Ly1^+$ or $CD5^+$, which has its origin in the peritoneal cavity of mice. B1-lymphocytes produce polyreactive natural antibodies to maintain normal intestinal microbiota. Another B2-lymphocyte residing in the lamina proparia of the gut produces specific IgA to exclude pathogens. Breast milk contains sIgA induced by the maternal mucosal immune system against microbial antigens to protect the gastrointestinal tract of newborns. Newborn infants are fed with IgA-rich preparation (70% IgA) since the anti-inflammatory property of IgA prevents necrotizing enterocolitis (Kerr 1990; Kraehenbuhl, Kerneis, Berdoz, *et al.* 1995).

4.11 EFFECTOR FUNCTIONS OF ANTIBODY

An antibody not only acts by itself on antigens and pathogens but also invokes various responses to eliminate pathogens and antigens; this is known as effector functions of antibodies. The constant regions of antibodies that interact with other cells and antigens play a crucial role in these functions. Effector functions of antibodies are essential for humoral response. The functions differ from one antibody to another due to their different functional properties in interacting with cell membrane receptors and antigens. There are four well-defined effector functions in antibodies, except in IgE in which there is the fifth unique function of activating basophiles, eosinophils, and mast cells.

4.11.1 Phagocytosis Activation

Phagocytosis is an important process through which macrophages and neutrophils remove antigens and pathogens after they invade the barrier of the immune defence system. The Fc region of the antibody binds to Fc receptors (FcRs) present on the surfaces of macrophages and neutrophils to form an antigen–antibody complex. The antigen–antibody complex is engulfed by the phagocyte cells after receiving signals that are generated after complex formation. Enzymatic digestion and killing of pathogens follow phagocytosis. The whole process is called *opsonization*.

4.11.2 Complement Activation

Complement or the complement system is a collection of serum glycoproteins that perforate cell membranes of RBCs and macrophages. The IgM and IgG subclasses of antibodies are able to activate the complement system. C3b protein is an important by-product of the complement-activation pathway, which can bind non-specifically to cell and antigen–antibody complexes near the complement-activation site. This effector function of antibody leads to the inactivation, killing, and removal of pathogens and other antigens in cooperation with the activated complement.

4.11.3 Activation of Cytotoxic Cells

In this effector function, at first the antibody binds to the cells infected with viruses called target cells or infected cells. The FcR-bearing cytotoxic cells, for example natural killer (NK) cells, bind to the Fc region of the antibody, which is attached to the target cells. Binding of cytotoxic cells to target cells causes the lysis of target cells. Since this type of cytotoxicity is mediated by antibody and cytotoxic cells, this whole process of killing of target cells is known as *antibody-dependent cell-mediated cytotoxicity* (ADCC). Examples of cytotoxic cells are NK cells, macrophages, monocytes, neutrophils, and eosinophils.

4.11.4 Transcytosis

Transcytosis is a process by which the antibody moves to the respiratory mucosal surfaces, urogenital mucosal tracts, and breast milk through epithelial layers. The Fc region of the antibody defines the capacity of transportation through the membrane. The antibody transported most abundantly is IgA in human and mice. Other two antibodies transported through the epithelial layer are IgM and IgG. IgG is transported from mother to foetus to trigger immune response. Such type of immunity is termed *passive immunization*. Passive immunization is defined as the acquisition of immunity by received antibodies (from other sources) rather than active synthesis of antibodies by antigen exposure.

4.12 DIFFERENT CLASSES OF ANTIBODIES BASED ON CONSTANT REGION

Antibodies are divided into various classes based on the amino acid sequences of the heavy chains present in the constant region. Since the structure differs from one antibody to another, their effector functions and biological activities also differ. The five major classes of antibodies are as follows.

4.12.1 Immunoglobulin G (or γ)

The molecular weight of IgG is 150,000. It always exists as a monomer and consists of two γ-heavy chains and two κ- or two λ-light chains. IgG makes up 80% of the total serum immunoglobulin. It is further divided into subclasses based on the structural differences of γ-chains (which shows 90%–95% homology with each other), size of the hinge region, total number of disulphide bonds between two chains, including its location and number based on its serum concentrations. The subclasses (Figure 4.3) are IgG1 (γ1), IgG2 (γ2), IgG3 (γ3), and IgG4 (γ4). Differences in the biological activities are due to specific amino acid sequences. The specific biological activities of the subclasses are as follows.

- Except IgG2, all other antibodies cross the placenta to immunize the foetus. The normal serum level in decreasing order is IgG1 (9 mg/ml) > IgG2 (3 mg/ml) > IgG3 (1 mg/ml) > IgG4 (0.5 mg/ml).
- Except IgG4, all other antibodies can activate the complement system. The efficiency of activation in decreasing order is IgG3 > IgG1 > IgG2.
- Affinity for phagocytic cell's FcRs is in the following decreasing order: IgG1 ≥ IgG3 > IgG4 > IgG2.
- Half-life in vivo in serum is 23 days except for IgG3 (eight days).

4.12.2 Immunoglobulin M (or μ)

Membrane-associated IgM is a monomer (180,000 MW) but is secreted into the serum as a pentamer (900,000 MW). Pentamer formation is due to the formation of disulphide bonds between the carboxyl terminal heavy chain domains Cμ4/Cμ4 with domains Cμ3/Cμ3 (Figure 4.4a). It constitutes approximately 7% of the total serum immunoglobulin with 1.5 mg/ml serum concentration.

4.12.2.1 J-chain

IgM contains an Fc-linked polypeptide chain called J-chain (J stands for joining chains). It is added just after secretion and is attached by two disulphide bonds to the Fc region in two different monomers at the cysteine residue of the carboxyl terminal.

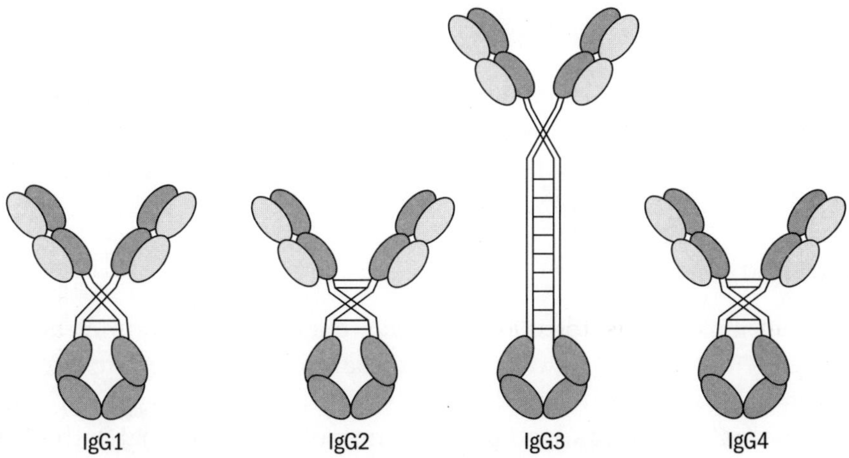

Figure 4.3 Various subtypes of IgG antibody

4.12.2.2 Important features of IgM

The important features of IgM are as follows.

- First immunoglobulin synthesized in neonates against antigen.
- Higher valency than other isotypes, hence has the highest efficiency; requires lesser amounts to agglutinate than IgG.
- It is a more efficient complement activator than IgG.
- It does not diffuse easily; hence, its amount in intercellular tissue fluids is very low.
- With the help of J-chains, it binds to secretary cell receptors to be transported across epithelial linings and enters the external environment (Hanson and Korotkova 2002).

4.12.3 Immunoglobulin A (or α)

IgA is found either as a monomer (150,000 MW) or as a polymer such as dimer, trimer or tetramer (600,000 MW). All polymeric forms contain J-chain polypeptides and are called secretory IgA. The production of secretory IgA in a day is more than the production of any other antibody. The plasma cells that synthesize this antibody are mostly present on the mucous membrane of the jejunum of small intestine. The secretory IgA is at least a dimer consisting of two molecules of IgA, covalently linked to each other by a J-chain (Figure 4.4b) and a secretory component (polypeptide chain). Secretory components are derivatives of epithelial mucosal cell membrane bound receptors and help in the transportation of IgA. The secretory components are five immunoglobulin-like domains and get attached to the dimeric IgA by the disulphide bond formed between the fifth domain and one of the IgA heavy chains. It constitutes approximately 12% of the total serum immunoglobulin with 3 mg/ml serum concentration (Hanson and Korotkova 2002; Hesslein and Schatz 2001). The half-life in vivo in serum is six days.

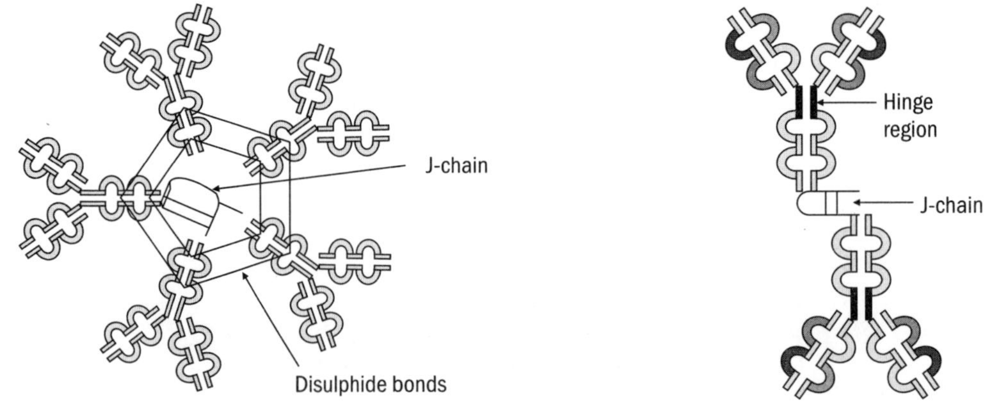

Figure 4.4 (a) IgM always present in pentamer forms, (b) IgA present as monomer, dimer, trimer or tetramer forms

4.12.3.1 Formation of secretory IgA

Dimeric IgA produced by plasma cells gets attached to the poly-Ig receptor of mucous membrane epithelial cells. The dimeric IgA-poly-Ig receptor complex is then internalized by receptor-mediated endocytosis into the *vesicle*. The vesicle formed in the epithelial cells gets transported to the luminal membrane where it gets fused with the plasma membrane. The poly-Ig receptor is enzymatically cleaved from the membrane to become secretory IgA, a complex of dimeric IgA-secretory component (Figure 4.5). The protease sensitive sites on secretory IgA are masked in the protease-rich mucosal environment to survive for longer periods of time (Hanson and Korotkova 2002; Hesslein and Schatz 2001; Kerr 1990).

- The secretory IgA binds to the surface of antigens of virus and bacteria, thus preventing them from attaching to the mucosal surface or colonizing at mucosal sites (Hesslein and Schatz 2001).
- Secretory IgA is present in mother's milk and helps protect the newborn from infection for a few months by boosting the immune response. Hence, breast feeding is always beneficial to the newborn.

4.12.4 Immunoglobulin E (or ε)

IgE is a monomer antibody (190,000 MW), named after the potent inducer, that is, E-antigen of ragweed pollen. The normal serum level is very low (0.0003 mg/ml) and the half-life in vivo in serum is 2.5 days. The antibody was first identified by K. Ishizaka and T. Ishizaka in 1966. The serum of an allergic individual was injected into rabbits to develop an anti-isotype antiserum. Using known human antibodies, the anti-isotype antibodies of rabbit antiserum was precipitated and removed. The unknown antibody that was left in the antiserum completely blocked the P–K reaction and was called IgE (Moiller 1978).

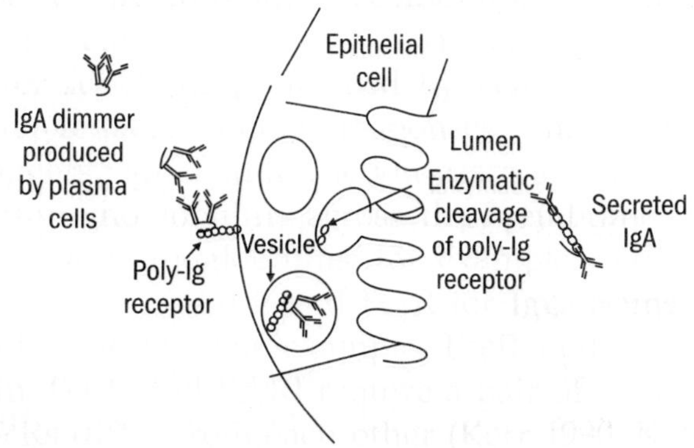

Figure 4.5 Mechanism of secretion of IgA into the lumen

Note First dimeric IgA binds to poly I-gA receptor followed by internalization into the vesicle and finally cleavage of poly-IgA receptor to secrete dimeric IgA into the lumen

4.12.4.1 P–K reaction

K. Prausnitz and H. Kustner in 1921 injected the serum of an allergic person into a non-allergic person intradermally. The second injection of appropriate antigen at the same site caused *wheal and flare* reaction (similar to hives). This reaction was called the *P–K reaction* after the names of its discoverers and used as a bioassay test for IgE. The attachment of IgE membrane-associated receptors of basophils and tissue mast cells causes release of granules (degranulation) to the outside environment. Degranulation causes allergic reaction, and mediators present in it act as defence tools against parasites (Moiller 1978).

4.12.5 Immunoglobulin D (or δ)

IgD is a monomer (150,000 MW). It was first discovered in the multiple myeloma protein taken from a multiple myeloma patient, which did not react with IgA, IgG, and IgM. Its normal serum concentration is 0.03 mg/ml with a half-life in vivo in serum of three days. Although IgD is associated with IgM in mature B-cell membrane, its effector function is still unknown (Moiller 1977).

4.13 EPITOPES OF IMMUNOGLOBULINS

Immunoglobulins are protein; thus they can act as immunogens to induce the synthesis of immunoglobulins in an animal injected with the original immunoglobulin. Synthesized immunoglobulins are called secondary immunoglobulins. The epitopes of immunoglobulin immunogens are of three types—isotypic, allotypic, and idiotypic.

4.13.1 Isotypic Determinants

Isotypic epitopes or determinants are normally present in the constant region of each heavy chain (Plate 5a). It is species specific, and members of the species harbour the same constant region gene or genes. The same species inherit the same constant region genes, while different species inherit different constant region genes. Similar genes will thus express similar isotypes, while different genes express different isotypes. The antibody of one species is foreign for another species and triggers synthesis of antibodies. For example, the IgG1 of mice differs from IgM.

4.13.2 Allotypic Determinants

Allotypic determinants are allele based. A species may have the same set of isotype genes, but their isotype gene set may differ from each other on the basis of alleles (some genes may have multiple alleles). The different allelic set encodes different amino acids called allotypic determinants (Plate 5b). There are four allotypes for four subclasses of IgG and one subclass for IgA. There are at least 25 different allotypes for γ-chain determined by γ-markers. When an antibody of a member of one species is injected into another member (of the same species) having different allotypic determinants, an antibody is synthesized due to immune response. For example, the IgG1 of strain-A differs from the IgG1 of strain-B and will trigger antibody synthesis in each other after cross injection.

4.13.3 Idiotypic Determinants

Idiotype is a variable domain (V_L and V_H) amino acids sequence-based determinant, which differs from each other due to exposure against different antigens (antigen-a or antigen-b) (Plate 5c). The idiotypic determinant region may be inside the binding site or outside the binding site. Each antibody presents multiple idiotypes.

4.14 RECEPTORS OF B CELLS

The receptors present on the membrane of B cells are called B-cell receptors (BCRs). They are heterodimeric transmembrane protein complexes of Igα-chain and Igβ-chain (Figure 4.6) linked together by disulphide bonds. BCRs were discovered by Michael Reth and his colleagues in 1990. Igα contains 61-amino-acid-long cytoplasmic tail, while Igβ contains 48-amino-acid-long tail (Reth 1995).

4.14.1 Fc Receptors and Binding Proteins

Fc receptors are glycoproteins present on many cell membranes. They have affinity for the Fc region of the antibody. Functionally, this is important for the movement of antibodies across the membrane, for example, movement of IgG across placenta to provide passive immunity to the foetus. Various cell types that acquire these receptors are neutrophils, mast cell macrophages, eosinophils, NK cells, B-lymphocytes, and T-lymphocytes. FcRs enable macrophages and neutrophils in the opsonization of antigens. Other effector functions performed by FcRs are ADCC, antibody-mediated cell differentiation, and reduced cellular responses. The binding of antigen–antibody complexes to FcRs triggers signal transduction cascades for opsonization, ADCC, and transport of various immunoglobulins across the membrane. There are specific FcRs for specific subtypes of immunoglobulins; for example, FcαR is for IgA, FcεR is for IgE, FcδR is for IgD, FcμR is for IgM, and FcγR for IgG. Some FcRs require accessory signal-transducing polypeptides. For example, FcεR and FcγRIIIA require a pair of γ-chains and a β-chain; FcαR and FcγRI require a pair of γ-chains only. The structures of antibody-specific FcRs differ from each other (Kerr 1990; Kohler and Milstein 2005). Normally these receptors have two parts—extracellular antigen–antibody complex

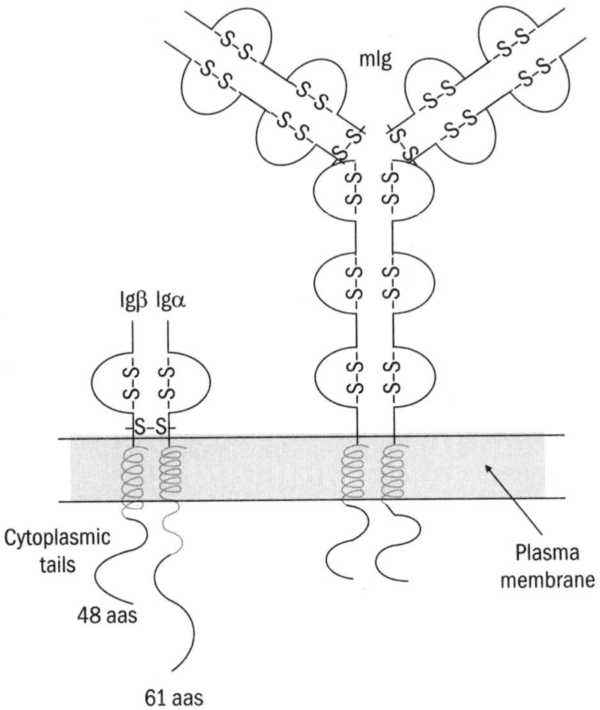

Figure 4.6 B-cells Fc receptor

Note B-cells receptor with Igα and Igβ chain with 61aas and 48aas long cytoplasmic tails with transmembrane domains. Igα and Igβ are associated with membrane associated receptor mIg which has transmembrane domains and cytoplasmic tail.

Source Based on Keegen and Paul (1992)

interacting part and intracellular signal-transducing part. The FcR responsible for translocation of IgG from the mother to the foetus is known as *neonatal Fc receptor* (FcR_N) (Hanson and Korotkova 2002). FcR_N requires an extra β2m assemblage as accessory signal-transducing polypeptides.

4.15 IMMUNOGLOBULIN SUPERFAMILY

The immunoglobulin family consists of membrane-bound glycoproteins. The carboxyl-terminal end of glycoproteins is attached with the membrane. It shows homology with the domains of the immunoglobulin. Homology might be due to derivation from the common primordial gene. These genes encode 110 amino acids polypeptides. Due to gene duplication or divergence, genes of various heavy and light chains may be created. Since these immunoglobulin proteins and respective genes are derived from common primordial genes, the term superfamily is used. Although they are called a superfamily, there are no genetic linkages or functions among these genes and they are evolved independent to each other. Examples of superfamily proteins are as follows.

- Igα/Igβ of BCR
- Poly-Ig receptor providing the secretory component to sIgA and sIgM
- Accessory proteins for T cells
- CD2, CD4, CD8, CD28, and γ-, δ-, and ε-chains of CD3
- Major histocompatibility complex (MHC) class I and class II molecules
- MHC class I-associated β2-microglobulin protein
- Cell adhesion molecules such as VCAM-1, ICAM-1, ICAM-2, and LFA-3
- Platelet-derived growth factor

All these proteins contain specific immunoglobulin domain (110 amino acids polypeptide sequences) intrachain disulphide bond to close loop of 50–70 residues. The immunoglobulin-fold structure does not bind with antigens; rather, it facilitates interactions with membrane-associated proteins.

4.16 MONOCLONAL ANTIBODY

The monoclonal antibody is derived from a single B-cell clone against a single epitope and, hence, is a homogeneous serum of a single-epitope-binding site. On the other hand, polyclonal antibodies are derived from many B-cell clones exposed against different types of epitopes and are a heterogeneous collection of binding sites. A polyclonal antibody is in fact a mixture of different antibodies and has various effector functions as mentioned earlier. Since polyclonal antibodies lose efficacy in vitro, monoclonal antibodies are preferred for research, diagnostic, and therapeutic purposes.

Georges Kohler and Cesar Milstein in 1975 were the first to discover a hybridoma technique, which could produce monoclonal antibodies. For this achievement, they were awarded the Nobel Prize. In this technique, antibody-producing B cells are fused with cancerous plasma cells (myeloma cells) to produce a hybridoma having the growth properties of myeloma cells and the antibody-secreting properties of B cells (Kohler and Milstein 2005). The clones of hybridoma can be cultured for an indefinite period of time with antibody-producing abilities (Plate 6).

4.16.1 Uses of Monoclonal Antibodies

4.16.1.1 Making diagnostic kits

Monoclonal antibodies are used to prepare diagnostic kits to test pregnancy, to measure level of drugs in the blood, to detect pathogenic infection, to detect histocompatibility antigens, and to detect tumour antigens. Monoclonal antibodies to breast cancer are labelled with I^{131} (iodine-131) to detect the spread of tumour antigens in lymph nodes.

4.16.1.2 Making immunotoxins

Immunotoxins can be made from monoclonal antibodies. They are prepared by attaching tumour-specific monoclonal antibody with the toxins (ricin, shigella, or diphtheria) that inhibit protein synthesis. These immunotoxins are used to kill monoclonal antibody-specific tumour cells by inhibiting protein synthesis.

4.16.1.3 Synthesis of abzymes

Abzymes are catalytic antibodies that have enzyme-like activity. Abzymes can cut peptide bonds at specific amino acid sites just like restriction enzymes that cut DNA at specific sites (Wentworth and Janda 1998). This property of abzymes is used for structural and functional analysis of proteins, to cleave viral glycoproteins at specific sites, to block viral ability to infect, and to dissolve blood clots.

Preparation of abzymes is done by synthesizing antibodies against hapten in a hapten–protein carrier complex. Hapten structurally resembles the transition state of ester, which is in the process of hydrolysis. This transition state of ester is used to immunize mice. The spleen cells of immunized mice are fused with myeloma cells to produce monoclonal anti-hapten monoclonal antibodies. The monoclonal antibodies produced increase hydrolysis of ester manifolds (approximately 1000 fold) (Wentworth and Janda 1998).

4.17 DIVERSITY OF ANTIBODY

Two models have been developed to explain the diversity of antibody.

1. *Germ line theories* The genome contributed by the sperm and the egg (that is, two germ cells) contains large numbers of immunoglobulin genes. The genome contains antibodies-coding regions, which provide the essential support for the survival of an organism with a good immune system.
2. *Somatic variation models* This model suggests that the genome contains a very small number of immunoglobulin genes. From these small numbers of genomic genes, a large number of specific antibodies are produced in the somatic cells due to mutation or recombination in the genes.

Both theories were unable to explain the main feature of the immunoglobulin structure, including the mechanisms involved in the generation of the variable region of heavy and light chain genes, keeping unchanged the constant region of each gene. The theories also found it difficult to explain the Til-myeloma protein amino acid sequence. The amino acid sequences of Til protein revealed that the identical variable region sequences of these proteins are associated with both γ- and μ-heavy chain constant regions. Evidence confirmed that sequences of a single variable region specific to a particular antigen could associate with many heavy chain constant region sequences (Schroeder and Cavacini 2010; Wedemayer, Patten, Wang, *et al.* 1997). We can say that different isotypes of antibodies, such as IgG and IgM, could be expressed by similar variable region sequences.

4.17.1 Two-gene Model or Recombination Model

The two-gene model was developed by W. Dreyer and J. Bennett in 1965. According to this model, two genes are responsible for the generation of one antibody. One gene encodes the variable region and another gene encodes the constant region of a single antibody. They proposed that variable region genes could be nearly thousands in number and that constant region genes could be a single copy for various classes and subclasses of genes.

This theory combined both the germ line theory and the somatic variation theory to give a single recombinational model according to which the single variable region combined with various constant regions to generate immunoglobulins. Dreyer and Bennett's two-gene theory for a single polypeptide was against the existing one-gene one-polypeptide theory and, hence, was hard to accept.

4.17.2 Immunoglobulin Genes Rearrangement Theory

In 1976, S. Tonegawa and N. Hozumi, for the first time, gave direct evidence for the two-gene theory proposed by Dreyer and Bennett in 1965. According to Tonegawa and Hozumi, two separate genes encode the variable and constant regions of a single immunoglobulin and rearrangement of genes takes place in the course of B-cell differentiation. Tonegawa was awarded the Nobel Prize for his work in 1987.

They isolated DNA from embryonic cells and adult myeloma cells at different stages of development. Using restriction endonucleases, they digested DNA to generate fragments of DNA. The DNA fragments were separated by size and hybridized with radiolabelled mRNA. The two embryonic DNA fragments hybridized with mRNA probe while only a single matured DNA fragment hybridized with the same mRNA probe. On the basis of this observation, they concluded that variable (V) and constant (C) genes are separated by a large segment of DNA that had a restriction site. During differentiation, the DNA segment gets eliminated bringing the V and C genes together. With the latest Southern blot technique, it is easy to prove the same thing without eluting the separated DNA restriction fragments from gel slices.

4.17.3 Multigene Organization Theory

The heavy chain and light chains (κ and λ) are encoded by separate multigene families of different chromosomes. These multigene families have several *coding sequences* or *gene segments* separated by non-coding sequences in germ line DNA. But in mature DNA, these coding sequences come together to generate functional mature immunoglobulin genes.

The κ-LC and λ-LC families have V, J, and C gene segments. The variable segment is at the 5′-end, while the constant segment is at the 3′-end. The VJ gene segments (J stands for joining; 39 base-pair sequence) encode the variable region of light chains while the C gene segment encodes the constant region. The variable segment is preceded by a small exon at the 5′-end that encodes a small signal or leader peptide. Leader peptide helps light chains and heavy chains to pass through endoplasmic reticulum.

The heavy chain family has V, D, J, and C gene segments. The variable segment is at the 5′-end, while the constant segment is at the 3′-end. The diversity (D) gene segment is the CDR because it contributes diversity to the antibody. The VDJ gene segment encodes the variable region of the heavy chain while the C gene segment encodes constant regions. Similar to light chains, the V gene segment is preceded by the 5′-end short exon that encodes leader peptides (Fugmann, Lee, Shockett, *et al.* 2000; Hesslein and Schatz 2001).

4.18 VARIABLE REGION DNA REARRANGEMENT

Recombination signal sequence (RSS) is the mechanism involved in the rearrangement of the variable region DNA in the antibody.

4.18.1 Recombination Signal Sequence

The RSS flanks each germ line V, D, and J gene segments. It is present at the 3′-end of the V gene segment, 5′-end of the J gene segment, and both 3′-end and 5′-end of the D gene segment. RSS provide signals for rearrangement. It comprises a palindromic heptamer and an AT-rich nonamer sequence separated by an intervening sequence (12 or 23 base pairs). Both are conserved sequences. The 12 base-pair intervening sequence corresponds to one turn of DNA helix, called one-turn RSS. While the 23 base-pair intervening sequence corresponds to two turns of DNA helix, called two-turn RSS.

Vκ has one-turn RSS, and Jκ has two-turn RSS. Vλ has two-turn RSS, and Jλ has one-turn RSS. V_H and J_H have two-turn RSS. D_H has one-turn RSS on either side. One-turn RSS joins only with two-turn RSS. This is called the one-turn/two-turn joining rule. This rule ensures that the V_L segment joins only to a J_L segment. Thus it sees to it that a particular segment joins in proper order and that same types of segments do not join each other. Gene segments are joined together by enzymes called V(D)J recombinase (Fugmann, Lee, Shockett, *et al.* 2000; Hesslein and Schatz 2001).

In 1990, David Schatz, Marjorie Oettinger, and David Baltimore discovered two recombination activating genes (RAG)—RAG1 and RAG2. RAG1, RAG2, and terminal deoxynucleotidyl transferase (TdT) are the only lymphoid gene products involved in the V(D)J rearrangement (Fugmann, Lee, Shockett, *et al.* 2000; Hesslein and Schatz 2001). Recombination occurs in the following steps.

- Recognition of RSS by recombinase enzyme RAG1/2 and synapsis.
- Single-stranded DNA cleavage at junctures and coding sequences by RAG1/2.
- The free 3′-OH group of cut DNA strand attacks the phosphodiester bond linking the opposite strand to the signal sequence and generating hairpin structure followed by double-stranded DNA break.
- Random cutting of the hairpin structure by endonuclease generates sites for the addition of P-region nucleotides, also trimming of a coding sequence by an ss-endonuclease.

AT A GLANCE

- Two identical light chains and two identical heavy chains of glycoproteins or immunoglobulin proteins joined together by disulphide bonds make a molecule of antibody. The amino-terminal ends of the heavy chains along with both the amino and carboxyl terminals of the light chains form the Fab region. The upper Fab region is variable and the lower region is constant; the carboxyl-terminal end of heavy chains forms the Fc region.
- Heavy chains are of five types constituting five isotypes—μ, γ, δ, α, or ε. One of these forms the constant regions of an antibody along with one of either κ or λ type of light chains.
- The type or class of antibody isotypes is determined by the types of heavy chains—μ for IgM; γ for IgG; δ for IgD; α for IgA; and ε for IgE constituting five antibodies.
- Each antibody class has different effector functions, serum concentrations, and half-lives.
- The variable domain of each heavy chain and light chain has three complementarity-determining regions (CDRs) specific to the antigen epitope against which it is raised.
- Antibodies could be secretory or membrane bound. Secretory antibodies from plasma cells remain in the soluble form. Membrane-bound antibodies associate with Igα/Igβ to form B-cell receptors (BCRs) on the B-cell surface.
- Antibodies have three effector functions: (1) opsonization, (2) complement activation, and (3) antibody-dependent, cell-mediated cytotoxicity (ADCC). Opsonization helps in phagocytosis by macrophages; complement activation helps in cell membrane perforation and lysis, and ADCC helps to kill target cells.
- Polyclonal antibodies are raised by many B-cell clones with heterogeneous multiple binding sites. Monoclonal antibodies are raised by single B-cell clones having homogeneous binding sites.

QUESTIONS

1. What is an antibody? Discuss the formation site of an antibody along with the composition of serum.
2. What is glycosylation of antibodies? Discuss its importance.
3. What do you mean by clonal selection? Discuss its importance for the immune system.

4. Draw a well-labelled diagram of an immunoglobulin. Discuss the enzymatic and chemical-based structures of an immunoglobulin.
5. On the basis of domains, how many types of immunoglobulin are there?
6. What do you mean by the variability of CDRs? Discuss the various constant region domains present in the Ig with their functions.
7. How many different classes of antibodies are present based on constant regions? Discuss their effector functions.
8. What are epitopes? Discuss their types and importance.
9. How many theories and models have been developed to prove the diversity of antibodies? Discuss any one of them.

OBJECTIVE QUESTIONS

A. Fill in the blanks.

a. Antibodies consist of _______ identical chains.

b. Two chains are linked by _______ bonds.

c. Each chain has _______ and _______ terminal end.

d. Variable region is present at _______ end.

e _______ isotypes of immunoglobulin are based on _______ region of _______ chains.

f. Each immunoglobulin molecule has a characteristic _______ protein structure called _______.

g. κ-chain is known as _______ chain while λ-chain is known as _______ chain and both chains are encoded by _______ separate multigene families located on _______ chromosomes.

h. V(D)J join by recombinase activating genes _______ and _______.

i. At the time of rearrangement, segments of genes flanked by _______ spacer join only to segments of genes flanked by _______ spacer to assure _______ and _______ joining.

j. Immunoglobulin gene rearrangement first occurs in _______ chain followed by _______ chain.

k. Gene rearrangement occurs in _______ manner.

l. After antigenic stimulation of B cells, there is, at first, expression of _______ antibody followed by _______ antibody.

m. Three regulatory sequences involved in immunoglobulin transcript generations are _______, _______, and _______.

n. In allelic exclusion, rearrangement of _______ parental chromosomal DNA occurs.

B. Match the following.

i. Secreted pentameric immunoglobulin	a. IgE
ii. Binds to mast cells surface receptor	b. IgA
iii. Immunoglobulin present on primitive B cells	c. IgM
iv. Most abundant immunoglobulin	d. IgE
v. Immunoglobulin present in saliva, tears, and breast milk	e. IgD
vi. Least abundant immunoglobulin	f. IgG
vii. CDR	g. Igα/Igβ
viii. Membrane-bound antibody	h. Variable region of immunoglobulin
ix. Secreted antibody	i. Myeloma cells
x. Monoclonal antibody	j. Plasma cells

CHAPTER FIVE 5

Antigen and Antibody Interaction

5.1 INTRODUCTION

The antigen–antibody interaction is mediated by many non-covalent bonds between the epitope (antigenic determinant) of antigen and the variable region (V_H/V_L) domain, also known as complementarity-determining regions (CDRs), of the antibody. In fact, CDRs are hyper-variable regions of the variable region domain. Antigen–antibody interactions are highly specific, and hence many types of immunoassays have been developed for diagnostic purposes to check the presence of antibody or antigen. An immunoassay is used to diagnose types of infections and diseases and to evaluate (qualitatively and quantitatively) molecules for biological and medical purposes.

5.2 TYPES OF BONDS IN ANTIGEN–ANTIBODY INTERACTION

Antigen–antibody interactions are due to non-covalent interactions or bond formations (Plate 1). Non-covalent bonds taking part in the interactions are as follows.

- Hydrogen bonds—between two electronegative atoms
- Ionic bonds—between opposite charges
- Hydrophobic bonds or interactions—water forces groups of hydrophobic atoms to come together
- Van der Waals interactions—between outer electron clouds of two or more atoms

Close proximity or complete fit is needed between antigen and antibody along with a large number of these weak bonds to give maximum strength to the interaction. Close proximity is required because these non-covalent bonds are effective only up to a very short distance (approximately 1×10^{-7} mm [1 Å]). It depends on the degree of complementarities between antigen and antibody. Complementarity is directly proportional to the strength, that is, maximum complementarity means high strength of interaction between antigen and antibody and least complementarity means weak interaction between antigen and antibody.

5.2.1 Interaction Strength (Affinity of Antibody)

The combined binding strength of non-covalent bonds between a single epitope of an antigen and a single binding site on an antibody is known as interaction strength. High-affinity antibodies will bind antigen epitopes strongly, while low-affinity antibodies will bind antigen epitopes weakly (Burton 1990; Conrad 1990). High-affinity antibodies will be in association with antigens for a longer period of time, while low-affinity antibodies will be in association with antigens for a shorter period of time and will dissociate immediately. Binding of an antibody with a monovalent antigen is described by Equation 5.1.

$$Ag + Ab \underset{k_{-1}}{\overset{k_1}{\leftrightarrow}} Ag - Ab \qquad \ldots(5.1)$$

where k_1 is the forward rate constant or association (in L/mol/s), k_{-1} is the reverse rate constant or dissociation (in 1/s), and the ratio k_1/k_{-1} is known as the association constant K_a (that is, $K_a = k_1/k_{-1}$).

K_a is equilibrium constant and a measure of affinity. It is denoted by Equation 5.2.

$$K_a = \frac{[Ag - Ab]}{[Ag]\ [Ab]} \qquad \ldots(5.2)$$

For small haptens, k_1 (forward rate constant) is very high, approximately 4×10^8 L/mol/s. This value is very close to a diffusion-limited reaction, which is 10^9 L/mol/s. For bigger size antigens, k_1 is very small, that is, 10^5 L/mol/s. The value of the antigen–antibody complex depends on the k_1 value, which varies in different antigen–antibody complexes. The affinity (K_a) of the anti-fluorescein antibody is 1000 fold higher than the affinity of the anti-DNP antibody. The k_1 for DNP is 1/5 of fluorescein, but k^{-1} is 200 times greater.

The dissociation of the antigen–antibody complex is denoted by Equation 5.3.

$$Ag - Ab \leftrightarrow Ab + Ag \qquad \ldots(5.3)$$

The dissociation constant K_d is given by Equation 5.4.

$$K_d = [Ab]\ [Ag]/[Ab - Ag] = 1/K_a \qquad \ldots(5.4)$$

K_d is a quantitative indicator of the stability of the antigen–antibody complex. The lower the value of K_d, the more stable the complex and vice versa. K_d is the reciprocal of the value of K_a.

5.2.1.1 Determination of affinity constant

The affinity constant K_a is determined by equilibrium dialysis. In equilibrium dialysis, a chamber is separated into two parts by a semipermeable membrane. The antibody is placed in one compartment and a small radiolabelled antigen, which is able to pass through the membrane, is placed in the other compartment. Small antigens could be haptens, oligopeptides or oligosaccharides. In the absence of antibodies, the antigens will be equally distributed in both sides of the compartment separated by the semipermeable membrane. In the presence of an antibody, some of the radiolabelled antigens will be bound to the antibody, and the rest of the free antigens will be distributed equally on both sides making the concentration of radiolabelled antigens more in the antibody-containing compartment. Thus antigens bound to the antibody make the two compartments different. If the affinity of the antibody is higher for the antigen, the formation of the antigen–antibody complex will be more (Figure 5.1).

The equilibrium concentration for the known concentration of antibody in a chamber is given by Equation 5.5.

$$K_a = \frac{[\text{Ab} - \text{Ag}]}{[\text{Ab}]\,[\text{Ag}]} = \frac{r}{c(n - r)} \qquad \ldots(5.5)$$

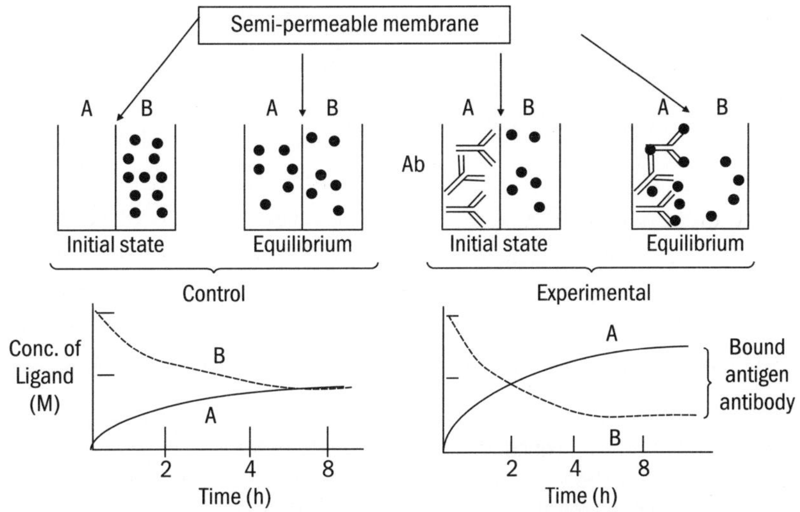

Figure 5.1 Equilibrium dialysis method for determining antibody affinity

Note Two dialysis chambers A and B are separated by semipermeable membrane. In chamber A, antibody is added and in chamber B, radiolabelled antigen is added. The concentration of antibody is measured in both chambers A and B after equilibrium. The concentrations of both antigens in each compartment are plotted, and the difference in the concentration of antigen denotes antigen bounded with the antibody.

where r is the concentration of the bound antigen with the antibody, c is the concentration of free antigen, n is the number of binding sites per antibody molecule. Thus the Scatchard equation (Equation 5.6) will be as follows.

$$\frac{r}{c} = K_{an} - K_{ar} \qquad \ldots(5.6)$$

The values of r and c are obtained by using the same antibody concentration but different antigen concentration. If K_a denotes antibodies with the same affinity, a Scatchard plot of r/c against r will be a straight line with $-K_a$ as the slope (Figure 5.2a). With an increase in the concentration of unbound antigens, c increases, r/c becomes 0, and r becomes n, the valency equal to the number of binding sites per molecule of antibody.

For a polyclonal antibody, K_a is a constant for heterogeneous antibodies and has a wide range of affinity. Therefore, the Scatchard plot of a heterogeneous antibody results in a curved line with constantly changing slope. Using this plot, the average affinity constant K_0 is determined by calculating the value of K_a when half the number of antigen–antibody complexes is generated (Figure 5.2b).

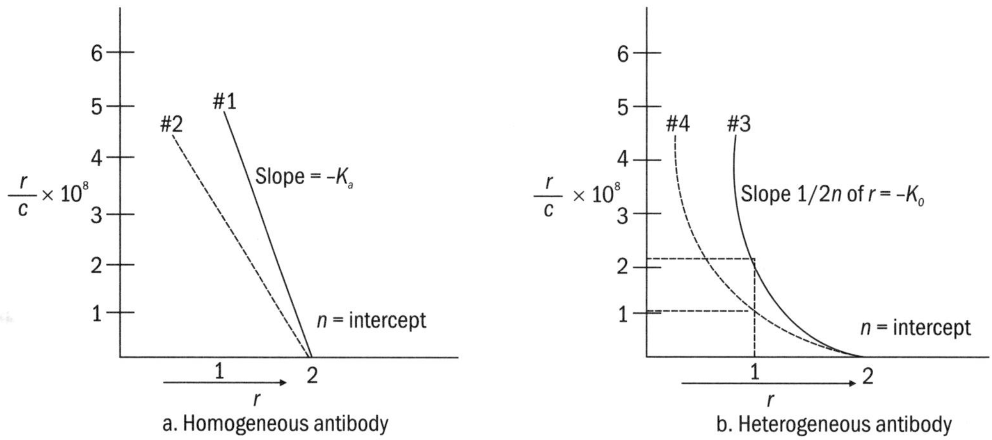

Figure 5.2 Scattered plots obtained by equilibrium dialysis with a constant antibody concentration and different concentration of ligand (antigen)

Note Here r is antigen per mole antibody, c is concentration of free ligand (antigen). The equilibrium constant (K_a) and the number of cites per antibody molecule (n) or valency could be found out. (a) Scatchard plot for the antibodies having same affinity will yield straight line with a slope of $-K_a$. The intercept (n) of antibodies 1 and 2 is 2 for IgG or other divalent immunoglobulins (n = 2). For IgA (dimeric) n = 4, and for IgM (pentameric) n = 10. The antibody 1 has higher affinity than antibody 2. (b) A curved line with constantly changing slope represents polyclonal antibody with a different range of affinities. K_a denotes half of the binding sites occupied, that is, r = 1, and this will give the value of average affinity constant (K_0). Here antibody 3 (K_0 = 2.4 × 108) has higher affinity than antibody 4 (K_0 = 1.25 × 108) for IgG (divalent).

5.2.2 Binding Capacity with Total Strength (Avidity of Antibody)

The strength of multiple interactions between a multivalent or polyclonal antibody and an antigen with many epitopes is called avidity. It is also the measurement of the binding capacity of the antibody with antigenic determinants residing on the virus or bacterial cell. The high avidity of an antibody can compensate for its low affinity. Avidity reflects the true strength of an antibody–antigen interaction because more than one site of the antibody and the antigen is involved in the interaction. IgM has lower affinity than IgG, but the avidity of IgM is more than that of IgG due to higher valency (Davies and Metzger 1984; Davies and Schulman 1989).

5.3 CROSS-REACTIVITY OF ANTIBODIES

If two different antigens have identical or similar epitopes, the antibody raised against one antigen will also bind with the second antigen. This property of antibodies is known as cross-reactivity.

Cross-reactivity is mostly observed in antigens such as polysaccharides (which contains similar oligosaccharide residues) and glycoproteins expressed on red blood cells (RBCs). ABO blood groups of RBCs are determined by the types of glycoproteins present on their walls. The terminal residues of the sugars of these proteins differentiate blood group antigens A and B of RBCs. The individual will have antiserum of the missing antigens of either A or B or AB. Antibodies are not due to the exposure of the antigens A or B of RBCs but due to the exposure of cross-reacting microbial antigens present on common intestinal bacteria. Thus the blood group antibodies are elicited by microbial antigens. Microbial antigens not matching with antigens present on RBCs will raise antibodies. The complementary antibody-recognizing self-antigens are weeded out during the developmental stage. A group-A individual has antiserum B, a group-B individual has antiserum A, a group-O individual (lacking both antigens A and B) has antiserum A and antiserum B, and a group-AB individual has no antiserum (Table 5.1).

The antibodies produced against the M-antigens of the cell wall protein of *Streptococcus pyogenes* show cross-reactivity with myocardial and skeletal muscle proteins. Similarly, the vaccine developed against *vaccinia* virus cross-reacts with the epitopes of *variola* virus (causes small pox).

Table 5.1 ABO blood groups

Blood type or group	Antigens on RBC	Serum antibody present
A	A	B
B	B	A
AB	A and B	None
O	None	Anti-A and Anti-B

5.3.1 Lattice Formation by Antibody

Antibodies interact with soluble antigen in aqueous solution to form a lattice, which is visible as a precipitate. The antibodies that aggregate soluble antigens are known as precipitins. A complex forms in a few minutes, but visible precipitation takes place in 24 h. Lattice formation depends on the valency of both antibody and antigen. In fact, both antigen and antibody require some essential features for lattice formation and visible precipitation after antigen–antibody complex formation.

Precipitation will not occur with monovalent Fab of antibody because for lattice formation and precipitation, the antibody and antigen must be bivalent or polyvalent. For example, myoglobin does not precipitate with monoclonal antibody (Figure 5.3b), but visible precipitation occurs with polyclonal antibody (Figure 5.3a). This is because myoglobin forms a cross-linked lattice structure with polyclonal antibody, but such a cross-linked lattice formation is not possible with monoclonal antibody (Burton 1990).

5.3.2 Precipitation Curve Formation in Fluids

Precipitation is a quantitative reaction. To analyse this reaction, a constant amount of antibody is taken in a number of tubes and an increasing amount of antigen is added to each tube. Centrifugation is done after precipitation, the pellet is recovered, and the amount of precipitate is measured. The amount of precipitate is plotted against an increasing amount of antigen concentration to get the precipitation curve (Plate 2). The maximum precipitation amount occurs at a point where the amount of antigens is equal to the constant amount of antibody; the area around this point is called the *equivalence zone*. Thus the equivalence zone is the region where the ratio of antigen to antibody is optimal. A large-size multi-molecular lattice is formed in the equivalence zone, which precipitates out of the solution. In case of excess antibody at the initial stage and excess antigen at the later stage, the lattice formation is not extensive and there is no complete precipitation (Plate 2). The sensitivity of precipitation reaction in fluids is 20–200 μg antibodies/ml, which depends on the valency of the antibody.

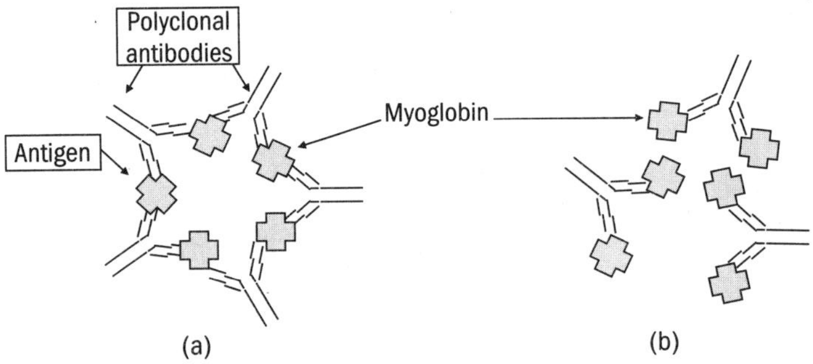

Figure 5.3 (a) Lattice formation by polyclonal antibodies to form visible precipitation line. (b) Lattice formation does not occur by monoclonal antibodies, and thus there is no visible precipitation line.

5.4 TECHNIQUES BASED ON ANTIGEN–ANTIBODY REACTIONS

Antigen–antibody interactions differ due to variation in the production of antibody against specific antigen because of gene rearrangement, organization, polymorphism, and their expression (Moiller 1993; Ravetch and Kinet 1991).

5.4.1 Immunodiffusion or Visible Precipitation Line in Gels

Immunodiffusion occurs due to diffusion of antibody and antigen towards each other, or when one is constant or fixed and the other is allowed to diffuse to form a precipitation line in the equivalence zone. Two types of immunodiffusion reactions are used to determine the relative concentration of antigen and antibody, either to compare two antigens or to determine the relative purity of an antigen preparation. One is radial immunodiffusion (Mancini method) and the second is double immunodiffusion (Ouchterlony method), both carried out in semi-solid agar medium.

In radial immunodiffusion (Mancini method), a constant concentration of antibody is mixed in a semi-solid agar medium, and the antigen is placed in a well and allowed to diffuse towards the antibody (Plate 3). A precipitation ring of precipitin is formed around the well. The precipitin formed is proportional to the concentration of antigen. The sensitivity of this reaction is 10–50 μg antibodies/ml. By measuring the precipitin areas of known concentrations of antigen, which give a standard curve, the concentration of unknown antigens can be obtained.

In the Ouchterlony double immunodiffusion method, both antigen and antibody are placed in two separate wells and allowed to diffuse towards each other. After reaching the equivalence zone, a visible line of precipitin appears (Figure 5.4). The sensitivity of this reaction is 20–200 μg antibodies/ml.

5.4.2 Immunoelectrophoresis

Immunoelectrophoresis is a technique in which antigens are first electrophoresed to separate different antigens on the basis of their charge. The antibody is then added in the trough formed in the direction of the electric field. Both antigens and antibodies are allowed to diffuse to form the precipitation line (Figure 5.5). The sensitivity limit of immunoelectrophoresis is 20–200 μg antibodies/ml. This technique is clinically used to detect the presence/absence or low/high production of specific proteins in the serum of an immunodeficient individual. It is highly qualitative and can detect only high concentrations of antibodies in quantitative abnormalities.

5.4.3 Rocket Electrophoresis

Rocket electrophoresis was first introduced by Laurell. It is also known as electro-immunodiffusion. In contrast to immunodiffusion, rocket electrophoresis is quantitative

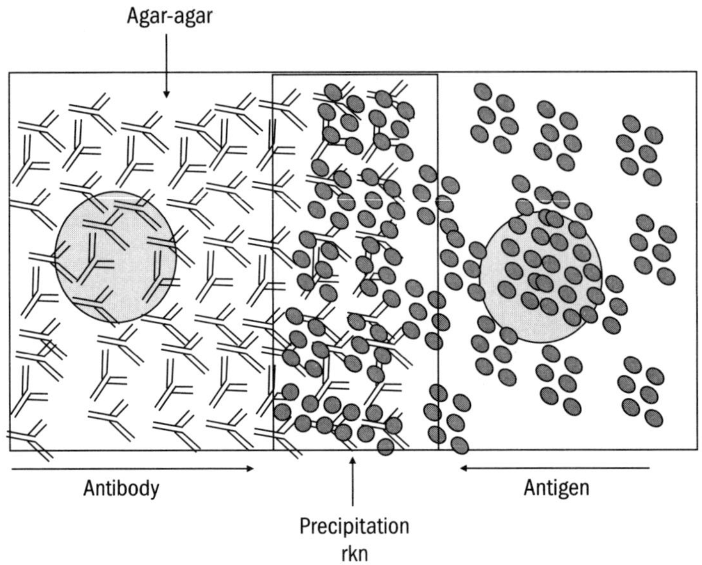

Figure 5.4 Double immunodiffusion

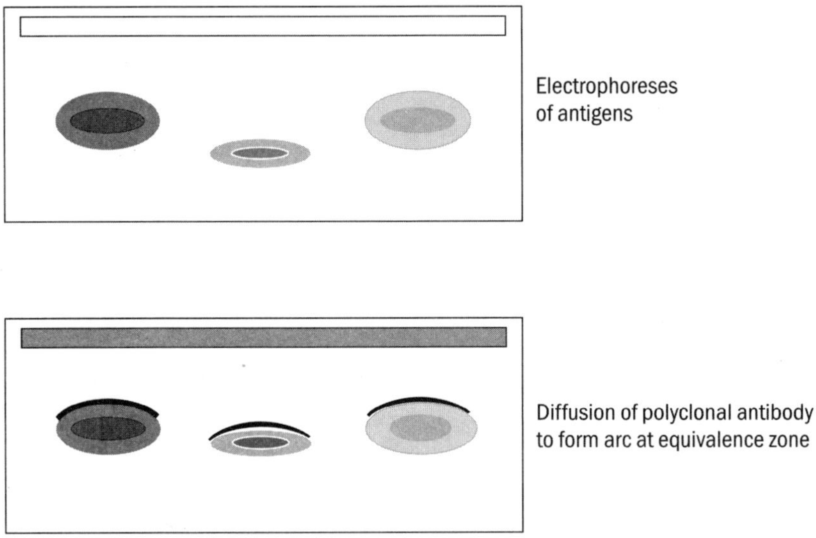

Figure 5.5 Immunoelectrophoresis of an antigen mixture

Note Protein is separated on the basis of charge and then detected by the antigen through diffusion of specific antibody, which forms an arch at the point of equivalence zone.

and measures the level of a specific protein in a protein mixture. Samples of unknown concentration and a series of diluted concentration of the protein are compared with a monospecific antibody. The samples are then loaded side by side in circular wells along the edge of an agarose gel. The agarose gel contains the monospecific antibody. Electrophoresis of the samples is done to enable interaction between antigen and antibody. As the proteins leave the well and enter the gel, the antigens bind to the antibodies. In the beginning, antigens are in excess compared to antibodies and no precipitation occurs. As the electrophoresis proceeds, more antibodies interact with antigens until equivalence is reached to form a line of precipitation. The precipitation line appears in the shape of a rocket. The precipitation amount is more at the head of the rocket, but fine precipitation lines are also formed up the side of the rocket by a small amount of antigen diffusing sideways as it passes through the gel. The diffused small amounts of antigens interact with a sufficient amount of antibodies to form a precipitation line. The greater the amount of antigen loaded in a well, the greater the distance the antigen will travel through the gel to interact with sufficient amount of antibodies to form precipitate. If the series of wells are loaded with increasing concentration of antigens, a series of rockets of increasing height will be produced with the area under the curve (rocket) being proportional to the amount of antigen in the well. The height of the rocket is also proportional to the area under the curve and hence antigen concentration (Figure 5.6). Rocket electrophoresis is, therefore, carried out by loading a series of samples of different antigen concentrations with one unknown antigen sample. The antigen sample is a complex mixture of proteins, for example, serum sample, tissue extract, and urine. As little as 1 μg/ml concentration of proteins can be measured, loading as little as 20 ng of protein in a well.

5.4.4 Agglutination Reactions

Agglutination is an interaction between a specific antigen and antibody resulting in visible clumping. The clump-forming antibodies are called agglutinins. The sensitivity of an agglutination analysis is 0.06–0.006 μg antibodies/ml. Like the precipitation reaction, excess of antibody also inhibits agglutination reactions. This inhibition is

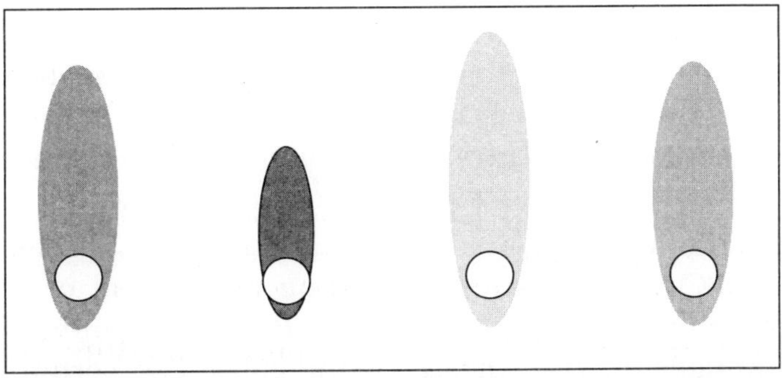

Figure 5.6 Rocket electrophoresis showing different concentration of proteins in each well

Note The containing well with high concentration has a longer rocket than the containing well with lower concentration protein. The shortest rocket has comparatively the lowest concentration of sample protein.

known as the *prozone effect*. The prozone effect is caused by high antibody concentration and the presence of incomplete antibodies. When antibody concentration is high, most antibodies, even though they are polyclonal or polyvalent, bind univalently instead of multivalently to the antigen, preventing cross-linking to each other and thus inhibiting the formation of lattices. In such a condition, no visible precipitation line will be present (Burton 1990).

Antibodies that bind to antigen without inducing agglutination are called *incomplete antibodies*. Incomplete antibodies are often of the IgG class, and in the case of high concentration, it occupies most antigenic sites, thus blocking access to these sites. IgM is known as a good agglutinin. Incomplete antibodies lack agglutinating activity due to their restricted flexibility in the hinge region, thus preventing them from cross-linking the epitopes of one or more antigens (Burton 1990; Davies and Schulman 1989).

5.4.5 Haemagglutination

Clumping of RBCs is called haemagglutination, and such a reaction is an example of agglutination. The substance that causes RBCs to clump is called haemagglutinin. Haemagglutinin could be viruses, toxins, or antibodies. Agglutination is a visible clump formation of antisera A or B with specific antigen A or B present on RBCs. Antisera A or B are mixed with RBCs. The visible clump or agglutinate is formed on the slide if specific antigens are present on the RBCs. For example, the antibodies that bind with blood group A are added. If agglutination occurs, the blood type is either A or AB. To distinguish between types A and AB, antibodies that interact with B group are then added to fresh blood. If agglutination is absent, the blood type is A. Agglutination reaction is a basis for blood transfusions and blood group typing.

Bacterial infection causes formation of bacteria-specific antibody. This formation is used to detect the infection of specific bacteria by agglutination reactions. The serum obtained from the infected individual is first serially diluted. The last dilution showing visible agglutination reflects the serum antibody titre of the patient and is called *agglutinin titre of the serum* of a specific patient. For example, if 1/540 dilution shows agglutination, but 1/1080 dilution does not, the agglutination titre of the diseased person is 1/540. The agglutinin titre is used to diagnose specific bacterial infection. For example, increase in agglutination titre of *Salmonella typhi* suggests typhoid fever in a patient. This is also an example of a *direct agglutination* reaction.

Passive haemagglutination or passive agglutination is used to detect soluble antigens. The antigen-coated RBCs are prepared by mixing tannic acid or chromium chloride treated RBCs with soluble antigens. Tannic acid or chromium chloride promotes adsorption of the antigen to the surface of the cells. The serum that contains the antibody is serially diluted in the wells of a microtitre plate. Antigen-coated RBCs are added to each well. The size of the characteristic spread pattern of RBCs on the bottom of the well suggests the pattern of haemagglutination reactions (Plate 4). Instead of RBCs, synthetic particles, such as latex and beads, can be coated with antigens for the agglutination reactions. Synthetic beads are more consistent, uniform, stable, and fast for the agglutination test. Its sensitivity is in nanograms. The sensitivity of direct agglutination is 0.3 μg antibodies/ml and that of passive agglutination is 0.006–0.06 μg antibodies/ml.

5.4.6 Agglutination Inhibition

Agglutination inhibition is useful to test unknown soluble antigens. The unknown antigen is mixed with a known agglutinin. The agglutinin cannot adhere to the cells or particles that carry the related antigen. The antigen is thus identified. Pregnancy is tested using agglutination inhibition techniques. Pregnancy test kits contain latex particles coated with human chorionic gonadotropin (HCG) hormone and the antibody to HCG. The addition of a pregnant woman's urine that contains HCG inhibits agglutination of the anti-HCG antibody-coated latex particle, which indicates pregnancy (Figure 5.7). The test's sensitivity is 0.006–0.06 μg antibodies/ml.

This technique is also used to identify individuals using cocaine or heroin drugs. The urine or blood sample of a suspected drug user is first incubated with antibody specific to the drug assumed to have been used. RBCs or latex coated with the drug are then added. The absence of agglutination by the antibody indicates that the sample contains an antigen already recognized by the antibody. This suggests that the individual is using illicit drugs. Some legal drugs have a chemical structure similar to the illicit drugs and hence can give false positive reaction (Moiller 1992).

5.4.7 Radioimmunoassay

The radioimmunoassay (RIA) technique was first developed by S. A. Berson and Rosalyn Yallow in 1960 to determine the levels of insulin–anti-insulin complex in diabetic patients. The sensitivity of this technique is 0.0006–0.006 μg antibodies/ml. The RIA technique is used to measure hormones, serum proteins, drugs, and vitamins. After the death of Berson, Yallow was awarded the Nobel Prize in 1977; this shows the importance of this technique.

The basic principle of RIA is the competitive binding of radiolabelled antigens and unlabelled antigen (test sample) to the antibody. First the labelled antigen is mixed with the antibody to saturate the antigen-binding sites of the antibody. Then the unlabelled antigen (of unknown concentration), which is the test sample, is added to the saturated antibody in larger amounts. The unlabelled antigen will replace the labelled antigen from the labelled antigen–antibody complex with increase in the amount of unlabelled antigens. The decrease in the amount of labelled antigen will determine the amount

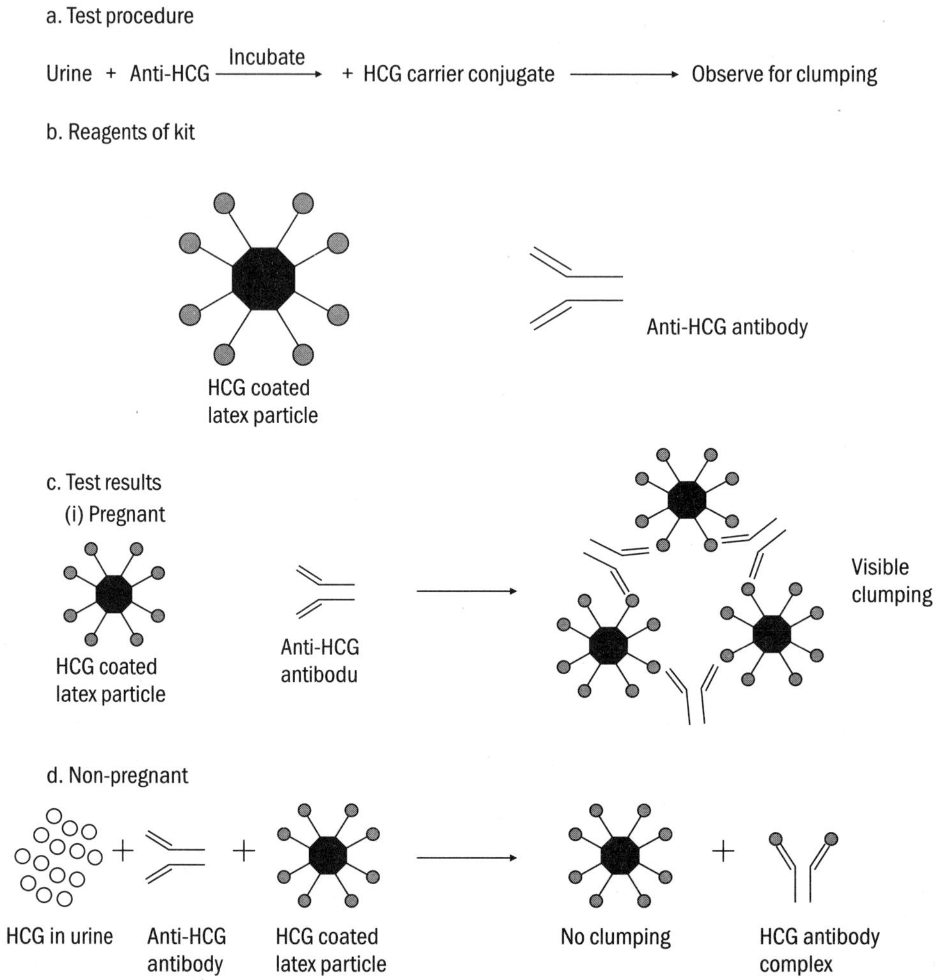

Figure 5.7 Use of agglutination inhibition to test pregnancy: (a) test procedure to check pregnancy; (b) reagents for kit used to test pregnancy; (c) visible clumping shows pregnancy; (d) no clumping after incubation of antibody with urine

of test antigen sample (Figure 5.8). Normally, antigen is labelled with ^{125}I, which emits γ-radiation. Tritium (^{3}H), which emits β-rays, can also be used in place of ^{125}I. The amount of labelled antigen in the antigen–antibody complex is precipitated, using either secondary anti-isotype antiserum protein A of *Staphylococcus aureus* or protein of formalin-killed *S. aureus*, to separate it from free-labelled antigen. The free-labelled antigen in the supernatant is measured in β- or γ-radiation counter. The amount of free-labelled antigen is subtracted from the total amount of labelled antigen added to give the amount of labelled antigen bound to the antibody.

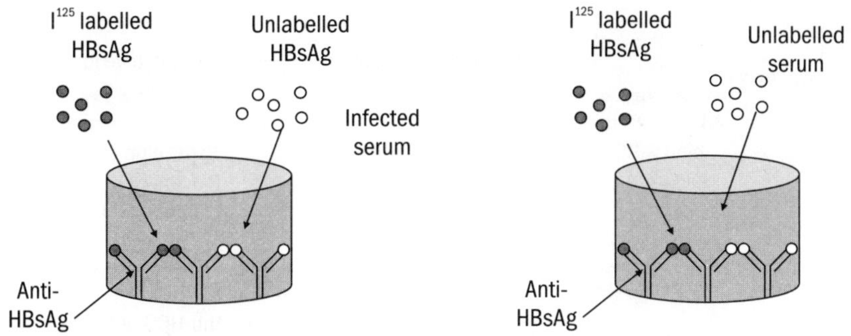

Figure 5.8 Radioimmunoassay procedure

Note Antibody (constant amount) is coated on microtitre wells to detect hepatitis B virus antigens HBsAg, unlabelled or labelled with ^{125}IHBsAg. Both labelled and unlabelled HBsAg are added, and radioactivity is measured. The infected samples will be having HBsAg and, hence, will compete with labelled HBsAg to read the amount of labelled HBsAg than the control.

Solid-phase RIAs can be used to separate the antigen–antibody complex from free antigen. The antibody is covalently linked to sepharose beads. Beads could be of polystyrene or polyvinylchloride. The antibody can also be immobilized on the walls of the well of the microtitre plate, which is used to determine the amount of antigen. This test is used to screen hepatitis B virus infection in blood.

5.4.8 Enzyme-linked immunosorbent assay

In Enzyme-linked immunosorbent assay (ELISA), enzyme-conjugated specific antibody is used in place of radioactively labelled antigen. A colourless *chromogenic substrate*, which produces colour when it comes in contact with the enzyme-conjugated antibody, is added to the sample. The enzymes used are alkaline phosphatase, horseradish peroxidase, and β-galactosidase. This technique is safer and cheaper than RIA, both qualitatively and quantitatively, to measure the presence of either antigen or antibody. Its sensitivity is less than 0.0001–0.01 μg antibodies/ml. It is of three kinds—indirect ELISA, sandwich ELISA, and competitive ELISA (Figure 5.9).

5.4.8.1 Indirect ELISA

Indirect ELISA is used to detect an unknown antibody quantitatively. The known antigen-coated microtitre well is incubated with a serum containing the unknown antibody (Ab1). The antibody present in the serum will attach to the antigen and free unattached antibody is removed by washing. Enzyme-conjugated secondary antibody (*E-SecAb2*) is added to bind with the primary antibody (Ab1). Free *E-SecAb2* is removed by washing. A chromogenic substrate is then added to develop colour, which can be measured by a spectrophotometric ELISA plate reader. This method is used to detect HIV infection in an individual whose serum will contain HIV-specific antibody.

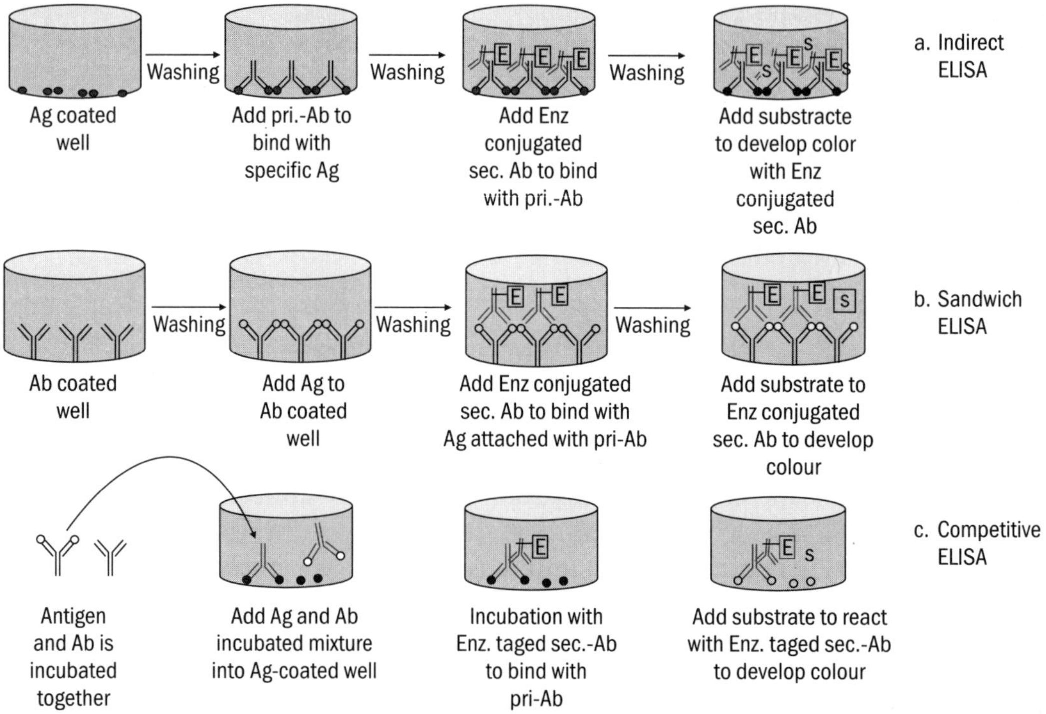

Figure 5.9 Types of enzyme-linked immunosorbent assay (ELISA): (a) indirect ELISA, (b) sandwich ELISA, and (c) competitive ELISA

5.4.8.2 Sandwich ELISA

In sandwich ELISA, a known primary antibody (Ab1) well coated microtitre plate is used to detect unknown antigens. The unknown antigen is added to the well and incubated. Unbound antigens are removed by washing. An enzyme-conjugated secondary antibody (*E-SecAb2*) specific to different epitopes of the same antigen is added and incubated to bind to the antigen. Free *E-SecAb2* is removed by washing. Chromogenic substrate is added to develop a coloured product, which can be measured in the ELISA plate reader.

5.4.8.3 Competitive ELISA

Competitive ELISA is used to measure unknown antigens. The unknown sample antigen is first incubated with a known antibody (Ab1) to form an Ag–Ab1 complex (Davies and Metzger 1984). The Ag–Ab1 mixture is incubated with the antigen-coated microtitre well for interaction. Free Ab1 will bind to the antigen. The amount of free Ab1 that will bind to the immobilized antigen will depend on the amount of antigen available in the unknown sample. An enzyme-conjugated secondary antibody

(*E-SecAb2*) specific to the primary antibody is added to the well to interact with Ab1. Free *E-SecAb2* is removed by washing. Substrate is then added to develop colour, which can be measured in the ELISA plate reader to know the amount of primary antibody bound with the antigen-coated well. The result shows that the higher the concentration of antigen in the original unknown sample, the less free the antibody and, hence, less absorbance will be showed.

5.4.8.4 ELISA using chemiluminescence

In this kind of ELISA, instead of a chromogenic substrate, a luxogenic (light-emitting) substrate is used. The emitted light is measured. In the presence of enhancing agents, the sensitivity is increased more than 200 folds. The advantage of this technique over ELISA using chromogenic substrates is that its detection limit is increased up to tenfold. Its sensitivity is less than 0.0001–0.01 μg antibodies/ml. The general reaction is as follows.

$$\text{Ab-HRP} + \text{Ag} \longrightarrow \text{Ab-HRP-Ag} \xrightarrow{\text{Luminінol} + H_2O_2} \text{Light}$$

5.4.9 ELISPOT Assay

The ELISPOT assay is an ELISA-type assay used for the quantitative determination of antibody-producing cells or specific antigen-producing cells using antigen-coated or antibody-coated plates, respectively. First the antigen-coated or antibody-coated plates are incubated with antibody-producing cells or specific antigen-producing cells, respectively. Next it is incubated with the enzyme-linked secondary antibody (*E-SecAb2*) specific to the secreted antibody or secreted antigen to bind with it. After washing off the unbound secondary antibody, chromogenic or chemiluminescence-producing substrate is added to locate the antibody- or antigen-producing cells at the coloured sites. Figure 5.10 shows the specific antigen (cytokines)-producing cells based on the colour development at particular sites and no colour development at non-producing cells site.

5.4.10 Western Blotting

Western blotting is a technique used to detect a specific protein in a mixture of proteins using a specific antibody. The name Western blotting is given to show its similarity to Southern blotting, which is used to detect DNA, and Northern blotting, which is used to detect mRNAs. First the protein mixture is electrophoretically separated using sodium dodecyl sulphate polyacrylamide gel electrophoresis (SDS-PAGE). Then the separated protein bands are transferred to a nylon membrane (nitrocellulose membrane) using the wet lot or semi-dry hot plate transfer method. After blocking the rest of the space on membrane with bovine serum albumin (BSA) protein or skimmed milk (5%), the

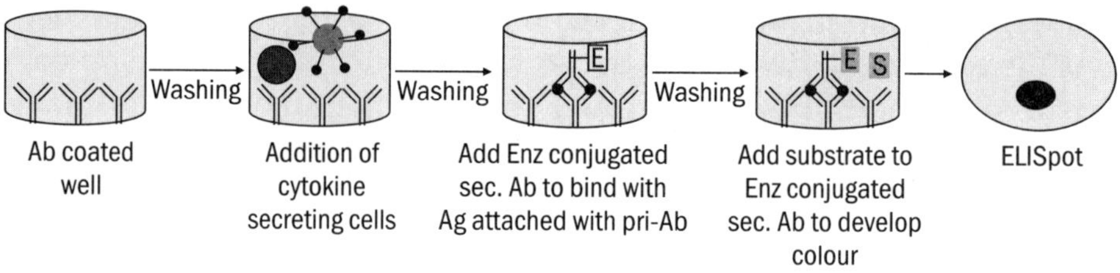

Figure 5.10 ELISPOT assay technique to detect specific cytokine-producing cells

Note Colour develops at the site of cytokine-producing cells, but no colour develops at the site of cells that do not produce cytokine. An antibody-coated plate is used in this assay method.

membrane is washed with a washing buffer and then flooded and incubated with a specific primary antibody, either polyclonal or monoclonal. After washing the membrane again with a washing buffer, it is flooded and incubated with *E-SecAb2* (enzyme-linked) or radiolabelled antibody. At the end, a chromogenic substrate is added to the *E-SecAb2*-used membrane to develop coloured bands or chemiluminescence over the part of the membrane containing the specific protein antigen. Radiolabelled antibody is exposed directly to determine the presence of the specific protein.

5.4.11 Immunoprecipitation

Selected monoclonal antibody is used to pull the specific antigen from a mixture of proteins and assayed by the Western blotting technique. The antibody attached to solid support/beads or secondary antibody is used to pull the specific protein from the mixture of proteins. Later the protein (antigen) and antibody complex is recovered as pellets by centrifugation (Moiller 1993). In some cases, the antibody is attached to magnetic beads. In such cases, magnetic beads attached to the secondary antibody are used to pull a specific antigen bound to the primary antibody. The immunoprecipitates are collected by placing a magnet against the wall of the tube (Figure 5.11).

Nowadays agarose beads-A, agarose beads-B or agarose beads-A+B are used for immunoprecipitation. Either of these beads is added to the prepared cell extract and incubated with the specific primary monoclonal/polyclonal antibody for 1–2 h in eppendorf tubes on a rotator. After centrifugation, agarose beads are collected as pellets and washed twice with buffer to remove unbound antigens. After two or three washes, the antibody-attached antigens are collected by centrifugation to recover purified proteins.

In growing cells, the protein synthesized is radiolabelled by placing the radioactively labelled amino acids such as leucine, cysteine, or methionine in the culture medium. These amino acids are resistant to metabolic modification and, hence, are normally used to radiolabel the protein during its synthesis. After the exponential growth phase, the cells are lysed using a lysis buffer and the specific primary antibody is

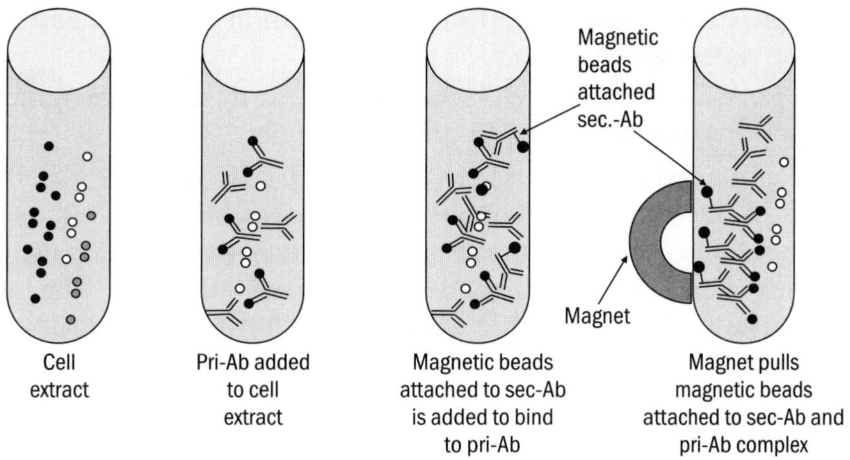

Figure 5.11 Immunoprecipitation

Note Cell extract is added with primary antibodies that bind to specific antigens. Magnetic beads or Sephadex beads or A or G beads or magnets are then used to collect the antigen–antibody complex, while antigens and antibodies that do not form a complex are discarded. The antigen can be dissociated for further study.

added for immunoprecipitation. The immunoprecipitate is washed to remove impurities and then counted in the scintillation counter to obtain the incorporated radioactive amino acid amounts for quantitative determination of the total amount of protein synthesized. The molecular weight of a specific protein synthesized is obtained by running an SDS-PAGE along with a molecular-weight marker and carrying out an autoradiography.

5.4.12 Immunofluorescence

Antibodies labelled with fluorescent dyes are used to detect specific biomolecules inside the cells. The labelling of antibodies with fluorescent dyes to detect an antigen-bound immunofluorescence using a fluorescence microscope is called immunofluorescence. This technique is used to measure and recognize the distribution of biomolecules in living cultured cells (Moiller 1993). Fluorescent dyes could be attached directly either to the primary antibody or to the secondary antibody used to recognize the primary antibody. Antibody is normally attached to the specific antigen or biomolecules (Plate 5). Its sensitivity is 1 μg antibodies/ml. The use of directly labelled primary antibody avoids cross-reactivity and high background of indirect immunofluorescence. This technique was used for the first time by Albert Coons in 1944. Fluorescent dyes or fluorochrome absorb light of one wavelength, called the excitation wavelength, and emit light having a different wavelength, called emission wavelength. Various types of fluorochrome are as follows.

- *Fluorescein (organic dye)* absorbs 490 nm wavelength of blue light and emits 517 nm wavelength of yellow-green light.
- *Rhodamine (organic dye)* absorbs 515 nm wavelength in the yellow-green range and emits a deep-red fluorescence of 546 nm wavelength.
- *Phycoerythrin* emits red fluorescence of wavelength 546 nm after absorbing light.

Immunostaining of the cell membrane with fluorescent antibody could be done directly by using fluorescent-labelled primary antibody or indirectly by using unlabelled primary antibody but fluorescent-labelled secondary antibody. The advantage of immunofluorescence is that it is highly qualitative, but it is also a disadvantage that it is not quantitative.

5.4.13 Flow Cytometry

Flow cytometry is used to count single intact cells in suspension by analysing and separating the cells stained with fluorescent antibody. Its sensitivity is 0.06–0.006 μg antibodies/ml. A flow cytometer uses a laser beam and a light detector to count the cells. Each time a cell passes the laser beam light, the light gets deflected and the interruptions of laser signal get recorded (Pascual and Capra 1991; Xing and Rao 2008). The cells that contain fluorescently tagged antibody bound to the antigens on their cell surface get excited by the laser and emit light, which is recorded by a second detector placed at right angles to the laser beam. On the basis of recorded data, instruments are able to sort cell populations into different containers according to their fluorescence profile (Plate 6). The determination of how many members of cell population bind fluorescently labelled antibodies is called *analysis,* and the placement of cells with different patterns of reactivity into different containers is called *cell sorting.*

Application of flow cytometer A flow cytometer (Ravetch and Kinet 1991) has the following multiple applications to know and determine the immunological data of patients.

- A flow cytometer is used to count the number of white blood cells (WBCs) and types of WBCs in the blood sample.
- Using T-cell-specific fluorescent antibody, the percentage of T cells in the total number of WBCs is determined and thus the T-cell fraction can be isolated from leukocytes.
- Same types of cells have the ability to express different types of antigens on the basis of developmental and physiological state. Using a flow cytometer, it is possible to determine the distribution of cell populations based on antigen densities.
- Based on light-scattering properties, it is possible to determine the size of cells.
- Using two different types of fluorescein-tagged and phycoerythrin-tagged antibodies specific to antigen of two different cell types (that is, T cells and B cells, respectively), it is possible to determine the percentages of B cells or T cells in a single analysis.

- The ratios and densities of T-cell subtypes, CD4, and CD8 antigens are determined using antigen-specific, labelled monoclonal antibodies. A decrease in the level of CD4 T cells is an indicator of infection in the patient by opportunistic pathogens. It is also an indicator of AIDS.
- A flow cytometer is used for typing of leukaemia, that is, acute lymphocytic leukaemia (ALL), acute myelogenous leukaemia (AML), chronic lymphocytic leukaemia (CLL), and chronic myelogenous leukaemia (CML).

5.4.14 Immunoelectron Microscopy

In this technique, an electron-dense label with either ferritin or colloidal gold is conjugated with the Fc part of a specific antibody (Moiller 1992; Ravetch and Kinet 1991) for direct staining of tissues components. It could also be conjugated to anti-immunoglobulin reagent for indirect staining. Since an electron-dense label has the ability to absorb electrons, an electron microscope can detect and visualize it as black dots in the tissues. Different sizes of gold/silicon or other metallic particles can be labelled with different antibodies to detect several antigens based on size (Xing and Rao 2008).

5.4.15 Nanotechnology and Quantum Dot Technology

Nanotechnology is currently in use for targeted drug delivery and diagnosis of many diseases, including cancer. Quantum dots are fluorophore semiconductor nanocrystals (Ravetch and Margulies 1994) with photophysical properties used in the detection of primary tumour of the ovary, breast, prostate and pancreas, regional lymph nodes, and metastasis. It has a diameter of 2–10 nm containing 200–10,000 atoms. Quantum dots have unique optical and electronic properties such as tunable light emission, improved signal brightness, resistance to photobleaching, and ability to simultaneously emit multiple fluorescence colours rather than synthetic organic dyes and fluorescent proteins. It is used for in vivo imaging, in vitro diagnosis, drug delivery, and multicolour imaging. Quantum dots attached to human epidermal growth factor receptor 2 (HER2) or prostate-specific membrane antigens (PSMAs) are used as tumour markers to detect ovarian cancer and prostate cancer, respectively. Quantum dot probes conjugated to PSMA monoclonal antibody are used under in vivo for multicolour fluorescence imaging of cancer cells. Single-chain antibody fragments (scFvs) conjugated with quantum dots have advantages such as solubility, activity, and ease of structure-dependent genetic engineering (Davies and Metzger 1984). Quantum dot-alpha-fetoprotein-antibody (QD-AFP-Ab) probes (binds to tumour cells) are used for targeting and imaging lung metastasis by using the confocal microscopic imaging technique. Fc monoclonal antibody attached quantum dots are able to attach to the specific target. The Fc antibody quantum dots are localized only in lungs (Plate 7b), while non-coated quantum dots get distributed all over the body (Plate 7a).

Archaeosome (Plate 8), a novel generation of liposome, which is a natural archaeal membrane lipid and/or synthetic lipid analogue, is used as a nanoparticle to deliver drugs and vaccines. Due to its ability to exhibit high stabilities at low or high temperatures, acidic and alkaline pH, oxidative conditions, high pressure, the action of phospholipases, and bile salts, nanotechnological applications are beneficial for gene delivery and drug delivery. Nanoparticles are also able to carry antigens and/or adjuvants leading to antigen-specific immune response (Ravetch and Margulies 1994).

AT A GLANCE

- There are four types of bond interactions present between antigens and antibodies: (1) hydrogen, (2) ionic, (3) hydrophobic, and (4) van der Waals.
- Scatchard analysis, used to determine affinity constant, gives the quantitative measurement of the interaction between the antigen's single epitope and the antibody's single binding site.
- Avidity is the sum of the strengths of all interactions between multivalent antibody molecules and multivalent antigen molecules.
- Multivalent antigen–antibody interactions result in precipitate when antigens bind in soluble form (liquid or gel) with multivalent antibodies. Precipitation visible in gel is known as immunoelectrophoresis due to the presence of immunoglobulin in gel electrophoresis.
- Visible clumping is produced by agglutinating antibody (agglutinin) and specific antigen in reaction called agglutination reaction. Agglutination is a simple, quick, and very sensitive immunoassay method to detect antigens.
- Radioimmunoassay (RIA) is a radioactively labelled antigen or antibody-based quantitative detection method of either antigen or antibody. It is a highly sensitive technique.
- Enzyme-linked immunosorbent assay (ELISA) is an enzyme-substrate reaction with coloured product (chromogenic reaction) or chemiluminescence-based detection using ELISA plate reader at a specific wavelength.
- Western blotting is a method in which antibody-specific proteins are detected after separation by gel electrophoresis and electrophoretic transfer on nitrocellulose membrane.
- Fluorescence microscopy is done with fluorescent-labelled antibody to detect or visualize a specific antigen on the surface of cells or within cells.
- Flow cytometry is a quantitative analysis or sorting of cells labelled with one or more fluorescent antibodies.

QUESTIONS

1. What are antigen–antibody interactions? Discuss various types of bonds present in these interactions.
2. What is affinity and avidity of antibodies? How do the two differ from each other?
3. Discuss the method to determine the affinity constant.

4. What do you understand by cross-reactivity? What is the mechanism in the formation of a precipitation curve? Discuss with a suitable example.
5. What is immunodiffusion? Discuss the Mancini method or Ouchterlony double immunodiffusion technique with a suitable diagram.
6. Write short notes on the following.
   - Immunoelectrophoresis
   - Rocket electrophoresis
   - Agglutination reaction
   - Radioimmunoassay (RIA)
   - Enzyme-linked immunosorbent assay (ELISA)
   - ELISPOT

OBJECTIVE QUESTIONS

A. Fill in the blanks.

a. Interaction of antigen and antibody depends on __________ interactions.
b. The interactions between antigen–antibody are __________, __________, and __________.
c. __________ reflects the overall strength between multivalent antibody and multivalent antigen.
d. __________ reflects interaction strength between epitope and paratope.
e. Immunoelectrophoresis is based on __________ reaction between antigens and antibodies in the gel medium.
f. Agglutination occurs due to interaction between __________ and __________.
g. Radioactively labelled antigen or antibody is used in __________ procedure.
h. Chemiluminescence reactions are __________ sensitive than chromogenic reaction.

B. Match the following.

i. Affinity constant	a. Antigen
ii. Epitope	b. Scatchard analysis
iii. Paratope	c. Agglutination
iv. Visible clumping	d. Generates colored product
v. RIA	e. Detect antigen on nitrocellulose membrane

vi. ELISA	f. Antibody
vii. Western blotting	g. Radioactive antibody
viii. Fluorescence microscopy	h. Sorting of cell population
ix. ELISPOT assay	i. Detect antigen inside the cells
x. Flow cytometry	j. Cell secreted antigens detection

CHAPTER SIX 6

Major Histocompatibility Complex

6.1 INTRODUCTION

Major histocompatibility complex (MHC) is a *cluster of genes* in animal cells (mammalian) whose products play an important role in the differentiation of self and non-self components and intercellular recognition. It also has a significant role in both cell-mediated and humoral immunity. T cells recognize only antigen–MHC molecule complexes, while antibodies have the ability to interact with antigens. Each individual-specific MHC molecule presents each antigen separately, to which specific T-helper cells (Th cells) and T-cytotoxic cells (Tc cells) might respond.

6.2 DISCOVERY OF MHC

The concept of MHC came from the work of Peter Gorer in 1930. He identified four groups of blood cell antigens designated as I to IV in mice. In the 1940s and 1950s, both Gorer and George Snell proposed that a group of genes, called II, takes part in transplanted tumours and tissues rejection. The tissue- or tumour-rejecting molecules reside on the cell surface known as *histocompatibility antigens.* Snell named the responsible genes as *histocompatibility genes* designated as histocompatibility-2 (H-2) genes to denote Gorer's group II blood antigens. Snell was awarded the Nobel Prize in 1980 for his work.

The MHC is a collection of genes in the continuous stretch of DNA on human chromosome-6 and mice chromosome-17. It is referred to as the human leukocyte antigen (HLA) complex in humans and the H-2 complex in mice (Bjorkman, Saper, Samraoui, *et al.* 1987a).

6.3 CLASSES OF MHC MOLECULES

Major histocompatibility complex genes encode three classes of molecules known as class I, class II, and class III (Bentley and Mariuzza 1996; Bjorkman and Parham 1990; Brown, Jardetzky, Gorga, *et al.* 1993; Powis and Trowsdale 1994; Stern, Brown, Jardetzky, *et al.* 1994; Stern and Wiley 1994; Zeng, Castaño, Segelke, *et al.* 1997).

1. *Class I MHC genes* encode glycoproteins expressed on the nucleated cell's surface. It presents peptide antigens to Tc cells.
   a. *Classical class I molecules* In mice, classical class I molecules constitute the K and D regions, while in humans they are encoded by A, B, and C loci. These molecules are expressed in the widest range of cell types.
   b. *Non-classical class I genes* Non-classical genes are additional genes within H-2 and HLA complexes to encode class I molecules. They are expressed only in specific cell types. The expression of class I HLA-G molecules protects the foetus from being recognized as foreign and rejected by the mother's Tc cells (Carosella, Dausset, and Kirszenbaum 1996).
2. *Class II MHC genes* encode glycoproteins expressed on antigen-presenting cells (APCs) (macrophages, dendritic cells, and B cells). They present processed antigenic peptides to Th cells.
   a. IA and IE regions encode two chains of class II MHC molecules in mice.
   b. DP, DQ, and DR regions encode two chains of class II MHC molecules in humans.

   Both class I and class II molecules are involved in antigen processing followed by presentation and have common structural features.
3. *Class III MHC genes* encode secreted proteins having immune functions (for example, complement components) and inflammatory molecules (for example, tumour necrosis factor [TNF] and heat shock proteins). The class III region is flanked by class I and class II regions. The molecules produced are completely different from the class I and class II molecules.

6.4 TYPES OF MICE STRAIN BASED ON MHC

6.4.1 Haplotype Strain

The loci of MHC are polymorphic in haplotype strain. Polymorphic means many alternative forms of genes or alleles reside at each locus. The MHC genes reside very close to each other and, hence, during inheritance, two sets of alleles (one from each parent) pass to the offspring. Each set of alleles is called a haplotype. The crossover frequency during mitosis for such genes is only 0.5% per 200 mitotic cycles. One haplotype is inherited from the mother and one from the father. When we say that alleles are *codominantly* expressed, it means that both parental alleles express in the same cells. In outbred populations, offspring are heterozygous, while inbred offspring are homozygous, that is, they have identical maternal and paternal haplotypes in each H-2 locus.

6.4.2 Prototype Strain

In prototype mice strains, the entire set of H-2 alleles is inherited without listing each allele individually. The same set of alleles, that is, similar MHC haplotypes, may be present in the same strain. Prototype strains are designated by arbitrary italic superscript, for example, H-2^a, H-2^b. The F1-generation of a cross of H-2^b and H-2^K strains will be H-$2^{b/k}$ (Figure 6.1). The F1-strain expresses MHC proteins of both strains on the cell membrane and, hence, will be histocompatible to accept grafts from either strain (Figures 6.2 and 6.3). But each parental strain will not be able to accept the grafts of F1-strain because half of the MHC molecule is foreign to the parent strain (Figures 6.2 and 6.3).

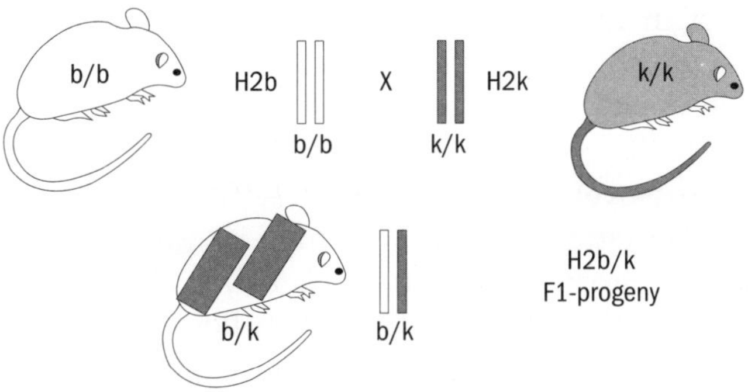

Figure 6.1 F1 progeny of two different MHC haplotypes mouse strains contain heterozygous chromosome

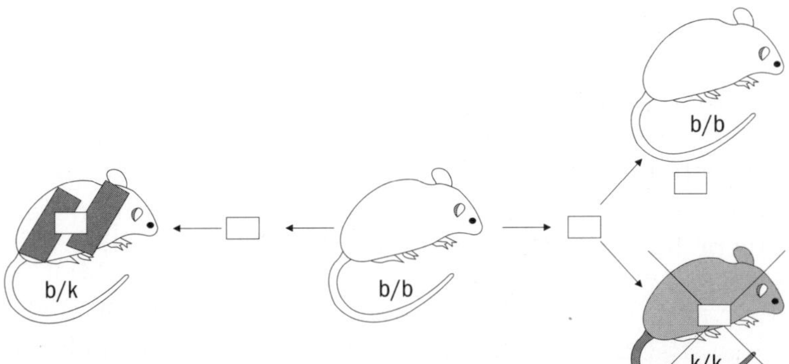

Figure 6.2 Graft from b/b acceptable in b/k and b/b while rejected in k/k

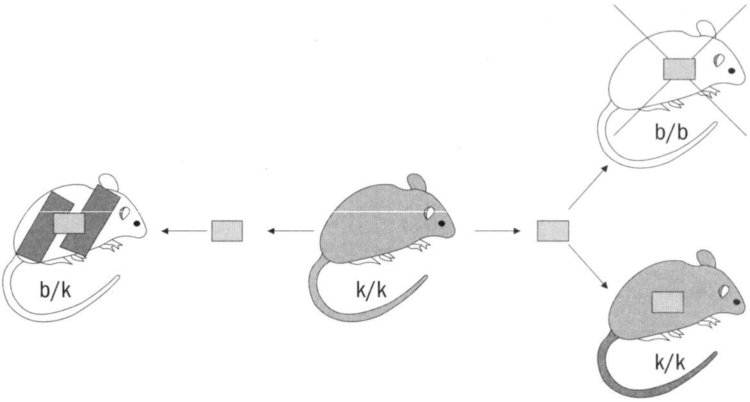

Figure 6.3 Graft from k/k acceptable in b/k and k/k while rejected in b/b

6.5 INHERITANCE OF HLA HAPLOTYPES IN HUMANS

The MHC complex, which is called HLA complex in humans, is highly polymorphic, and each locus contains genes from both class I MHC and class II MHC molecules. Similar to mice, the MHC loci in human are linked and inherited together as a haplotype. One in four siblings will inherit the same paternal and maternal haplotypes to become histocompatible, while none of the offspring will be histocompatible to either mother or father (Plate 1).

Recombination by crossover (rate is very low) is present within the HLA complex to generate new allelic combinations. In this way, high numbers of intervening generations are produced (Bjorkman, Saper, Samraoui, *et al.* 1987a; Garrett, Saper, Bjorkman, *et al.* 1989; Madden, Gorga, Strominger, *et al.* 1991; Salter, Benjamin, Wesley, *et al.* 1990). Thus two unrelated individuals rarely have sets of identical HLA genes.

6.5.1 Congenic Mice Strains

Two strains are congenic if they are genetically identical except at one genetic locus or region. This non-identical region is responsible for the phenotypic differences between the two strains. For example, if the H-2 complex of a homozygous strain-B is introduced into the background genes of a homozygous strain-A, a congenic strain-A.B. will be produced. The first letter of the congenic strain denotes the strain that provides the genetic background, and the second letter denotes the genetically different MHC region of that strain. Except for the MHC locus of strain-A.B., strain-A will be genetically identical to strain-A.B.

6.5.2 Recombinant Congenic Strain

Sometimes, due to crossover within the H-2 complex, the resultant recombinant strain differs from the parental strains of the congenic strain. Such recombinant strains are used to analyse the MHC by comparing the functional differences between the two strains.

6.6 STRUCTURE OF CLASS I MOLECULES AND CLASS II MOLECULES

6.6.1 Structure of Class I Molecules

A class I molecule (Plate 1) has a glycoprotein heavy chain and a small protein light chain. It is composed of a 45 kDa α-chain and a 12 kDa β2-microglobulin molecule. The β2-microglobulin molecule is associated with the α-chain by a non-covalent bond. The α-chain is a glycoprotein anchored in the membrane by a hydrophobic head segment and a cytoplasmic hydrophobic tail. It is encoded by the polymorphic genes residing in the A, B, and C regions of the HLA complex. The β2-microglobulin is encoded by highly conserved genes located on a different chromosome. Both chains are essential for the expression of class I MHC molecules on the membrane of respective cells.

The external segment of the α-chain has three external domains—α1, α2, and α3. As mentioned before, it is non-covalently attached with the β2-microglobulin. A membrane distal pair is made up of α1- and α2-domains, while the membrane proximal part is made up of α3-domain and β2-microglobulin. Each domain is composed of 90 amino acids. The external domains are followed by a transmembrane part consisting of 25 hydrophobic amino acids. This transmembrane part is followed by a short hydrophilic amino-acid stretch and a 30-amino-acid cytoplasmic tail. The size and organization of the β2-microglobulin are similar to the α3-domain. The β2-microglobulin does not contain a transmembrane segment and is only non-covalently attached to class I glycoprotein. The sequence of the α3-domain is also similar to the sequence of β2-microglobulin and the constant domains of immunoglobulin. The extracellular domain is separated by papain at 13 residues proximal to the transmembrane.

A platform of eight antiparallel β-strands is formed in class I MHC by spanning α1- and α2-domains with the help of the two long α-helical regions. The long α-helices form a 25 Å × 10 Å × 11 Å deep groove or cleft, and the β-sheet made up of β-strands forms the bottom. The cleft is large enough to bind an 8–10 amino-acid-long peptide. A small peptide that might be processed antigen and self-peptides was found in the cleft of class I MHC molecules using X-ray crystallography.

The β2-microglobulin and α3-domain are organized to form two β-pleated sheets. Each β-pleated sheet is made up of antiparallel β-strands of amino acids. The immunoglobulin fold formed in such a way is characteristic of domains of immunoglobulin. Due to their structural similarity with the constant regions of immunoglobulin, class I MHC molecules and β2-microglobulin are placed in the immunoglobulin superfamily list. Compared to other domains, the α3-domain is highly conserved and it interacts with Tc cells membrane-bound CD8 molecules.

The interaction of β2-microglobulin with α-chains is essential for the class I molecule to give folded conformation. The dimmer formed due to the initial interaction of β2-microglobulin with α-chains is later stabilized by an appropriate peptide to form a native trimeric class I structure of α-chain, β2-microglobulin, and peptide. The trimeric complex is then transported to the cell membrane surface. The expression of class I molecule is impossible without β2-microglobulin, such as *Daudi* tumour cells.

6.6.2 Structure of Class II Molecules

A class II molecule (Plate 1) has two non-identical glycoprotein chains—α-chain (33 kDa) and β-chain (28 kDa)—associated by non-covalent interactions. It contains three parts—external domain, transmembrane part, and cytoplasmic anchor part. There are two external domains: (1) α1- and α2-domains containing chain and (2) β1- and β2-domains containing chain. The α2- and β2-domains are proximal portions from the membrane and bear similar sequence. They are included in the *immunoglobulin superfamily*. The α1- and β1-domains are distal portions from the membrane and have a cleft-like structure to bind to processed antigens.

The class I and class II molecules are superimposed structures as revealed by X-ray crystallography. The peptide-binding cleft is composed of eight antiparallel β-strands and antiparallel α-helices in the side. Conserved residues that bind to short peptide terminal residues are absent in class II molecules but are present in class I molecules. Class II molecules form open-ended grooves, while class I molecules form sockets or open pockets. The differences between the two molecules are given in Table 6.1.

Table 6.1 Difference between class I and class II molecules

Property	*Class I molecules*	*Class II molecules*
Domain involved in peptide binding	α1/α2	α1/β1
Nature of cleft	Closed at both ends of chain	Open at both ends of chain
Peptide size fit in cleft	8–10 amino acids	13–18 amino acids
Peptide motifs interacting with MHC molecules	Conserved anchor residues at both ends of peptide, but carboxyl terminal is preferred	Lack conserved anchor residues, and binding site is throughout peptide
Nature of peptide	In extended peptide, both ends interact with cleft, but the middle part arches up away from contact with the MHC molecule	An extended structure of peptide held at constant elevation above the floor of the MHC cleft

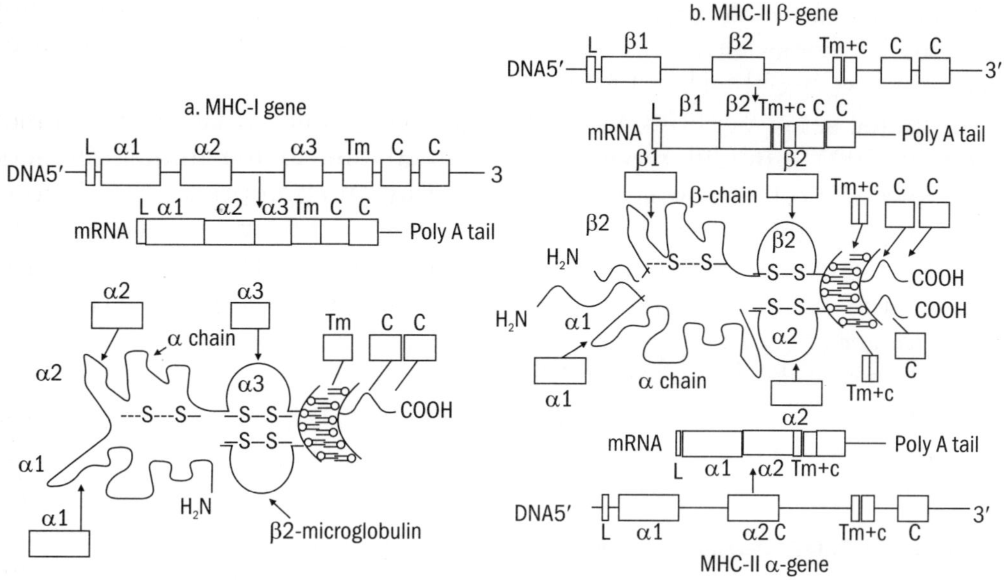

Figure 6.4 Organization of exons and introns in (a) class I and (b) class II MHC gene molecules showing mRNA transcripts, protein molecules

Note *L* is a leader exon; *Tm+c* encodes both transmembrane and cytoplasmic protein; *Tm* encodes transmembrane protein; and *C* encodes cytoplasmic protein.

The class II molecule for human DR1 is a *dimer of dimers,* that is, a dimer of αβ-heterodimer. It has two sites: (1) a peptide-binding cleft in which two dimers face each other in opposite directions and (2) a CD4-binding site on the opposite side of the class II molecule to stabilize class II dimers.

6.6.3 Exon and Intron Arrangement

In both class I and class II molecules, there are separate exons to encode a separate domain of the MHC molecule. In class I molecules, the 5′-leader exon that encodes a short signal peptide is followed by five to six exons encoding the α-chain (α1, α2, and α3) and the transmembrane (Tm) region, and two exons for the cytoplasmic domain (C) at the 3′-terminal end. The signal peptide facilitates the insertion of α-chain into the endoplasmic reticulum (ER). After completion of translation, it is removed by proteolytic activity. In class II molecules, both α- and β-chains are encoded by a leader exon, an α1- or β1-exon, an α2- or β2-exon, a transmembrane exon, and one or more cytoplasmic exons (Figure 6.4).

6.6.4 Class I MHC Peptide Interaction

Class I MHC molecules associate with peptides to present to $CD8^+$ T cells. Peptides are derived from endogenous intracellular proteins processed in the cytosol. The processed peptides are transported to the cisternae of the ER to interact with class I

MHC molecules. The whole process is known as *cytosolic processing*. A single nucleated cell expresses 10^5 copies of each class I molecule; different specific peptides will be expressed on the surface by class I MHC molecules. Each type of class I MHC molecule binds to a unique set of peptides. Since there are 10^5 copies of class I MHC molecules, approximately 2000 different types of peptides are presented with a frequency of 100–4000 copies per cell. Approximately, 100 peptide–MHC complexes are required to target a cell for recognition by Tc-lymphocyte. The class I MHC binding peptide must be nine-amino-acids long and contain specific amino-acid residues (Carosella, Dausset, and Kirszenbaum 1996). The binding affinity of non-americ peptides is 100–1000 fold higher than longer or shorter peptides. The amino-acid residues that anchor the peptide into the MHC molecule groove are called *anchor residues*. The binding sites lining the amino acid residues vary in different class I variants to specify the anchor residues. All binding peptides contain a carboxyl terminal anchor and have hydrophobic residues such as leucine and isoleucine (Bjorkman, Saper, Samraoui, *et al.* 1987b).

6.6.5 Class II MHC Peptide Interaction

The peptides associated and presented with class II MHC molecules bind to $CD4^+$ T cells. They are derived from either membrane-bound proteins or vesicle peptides formed due to the endocytic processing pathway. Membrane-bound proteins are internalized by receptor-mediated endocytosis and get processed. Class I MHC molecules processed peptides are also bound by class II MHC molecules. The cleft of class II molecules is open at both ends to allow the binding of longer peptides (13–18 amino-acid residues). Bound peptide maintains constant elevation on the cleft floor. They have conserved *motifs* but lack conserved anchor residues (Brenner, McLean, Dialynas, *et al.* 1986; Brown, Jardetzky, Gorga, *et al.* 1993).

6.6.6 Diversity of Class I and Class II Molecules

Major histocompatibility complex molecules are available in a wide diversity, which influences the diversity of antibodies and T-cell receptors (TCRs) (Garcia, Teyton, and Wilson 1999). Diversity in antibodies and TCRs is generated by gene rearrangement and somatic mutation of rearranged genes (Lefranc and Rabbitts 1989; Leiden 1993). Thus the generation of B cells and TCRs changes with time in an individual, but the generation of MHC molecules is fixed and does not change with time. The diversity of MHC molecules is determined not only by the alleles of each gene but also by duplicated genes with overlapping functions. Thus MHC molecules are governed by a polygenic system. In humans, HLA class I genes have 240 A alleles, 470 B alleles, and 110 C alleles. In HLA class II, the DRA chain is highly conserved with two different alleles, while DRB has 350 alleles. Hence, there could be $240 \times 470 \times 110 = 12 \times 10^6$ (12 million) different types of class I haplotypes in a population. In class II, there are five DRB genes, B1 to B5, having 304, 1, 35, 11, and 15 alleles, respectively. DQA1 and B1 have 22 and 49 alleles, respectively, and DPB1 has 96 alleles. Thus the total number of different haplotypes in class II MHC is 1.8×10^{11} (= $304 \times 35 \times 11 \times 15 \times 22 \times 49$) due to different combinations. Each haplotype has both class I and class II genes; thus the total combinations of these class I and class II alleles are 2.25×10^{18} (Garcia, Degano, Stanfield, *et al.* 1996).

6.6.7 Non-peptide Presentation

Both class I and class II molecules are able to present only peptide antigens. However, there are molecules structurally similar to class I MHC molecules that present non-peptide antigens, that is, glycolipids or bacterial lipids to T cells (Burdin and Kronenberg 1999; Porcelli, Segelke, Sugita, *et al.* 1998). Such molecules are known as CD1 molecules. CD1 molecules present glycolipids or lipids derived from bacteria. CD1 molecules are not encoded within MHC molecules; rather, they are located on chromosome-1 (Bentley and Mariuzza 1996).

6.7 LINKAGE DISEQUILIBRIUM

The actual diversity of random combinations of alleles is less than the theoretical diversity in HLA haplotypes; this is known as linkage disequilibrium. It is the difference between the frequencies observed and the frequencies expected for a particular combination of alleles. The expected frequency is calculated by multiplying the frequencies of the two alleles.

For example, HLA-A1 occurs in 16% of individuals in a particular population (frequency = 0.16) and HLA-B8 in 9% of the same group (frequency = 0.09). The expected frequency of both alleles will be 1.4% ($0.16 \times 0.09 = 0.014$). Yet, data suggests that HLA-A1 and HLA-B8 are present in 8.9% individuals. The difference between the two is the linkage disequilibrium between the two alleles of class I MHC molecules genes.

Uses of linkage equilibrium Certain allelic combinations might be resistant to certain diseases enabling them to be selected and over-represented.

Harmful effects of linkage equilibrium The allele might be susceptible to autoimmune diseases and can go for negative selection. Crossover might be present in certain DNA sequence sites, and the presence and absence of this site could be responsible for the crossover (*hotspots*) between alleles.

6.8 RELEVANCE OF MHC POLYMORPHISM

Major histocompatibility complex polymorphism is due to the sequence variance among MHC molecules. The sequence variance is clustered in short stretches in the α1- and

α2-domains of the class I MHC molecule distal to the membrane region of the MHC, and α1- and β2-domains of class II MHC molecules.

In humans, many proteins are present in more than one genetic variant. The differences are found between those proteins that are designated as *polymorphisms*. Classical MHC molecules are present in 500 and more different variants. Polymorphism reflects selection to disease under pressure from highly mutable pathogens. Thus diversity in the binding peptide is due to the diversity and mutability of the pathogens.

Major histocompatibility complex polymorphism is driven by the need to maximize peptide-binding diversity. This view is supported by two evidences. First, polymorphisms are concentrated in the amino acids forming the peptide-binding groove. Second, each variant of MHC molecule binds with a wide range of different peptides. Polymorphism in MHC is promoted by recombination, which generates new variants.

Why is it difficult to find tissue matches for transplant patients? This is due to MHC polymorphism. The chain of class I MHC molecule is encoded by three genes, that is, class I MHC α-chain is *polygenic* and each individual inherits six α-chains—three from the mother and three from the father (Figure 6.4a). Each α-chain gene has 90–500 different variants or alleles.

The chances of an individual inheriting identical copies of any of the three MHC class I genes from both father and mother are very low, and cells of most individuals express six different class I MHC molecules (Figure 6.4b). Thus there is very less chance that two individuals will have the same six MHC molecules, and hence has problem in tissue matching.

Major histocompatibility complex molecule polymorphism also has a role in disease resistance because two individuals will always be heterozygous for each MHC molecule. The most important advantage of *heterozygosity* is resistance to HIV or AIDS.

Class II MHC molecule polymorphism is more complex than class I MHC molecule polymorphism but both have the same effect. There are three DP, DQ, and DR class II MHC molecules in humans. The β-chain is much more polymorphic than the α-chains. The reason is not known. DR is more complex. The DR α-chain is not polymorphic but binds to the DR β-chain. The DR β-chain is encoded by DRβ1, DRβ3, DRβ4, and DRβ5. Chromosome-6 contains DRβ1 gene (with more than 300 alleles) and one of the three other DRβ genes. The kind of variation in which one gene may be present in one individual and absent in another is called *presence/absence polymorphism.*

6.9 MHC MOLECULE DISTRIBUTION IN VARIOUS CELL TYPES

The highest levels of class I molecules express in lymphocytes, approximately about 5×10^5 molecules/cell. Other cell types such as fibroblasts, muscle cells, liver hepatocytes, and neural cells express very low levels of class I MHC molecules. Low levels of MHC on liver cells reduce the chance of graft rejection by Tc of recipient cells. At certain stages of differentiation, cell types such as neurons and sperm cells lack class I MHC molecules. An F1 mouse expresses the K, D, and L from each parent (six different class I MHC molecules) on each of its nucleated cells (Plate 2).

6.10 IMMUNE RESPONSIVENESS TO MHC

B. Benacerraf immunized guinea pigs with synthetic antigens and showed that the ability of an animal to raise an immune response by producing antibodies is determined by MHC haplotypes. Later, H. McDevitt, M. Sela, and their colleagues mapped the control of immune responsiveness to class II MHC molecule genes by using congenic and recombinant congenic mouse strain. The genes responsible for phenotype were designated as Ir or *immune response genes*, and hence, the mouse class II MHC genes products are called IA and IE.

6.11 VARIABILITY IN IMMUNE RESPONSIVENESS

Two theories have been given for the variability in immune responsiveness in haplotypes.

- *Determinant selection model* Different class II MHC molecules have different binding ability to process antigens. Thus MHC polymorphism in a species generates a binding diversity and specificities along with different responsiveness patterns to antigens. For example, radiolabelled peptides bind either class II IA or IE molecules of $H\text{-}2^d$ or $H\text{-}2^k$ haplotype. According to this model, the haplotypes (class II MHC) showing maximum affinity for a specific peptide show the same affinity for the same specific peptide for the haplotype responder strains.
- *Holes in the repertoire model* This is an alternative theory. According to this model, TCR recognizing antigens (Brenner, McLean, Dialynas, *et al.* 1986; Davis and Bjorkman 1988) that resemble self-antigens are eliminated during processing in thymus. Since TCR binding with antigen is tripartite and also includes MHC, this theory along with the above theory might be correct (Manning and Kranz 1999; San José, Sahuquillo, Bragado, *et al.* 1998).

The absence of MHC molecules that can bind and present peptide or TCR that recognize the MHC–antigen complex will result in the absence of immune responsiveness. This shows correlation between MHC haplotypes and immune responsiveness to exogenous antigens. The λ-repressor is an exception to the general rule that has maximum binding correlates with the MHC restriction of high responding strains. Here the Th cell specific for the λ-peptide–IE^d complex has been deleted. This is an example of the model.

6.12 MHC GENE REGULATION

Major histocompatibility complex genes are localized on mouse chromosome-17 and human chromosome-6. In humans, class I MHC is 2000 kb, class II is 1000 kb, and in between the two, class III is 1000 kb. It makes a total of 4000 kb HLA gene complex. In mice, class I MHC is 50 kb, followed by class II of 400 kb, and class III of 1050 kb—a total of a 1500 kb long H-2 gene complex. The 5'-end of both class I and class II MHC genes contains promoter sequences that bind with specific transcription factors. MHC genes are regulated by both positive and negative elements. Class II MHC gene binding promoter has been identified as CIITA and RFX. Defects in these transcriptions result in *bare lymphocyte syndrome*. Patients having this disorder will have no class II MHC molecules on their cells and consequently suffer severe immunodeficiency because class II MHC molecules have a major role in the activation and maturation of T cells. Many cytokines are involved in the upregulation and downregulation of transcription factors that bind with the promoter region of the genes of class I MHC molecules to regulate them. Such cytokines are IL-1 α, β, γ, TNF, and IFNγ. Many viral infection and their products affect the regulation of transporter genes (*TAP1* and *TAP2*). TAP gene products are responsible for the transport of peptide from the cytoplasm into the rough ER. If the transport of peptide is blocked, class I MHC molecules will not assemble with β2-microglobulin and thus will not be transported to the membrane. This will lead to the presentation of a viral peptide–MHC complex to be targeted by the CTL to destroy.

6.13 TCR–CD3 COMPLEX

Like B-cell receptors (BCR), which form a complex with Igα/Igβ-heterodimer, TCRs associate with CD3 forming a TCR–CD3 complex (Figure 6.5). These accessory molecules do not participate in the interaction with antigen; rather, they help in signal transduction after interaction. CD3 is closely associated with $\alpha\beta$-heterodimer and also help in the expression of $\alpha\beta$- and $\gamma\delta$-TCRs to form the TCR–CD3 complex on the cell membrane. Loss in either $\alpha\beta$- or $\gamma\delta$-chain of TCRs of CD3 results in the loss of the molecular complex from the membrane.

The CD3 complex consists of two zeta homodimer chains ($\zeta\zeta$), zeta and eta heterodimer chains ($\zeta\eta$), gamma and epsilon heterodimer chains ($\gamma\varepsilon$), and delta and epsilon heterodimer chains ($\delta\varepsilon$). The ζ and η chains differ from each other in their carboxyl-terminal ends although encoded by the same gene. Thus the TCR complex consists of four dimmers—$\alpha\beta$- or $\gamma\delta$-TCR heterodimer, which provides ligand binding specificity, while the CD3 dimers ($\gamma\varepsilon$, $\delta\varepsilon$ and $\zeta\zeta$ or $\zeta\eta$) help in membrane expression of TCR and in signal transduction (Chien, Jores, and Crowley 1996; Davis, Boniface, Reich, *et al.* 1998; Haas, Pereira, and Tonegawa 1993; Raulet 1994).

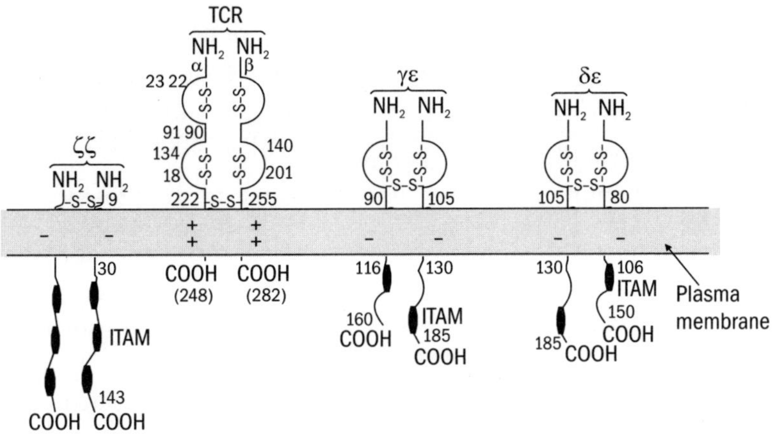

Figure 6.5 TCR–CD3 complex for antigen binding

Note The CD3 complex constitutes ζζ homodimer (alternately α ζ η heterodimer) along with γε and δε heterodimers. The cytoplasmic tails with transmembrane domains contain ITAM that takes part in signal transduction.

The cytoplasmic part of the CD3 chains taking part in signal transduction is called immunoreceptor tyrosine-based activation motif (ITAM). They get activated after phosphorylation by tyrosine kinases. A single copy of ITAM is also present on γ-, δ-, and ε-chains. The ζ-chain and η-chain contain three copies of ITAM (Sloan-Lancaster and Allen 1996; Weiss and Littman 1994).

AT A GLANCE

- Major histocompatibility complex (MHC) is a group of tightly linked genes that present antigens to T-lymphocytes for intercellular recognition.
- The groups of linked MHC genes that are inherited from parents to offspring are known as haplotypes.
- For each MHC gene, there are a large numbers of alleles that provide them with polymorphic features. Since there are a number of different MHC genes, they can be called polygenic.
- A class I MHC molecule has three extracellular domains containing a large glycoprotein with a transmembrane segment and a single domain β2-microglobulin protein.
- A class II MHC molecule is a glycoprotein having one α-chain and one β-chain, which are non-covalently attached to each other. The α-chain and β-chain are encoded by two separate genes.
- The distal regions of class I and class II MHC molecules form an invaginated membrane, which has sites for peptide-binding clefts as revealed by X-ray crystallography.
- Class I and class II MHC molecules present two different types of antigens after processing—class I molecules process and present endogenous antigens to $CD8^+$ T cells called Tc cells, while class II molecules process and present exogenous antigens to $CD4^+$ T cells called Th cells.
- Almost all nucleated cells express class I MHC molecules, while class II molecules are exclusively expressed by macrophages, dendritic cells, and B cells.

- Class III MHC regions express groups of proteins that have no role in antigen presentation.
- The MHC gene map of both humans and mice reveals the presence of antigen-processing proteasomes and transporter protein genes.
- The MHC haplotype is responsible for antigen presentation and immune responsiveness as deciphered by transgenic mouse strain experiments.
- Susceptibility to autoimmune and other diseases is due to the presence of specific MHC alleles.

QUESTIONS

1. What do you understand by MHC? Why is MHC called as H-2? How many classes of MHC molecules are present? Discuss any one in brief.
2. How many types of mice breeds are present on the basis of MHC? Discuss their important features.
3. Discuss the structure and properties of class I and class II MHC molecules. How do they structurally and functionally differ from each other?
4. With a suitable diagram discuss the exon and intron arrangement in the genes of class I and II molecules.
5. What is MHC polymorphism? How it is generated? Discuss its importance.
6. Discuss the experimental plan to differentiate between two identical inbred mouse strains having identical MHC haplotypes.
7. Are the RBCs of human expressed MHC molecules or not? Discuss its importance in blood transfusion.
8. What do you understand by immune responsiveness to MHC? Discuss variability in immune responsiveness.

OBJECTIVE QUESTIONS

A. Fill in the blanks.

a. _______ is associated with intercellular recognition and antigen presentation to T cells.
b. Groups of genes inherited together are called _______.
c. MHC genes are _______ because there are many alleles for each gene.
d. MHC genes are _______ because there are many different genes.
e. The α-chain and β-chain in class II MHC are encoded by _______ MHC genes.

f. Peptide-binding clefts in class I and class II MHC are located at _______ end revealed by _______ analysis.
g. Class I MHC molecules are expressed mainly on _______ cells.
h. Class II MHC molecules are expressed on _______ or _______ and _______ cells.
i. MHC genes can perform functions like _______, _______, and _______ .

B. Match the following.

i. Congenic mice strains	a. Many alleles for each gene
ii. Class III region of MHC	b. Number of different MHC genes
iii. Polymorphic	c. Distal cleft
iv. Polygenic	d. Single domain
v. β2-microglobulin	e. Three extracellular domains and a single domain
vi. Class I MHC molecules	f. Two non-covalently associated glycoproteins
vii. Class II MHC molecules	g. Strains differ in single genetic loci
viii. Peptide binding	h. No role in antigen presentation
ix. Linked MHC genes	i. Strain identical in all genetic loci
x. Syngenic mice strains	j. Inherited as unit

CHAPTER SEVEN 7

Complement System

7.1 INTRODUCTION

The complement system is a large complex group of proteins. It is the major group of effector molecules of humoral immunity. The complement system was discovered in 1890 by Jules Bordet at the Pasteur Institute of Paris. He found that unheated sheep antiserum to *Vibrio cholerae* causes lysis of bacteria (bacteriolytic) but heated antiserum does not. Adding fresh serum that contains no antibodies against the bacterium restored the bacteriolytic activity of the heated serum. This means that the bacteriolytic activity of serum depends on two components: (1) the heat-resistant component, which is the antibody that forms a coat around the bacterium, and (2) the heat-sensitive component having bacteriolytic activity. The term *complement* was coined by Paul Ehrlich who defined the complement as *"the activity of blood serum that completes the action of antibody"*.

7.2 SYNTHESIS AND COMPONENTS OF THE COMPLEMENT SYSTEM

The complement system is a multi-enzymatic cascade or group of proteins and glycoproteins. After activation, this multi-enzymatic cascade either leads to an assembly of membranolytic complex proteins or generates protein fragments that trigger inflammation and enhances phagocytosis. The cascade also interacts with the B-cell surface receptors to control the activities of B cells, thus providing a more effective acquired immune system (Carroll 2000). These proteins constitute 5% of the total serum globulin fraction. They are synthesized by liver hepatocytes and also by blood monocytes, tissue macrophages, and the epithelial cells of the gastrointestinal and genitourinary tract. The circulating serum complement components are functionally inactive due to the attachment of an inhibitory fragment. They circulate in the form of *zymogens* and proenzymes that become active after enzymatic removal of the inhibitory fragment. A cascade of enzymes is involved in the initiation of the complement reaction sequence. The components are denoted by numerals (C1–C9) in order of their discovery, before their functional roles were determined. Hence, the number does not reflect the reaction order. The active peptide fragment of the components is denoted by small letters. The smaller fragment is denoted by letter "a" and the larger fragment by letter "b", that is, C3a, C3b. The exception is C2a, which is the larger fragment. Larger fragments attach near the site of activation, and smaller fragments diffuse inside to initiate inflammation by binding to complement receptors (Ahearn and Fearon 1989; Molina and Holers 1996). The complement fragments attach to each other to form functional complexes. Enzymatic components are designated with a bar over the number or symbol, for example, $\overline{C4bC2a}$, $\overline{C3bBb}$.

7.3 ACTIVATION OF COMPLEMENT SYSTEM

The complement activation pathways work in three routes or pathways (Nielsen, Fischer, and Leslie 2000; Nonaka 2000; Pickering and Walport 2000). Each one culminates with the formation of C5b. The three pathways are (1) classical pathway or antibody-dependent pathway, (2) alternative pathway or antibody-independent pathway, and (3) lectin pathway. C5b leads to the formation of a membrane attack complex (MAC), which is similar in all three pathways.

7.3.1 Classical Pathway or Antibody-dependent Pathway

This pathway is a major element of innate immunity against infection. It is also involved in immune tolerance, graft rejection, and various pathogenesis. It is triggered by C1, a multimolecular protease (Figure 7.1), in following steps (Plate 1).

- Formation of an antigen–antibody (IgM or IgG subtypes IgG1, IgG2, and IgG3) complex either of soluble antigen or bacterial cell wall antigen.
- Complex formation causes a conformational change in the Fc part of the IgM molecule making a site for the binding of C1 component. As seen in Plate 1, C1 is a macromolecular complex of one molecule of C1q (hexameric protein), a recognition subunit and modular serine proteases, two molecules of C1r and two molecules of C1s (C1qr2s2) held together by a Ca^{2+}-dependent tetramer (C1s-C1r-C1r-C1s).
- C1q consists of 18 polypeptide chains arranged in six collagen-like triplets, the globular tips of which bind to at least two C_H2 domains (Fc region) of two antibody molecules to make a stable interaction.
- C1q is capable of enhancing phagocytosis, stimulating oxygen radical generation and stimulating immunoglobulin secretion. A C1q-deficient individual develops immune-complex-related diseases, such as lupus-like disorders or severe bouts of infection.

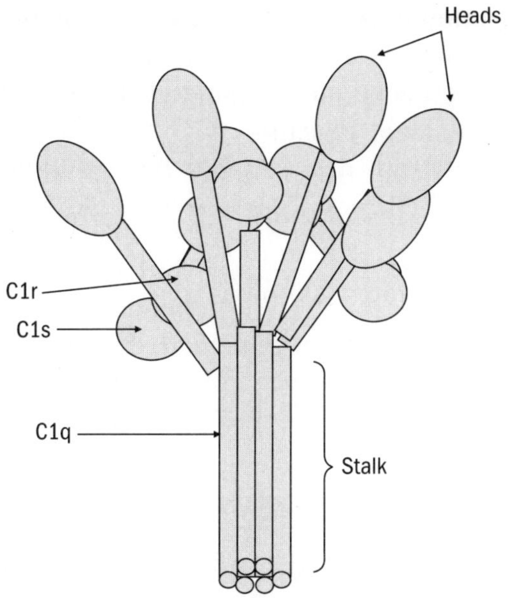

Figure 7.1 C1 macromolecular protease complex structure with C1q hexameric molecule, C1r bi-molecules, and C1s bi-molecules (C1qr2s2)

- C1r and C1s are monomers having a catalytic domain (enzymatic activity) and an interaction domain to facilitate binding with C1q or with each other.
- The antigen–antibody complex of IgM forms a staple-like configuration, exposing at least three binding sites. The planer configuration formed by a circulating antibody has no binding sites exposed for C1q binding and is thus unable to activate the complement cascade.
- The IgG molecule has a single C1q-binding site, which is of Fc-C_H2 domain. Hence, two IgG molecules are required to give two binding sites for C1q.
- Interaction of a single molecule of IgM with red blood cells (RBCs) activates the classical pathway and lyses the cells, while 1000 molecules of IgG are required to activate the same classical pathway.
- C1q causes conformational changes in C1r to make it an active serine protease enzyme $\overline{\text{C1r}}$.
- $\overline{\text{C1r}}$ cleaves C1s to form the active enzyme $\overline{\text{C1s}}$ that has two substrates C4 (glycoprotein of α, β, and γ polypeptide chains) and C2.
- $\overline{\text{C1s}}$ hydrolyses C4a (smaller fragment) of the C4 molecule from its amino-terminus of α-chain to expose a binding site on C4b (larger fragment).
- The exposed binding site of C4b attaches C1 and C2 proenzymes, which is cleaved by $\overline{\text{C1s}}$ to form C2a (larger fragment) and C2b (smaller fragment). The smaller fragment C2b diffuses away leaving complex $\overline{\text{C4b2a}}$.

- $\overline{\text{C4b2a}}$ complex is also known as C3 convertase because it converts C3 into its active form.
- C4a (smaller fragment) of C4 is known as anaphylatoxin or a mediator of inflammation and has no role in complement cascade.
- Anaphylatoxin includes smaller fragments of C4, C3, and C5.
- C3 has two polypeptide chains α and β. C3 convertase generates C3b from C3 by hydrolysing C3a (shorter fragment) from the amino-terminus of the α-chain of C3.
- A single C3 convertase generates 200 molecules of C3b (larger fragment).
- C3b binds to $\overline{\text{C4b2a}}$ to form $\overline{\text{C4b2a3b}}$, a trimolecular complex known as C5-convertase.
- C3b of complex $\overline{\text{C4b2a3b}}$ binds to C5 to make conformational changes in C5 and $\overline{\text{C4b2a}}$ cleaves C5 into C5a and C5b.
- C5a diffuses away and C5b binds to C6 to form a membrane attack complex (MAC).
- In some cases, C3b does not bind to $\overline{\text{C4b2a}}$; rather, it coats antigens and immune complexes to work as opsonin. C3b also has the ability to bind directly to cell membranes.

7.3.2 Alternative Pathway or Antibody-independent Pathway

The alternative pathway does not involve the generation of an antigen–antibody complex for initiation and generation of C5b. Hence, it is a component of the innate immune system. It involves four serum proteins: C3, factor B, factor D, and properdin. This pathway is activated by foreign surface cell wall constituents of the Gram-positive or Gram-negative bacteria or viral particles (Turner 1998). It occurs in the following steps (Plate 2).

- Hydrolysis of the unstable thioester bond of serum C3 to C3a and C3b, which is a slow and spontaneous process. C3b binds to foreign microbial surface antigens (bacterial, yeast or viral), including own cells.
- The sialic acid present in high amounts on host cells causes inactivation of C3b. Since sialic acid is present in very low amounts on microbial cell surfaces, C3b binds to them and remains active for longer time periods.
- There is an Mg^{2+}-mediated binding of factor-B to C3b on the foreign cell surface. The binding exposes the surface of factor-B as a substrate for another serum-active enzyme protein called factor-D that cleaves away the smaller fragment (Ba) leaving the larger fragment Bb attached with C3b of factor-B, generating the complex $\overline{\text{C3bBb}}$.
- $\overline{\text{C3bBb}}$ of the alternate pathway is analogous to $\overline{\text{C4b2a}}$ of the classical pathway, and both have similar C3 convertase activity.

- The half-life of $\overline{\text{C4b2a}}$ is analogous to $\overline{\text{C3bBb}}$. It is 5 min, but in the presence of properdin (serum protein), it extends to 30 min.
- $\overline{\text{C3Bb}}$ converts C3 into C3b automatically to deposit more than 2×10^6 molecules of C3b on the antigenic surface.
- $\overline{\text{C3Bb}}$ forms a complex with C3b to generate $\overline{\text{C3bBb3b}}$ complex with *C5 convertase* activity similar to $\overline{\text{C4b2a3b}}$ of the classical pathway.
- C3b component of $\overline{\text{C3bBb3b}}$ binds to C5 and Bb components and hydrolyses C5 to C5a and C5b. C5b binds to the antigenic microbial surface (Plate 2).

7.3.3 Lectin Pathway or Host Protein or Microbial Surface Interaction Pathway

Lectins are carbohydrate-binding proteins, which specifically bind to mannose residues. Hence, they are normally designated as mannan-binding lectin (MBL) pathway (Turner 1998). This pathway is also an antibody-independent activation pathway like the alternative pathway. Since it initiates and proceeds through the activation of C4 and C2 to generate C5 convertase, it is more or less similar to the classical pathway. It occurs in the following steps (Plate 2).

- Binding of MBL protein to mannose residue on glycoproteins or carbohydrates of microbial cell wall leads to the activation of this pathway.
- MBL is functionally similar to an acute phase protein and structurally similar to C1q.
- Later on, the MBL-associated proteins MASP-1 and MASP-2 bind to form an active complex, which cleaves and activates C4 and C2 to form C5 convertase.
- Both MASP-1 and MASP-2 show structural and functional similarity to C1r and C1s leading to the generation of *C5 convertase.*
- The generation of C5 convertase is without the help of antibodies. Thus it shows similarity to the alternative pathway, which helps in innate defence mechanism. The lectin pathway uses the classical pathway elements, except C1 proteins.

7.3.4 Formation of Membrane Attack Complex

The process of MAC is described in the following steps.

- All the three pathways of the complement activation system end with the production of C5 convertase, which causes the C5 (made up of two proteins α- and β-chains) conversion to C5b (C5a diffuses away). C5b binds on the cell surface providing a site for other components C6, C7, C8, and C9 to combine and form MAC (Nielsen, Fischer, and Leslie 2000).

- The MAC forms a pore in the target cell membrane for the diffusion of ions and small molecules. The C5b component gets inactivated within 2 min if C6 does not bind and stabilize its activity.
- Up to C5b6 complex formation, the surface is hydrophilic. As soon as C7 binds to the complex, it changes from hydrophilic to amphiphilic exposing hydrophobic regions, which is a binding site for phospholipids of the membrane to anchor the complex C5b67.
- This reaction enables the C5b67 complex to enter into the phospholipid bilayer.
- On non-activating cellular surface, the hydrophobic region is unable to hold the complex and it is released to insert itself into the nearby cell membrane where it mediates *"innocent-bystander"* lysis.
- The innocent-bystander lysis is stopped by regulator protein. But in case of disease and disorder due to lack of regulator protein, cells and tissues are damaged by this form of lysis.
- C8 component after binding with C5b67 form the complex C5b678. This complex undergoes hydrophilic–amphiphilic structural transition to expose a hydrophobic site, which can interact with plasma membrane.
- The C5678 complex makes a 10 Å diameter pore, which causes the lysis of only non-nucleated cells such as RBCs.
- The binding of C9 component enables the formation of MAC complex where polymerization of 10–17 molecules of C9 takes place by a single C5b678 complex.
- Due to the hydrophilic–amphiphilic transition of C9 component, the complex C5b6789 is inserted into the membrane.
- The completed MAC is a tubular pore of size 70–100 Å made of C5b678 surrounded by a poly-C9 complex.

7.4 REGULATION OF THE COMPLEMENT SYSTEM

The ability of components of the complement system to attack healthy cells of host, foreign cells, and microbes must be regulated by a regulatory mechanism to specify its targets and make it more precise (Lindahl, Sjöbring, and Johnsson 2000). Many regulatory proteins are involved in the checking of unwanted activity of components of the complement system (Table 7.1). The activity mechanism is discussed as follows.

- The components are involved at various stages of the pathway. They are labile proteins that need to be protected from spontaneously becoming inactive by binding with other components.
- A series of regulatory proteins are involved to inactivate various complement components, for example, C1 inhibitor (C1inh) binds with C1r2s2 leading to the dissociation of C1r2s2inh complex from C1q, thus ceasing the activation of C4 or C2.

Table 7.1 Regulatory proteins and their effect on the complement system

Regulatory proteins/location	*Effect on the complement system*	*Affected pathway*
C1 inhibitor (C1inh)/soluble	Forms a complex with $C1r_2s_2$ to dissociate it from C1q	Classical
C4bBP (C4b-binding protein/soluble)	Binds to C4b thus preventing the formation of C3 convertase and acts as a cofactor for factor-I for the cleavage of C3b	Classical and lectin
Factor-H/soluble	Binds to C3b thus preventing the formation of C3 convertase and acts as a cofactor for factor-I for the cleavage of C3b	Alternative
Complement receptor type 1 (CR1); membrane cofactor protein (MCP)/ associated with membrane	Binds to C3b or C4b thus preventing the formation of C3 convertase; acts as a cofactor for factor-I for the cleavage of C3b or C4b	Classical, alternative, and lectin
Decay-accelerating factor (DAE) or C55/ associated with membrane	Dissociation of $\overline{C4b2a}$ and $\overline{C3bBb}$ gets accelerated	Classical, alternative, and lectin
Factor-I/soluble	Serine protease enzymatic activity; cleaves C4b or C3b in the presence of cofactor C4bBP, CR1, factor-H, DAE or MCP	Classical, alternative, and lectin
S-protein/soluble	Forms complex with C5b67 and prevents insertion into target cell membrane	Terminal
Homologous restriction factor (HRF); membrane inhibitor of reactive lysis (MIRL) or CD59/membrane associated	Forms complex with C5b678 on cells preventing binding of C9	Terminal
Anaphyla toxin inactivator/soluble	Removes carboxypeptidase N of C-terminal Arg of C3a, C4a, and C5a, thus inactivating anaphylatoxin activity	Effector

- The product C3b, a component of a major amplification step in the complement activation pathway, has the ability to opsonize healthy normal cells, thus killing C3b receptor-bearing phagocytic cells. Such killing is prevented by spontaneous hydrolysis of C3b if it is diffused at 40 nm of distance from enzyme $\overline{C4b2a}$ or $\overline{C3bBb}$ convertase, thus stopping the binding of C3b to $\overline{C4b2a}$ or $\overline{C3bBb}$.
- The regulator of complement activation (RCA) proteins, which are short consensus repeats (SCRs) of 60 residues, prevent the assembly of C3 convertase. RCA genes are located on chromosome-1 in humans.
- RCA proteins do not allow the assembly of C3 convertase by binding to C4b, thus preventing its binding to C2a in classical and lectin pathways. RCA proteins, which

act as regulatory proteins, include soluble C4b-binding protein (C4bBP), complement receptor type 1 (CR1) (Ahearn and Fearon 1989; Muller-Eberhard 1988) and membrane cofactor protein (MCP).

- After the formation of the C3b–RCA proteins complex, soluble factor-I regulatory protein cleaves C3b to C4d (bound) and C4c (soluble).
- In the alternative pathway, CR1, MCP, and soluble factor-H (regulatory component) bind and form a complex with C3b where soluble factor-I cleaves C3b to iC3b (bound) and C3f (soluble) fragments.
- Again soluble factor-I cleaves iC3b to C3dg, which remains bound to the membrane and releases C3c.
- RCA proteins such as C4bBP, CR1, and factor-H can also dissociate C3 convertase complex.
- Decay accelerating factor (DAF) or CD55, a glycoprotein anchored covalently into membrane, can dissociate C3 convertase complex.
- Each component of RCA proteins has an ability to dissociate C3 convertase by removing either C2a or Bb from C4b, or C3b cell bound components.
- After dissociation, factor-I cleaves the rest of C4b or C3b membrane-associated components, permanently inactivating the convertase activity.

7.4.1 Regulation at MAC Level or Terminal End of Pathway

The S-protein of serum binds and induces hydrophobic transition to C5b67, thus preventing its insertion into the membrane of nearby cells and stopping the threat of innocent-bystander lysis of normal healthy cells.

The homologous restriction factor (HRF) and membrane inhibitor of reactive lysis (MIRL) proteins or CD59 present on many cell membranes bind to C8 component and prevent poly-C9 assembly and insertion into plasma membrane. This type of inhibition occurs if complement components are from the same species. That is why this type of inhibition is called homologous restriction. It poses a barrier in transplantation of organs from different species.

7.5 EFFECTS OF COMPLEMENT ACTIVATION ON BIOLOGICAL SYSTEM

The end product of the complement system MAC causes cell lysis, while other components or split elements have a role in inflammation, antigen opsonization and phagocytosis, neutralization of viruses, and removal of unwanted immune complexes (Table 7.2).

Table 7.2 Complement products with their biological effects and importance

Biological effect	Affected cell and body part	Binding receptor	Complement product
Cell lysis	• Target cells		C5b-9, MAC
	• Inflammation mast cells degranulation	C3a, C4a, and C5a	C3a, C4a, and C5a
	• Basophils degranulation	C3a, C4a and C5a	C3a, C4a, and C5a
	• Eosinophils degranulation	C3a and C5a	C3a, C5a
	• Extravasations and movement of leukocytes to inflammation sites	C3a, C5a, and C5b67	C3a, C5a, and C5b67
	• Platelets aggregation at infection site	C3a and C5a	C3a and C5a
	• Migration of monocytes and inhibition of macrophage		Bb
	• Release of neutrophils from bone marrow	C3c	C3c
	• Release of hydrolytic enzymes by neutrophils	C5a	C5a
	• Increase of expression of CR1 and CR3 on neutrophils	C5a	C5a
Opsonization	• Antigens for better phagocytosis	CR1 (CD35), CR3 (CD11b/18), CR4 (CD11c/18)	C3b, C4b, iC3b
Neutralization	• Viral particles	CR1(CD35)	C3b, C5b-9 (MAC)
Removal and clearance	• Immune complexes	CR1(CD35)	C3b

7.6 PROTECTION OF MICROBES AGAINST COMPLEMENT-MEDIATED DAMAGE

Many microbial species such as *Escherichia coli* and *Salmonella* have long-side polysaccharide chains in the cell wall. The LPS side chains prevent insertion of MAC into bacterial membrane by protecting them from complement-mediated lysis. The outer membrane proteins of some bacteria interact with MAC and prevent its insertion into the bacterial membrane, such as *Neisseria gonorrhoeae*. Gram-negative bacterial cell wall contains elastase, which inactivates anaphylatoxins C3a and C5a. A thick layer of peptidoglycan present in Gram-positive bacterial (such as *Streptococcus*) cell wall prevents the insertion of MAC into the membrane, thus preventing the damage of cells. The capsule present on bacterial cell wall in most cases does not allow the deposited C3b on the membrane to interact with the CR1 present on the phagocytic cells. The exceptions are *Streptococcus pneumoniae*, which is encapsulated, and also C3b, which interacts with CR1 of phagocytic cells. Some microbes (such as viruses, fungi, protozoa, and bacteria) have protein that has an ability to mimic the regulatory protein of complement cascade, thus inhibiting the cascade to proceed and cause complement-mediated damage to the cells (Table 7.3) (Rautemaa and Meri 1999).

Table 7.3 Microbes protection mechanism against complement-mediated damage

Type of bacterial strains	Protecting microbial component	Mechanism of protection
Gram-negative strains	LPS, OMP Elastase	Prevents insertion of MAC; inactivates C3a and C5a
Gram-positive strains	Peptidoglycan of cell wall, capsule	Prevents insertion of MAC; has a barrier to C3b (target cell) and CR1 (phagocytic cells) interaction
Other microbes	Complement regulatory mimic proteins	Inhibits complement cascade of microbes

7.7 DEFICIENCIES OF COMPONENTS OF THE COMPLEMENT SYSTEM

Homozygous deficiencies in each classical pathway component (C1q, C1r, C1s, C4, and C2) can cause immune complex diseases such as systemic lupus erythematosus, glomerulonephritis, and vasculitis (Lokki and Colten 1995, Rautemaa and Meri 1999). These classical pathway components are important to generate C3b, solubilize C3b, and remove the immune complexes formed. A complement-deficient person is infection prone to *Streptococci, Staphylococci,* and other pyogenic Gram-positive bacterial strains. The deficiency of factor-D of alternative pathway provides an opportunity to *Neisseria* infections. C3 deficiency causes frequent severe bacterial infections along with immune complex diseases. Deficiency in MAC-forming components in individuals makes them prone to meningococcal and gonococcal infections by *Neisseria* sp. Immune complex diseases are not found in MAC-deficient individuals suggesting that C3b is present in enough amounts to remove the immune complexes. Deficiency in C1inh, which regulates the classical pathway, gives rise to hereditary angioedema (Laurent and Guinnepain 1999). Angioedema is a localized oedema of the tissue after trauma. It causes pain in the upper respiratory tract or in the abdomen. The effect of the deficiency of complement components shows the broad spectrum of the role of each component in the immunity of an individual.

AT A GLANCE

- The complement system is a group of proteins present in the serum in active or inactive forms.
- The activation of the complement occurs in three pathways: (1) classical, (2) alternative, and (3) lectin. These pathways get activated independent to each other, generating and forming a molecular complex in a common sequence of events.
- The classical pathway is initiated or activated by the binding of IgM and subclasses of IgG antibody to target cells. Alternative and lectin pathways are initiated or activated by the interaction of complement proteins with microbial surface molecules on the antibody in an independent manner.

- Apart from cell lysis, the complement system mediates opsonization of bacteria, inflammation, and removal of immune complexes. Both innate and adaptive immune responses are controlled by complement proteins; by the interaction its subunits with the receptors present on immune system cells.
- The complement system has active and passive regulatory mechanisms to kill the target cells or microbes.
- Deficiency in the inheritance of complement system leads to the increase in susceptibility to infection in the host.

QUESTIONS

1. What do you understand by complement system? Discuss the different pathways of activation of complement system.
2. What is MAC? How it is formed? Discuss its importance in the immune system.
3. What kind of regulatory mechanisms is present in the complement system? Discuss the effect of activation of the complement on the biological system.
4. What do you understand by the following?
   a. Deficiencies of components of complement system
   b. Lectin pathway of complement system
   c. Classical pathway of complement system.
5. Define the complement system. Discuss the differences between classical pathway and lectin pathway.

OBJECTIVE QUESTIONS

A. Fill in the blanks.

a. Complement is a group of serum _______ present in _______ forms.
b. Three pathways of complement activation are _______, _______, and _______ .
c. Classical pathway is initiated by _______ and _______ subclasses of antibodies.
d. Complement system activation leads to the following four functions: _______, _______, _______, and _______ .
e. Viral neutralization is done by _______ subtypes of complement in the presence of antibody.
f. Anaphylatoxins are _______, _______, and _______ complement fragments.

g. Multiple membrane attack complexes (MAC) associated tumour antigen specific antibody is not able to kill cancer cells due to _______ of the MAC.

h. Complete MAC is a _______ complex surrounded by _______ .

i. MAC prevents _______ stability in the cells and lysis of cells occurs due to _______ of water and electrolytic _______

B. Match the following.

i. Classical pathway	a. Antibody independent
ii. Lectin pathway	b. Opsonin
iii. C3b	c. Antibody dependent
iv. Carboxypeptidase N	d. Multiple MAC
v. Lysis of RBC	e. Prevents C2a association
vi. Lysis of nucleated cells	f. Binds to C3b
vii. C4bBP	g. Inactivates anaphylatoxins activity
viii. Factor-H	h. Single MAC
ix. C5b6789	i. Properdin
x. $\overline{C3bBb}$	j. Perforin-like molecule

Colour Plates

Chapter 1
Basics of Immunity

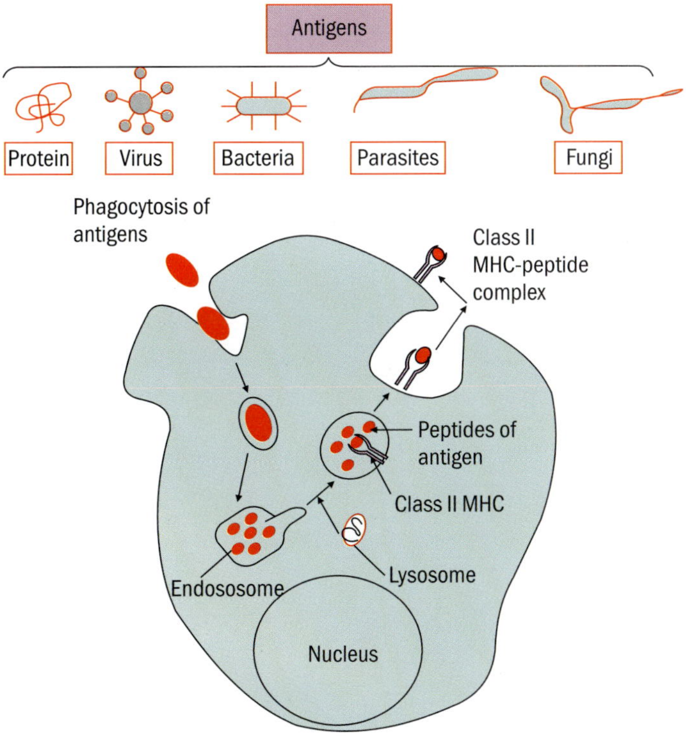

Plate 1 Processing and presentation of exogenous proteins by antigen-presenting cells
Note Class II MHC peptide association occurs in the vesicles and are present on the cell surface

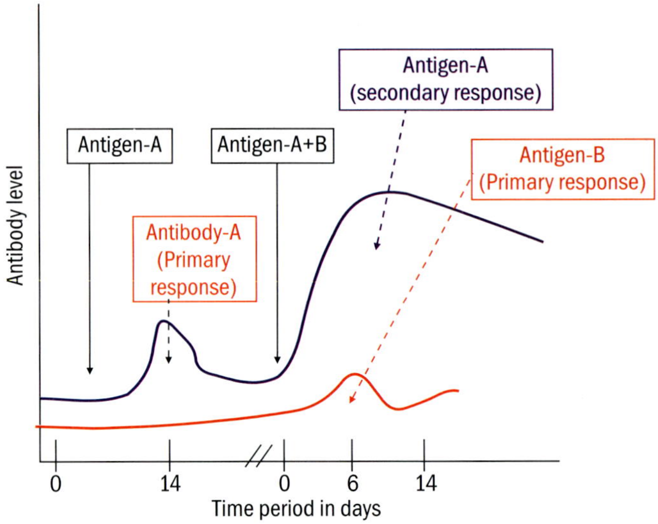

Plate 2 Plot of antibody level versus time period (in days)

Note Blue line represents primary and secondary response upon first-time interaction and second-time interaction with antigen-A and subsequently production of antibody in both response. Red line represents only primary response upon first-time antigen interaction and production of antibody in response.

Chapter 2
Composition of Immune System

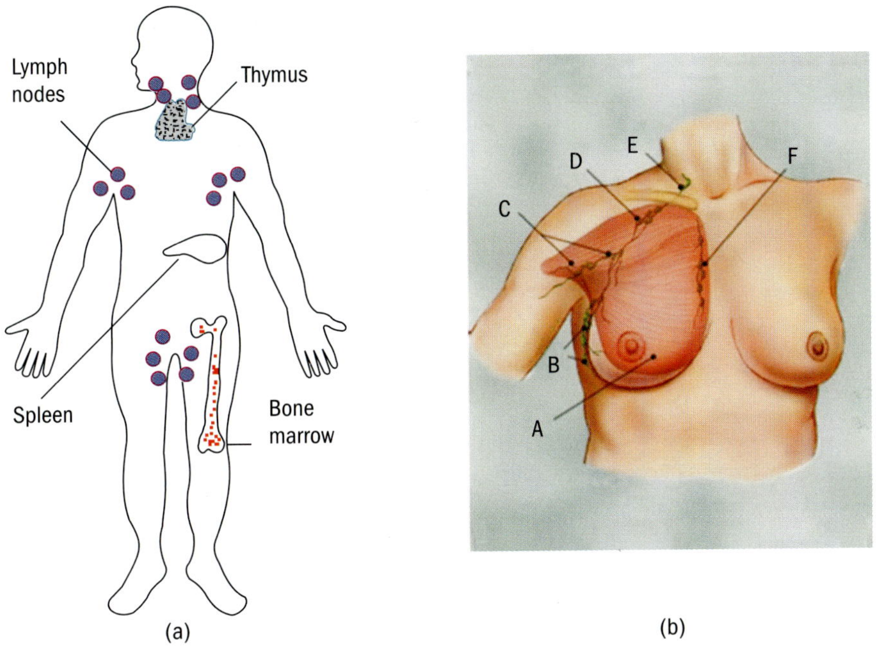

Plate 1 Lymphoid organs

Note (a) Shows both primary and secondary lymphoid organs, while (b) shows only secondary organs. A to F show lymph nodes along with lymph vessels. Primary organs include thymus and bone marrow and secondary organs include spleen, tonsils, and lymph vessels.

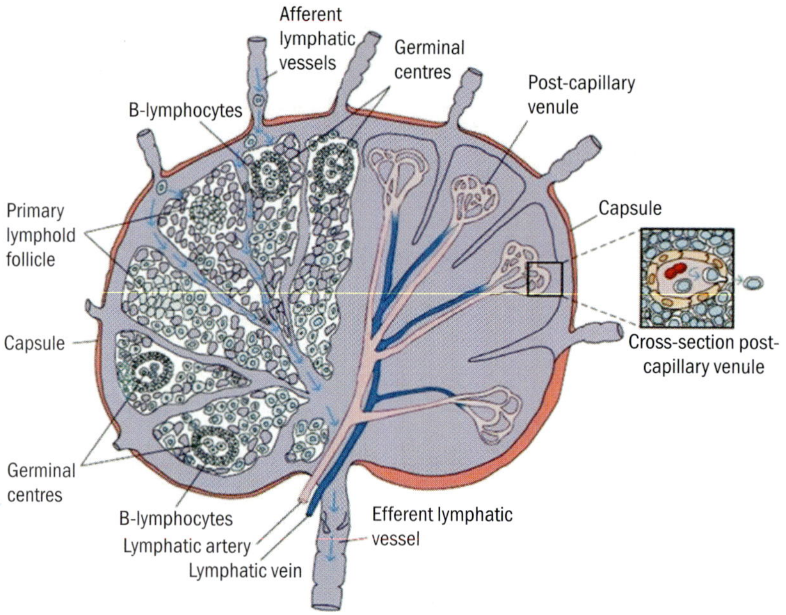

Plate 2 Structure of lymph nodes

Note Lymph nodes are made up of three zones: cortex, paracortex, and medulla. Afferent lymphatic vessels collect the lymph into lymph node, while efferent lymphatic vessels leave the lymph node to drain lymph into blood vessels.

Chapter 3
Antigens and Immunogens

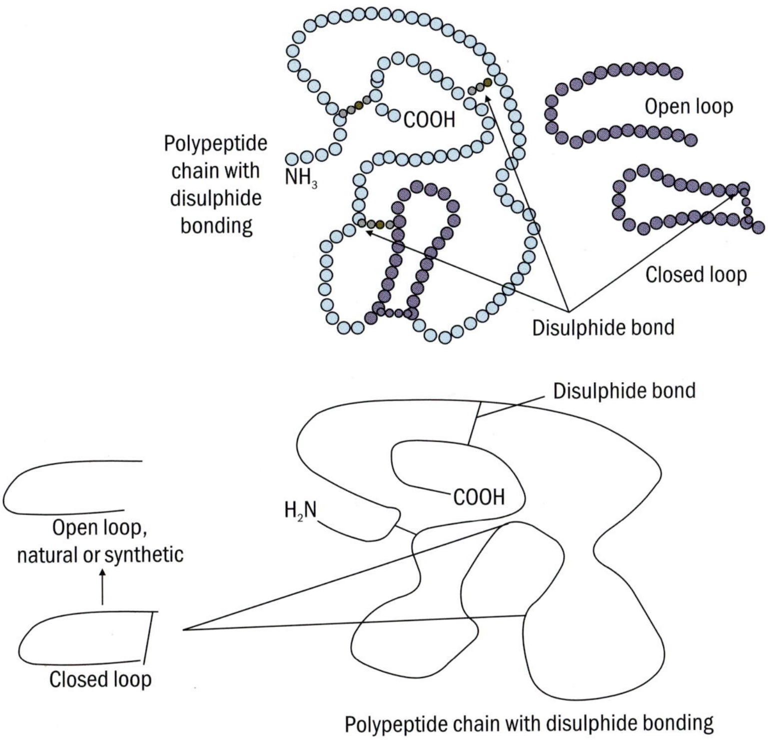

Plate 1 Determination of epitope

Note Antigenic behaviour is shown only by the closed loop in either (a) or (b), which determines the epitope structure due to bisulphide bonding, but not by the open loop.

Chapter 4
Antigen Determinants

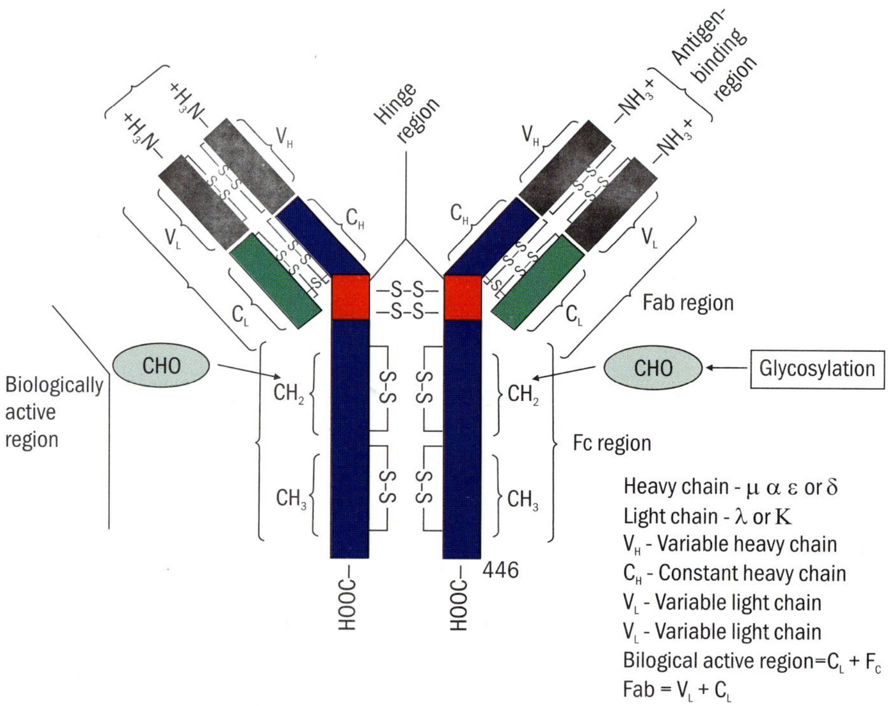

Plate 1 General structure of IgG

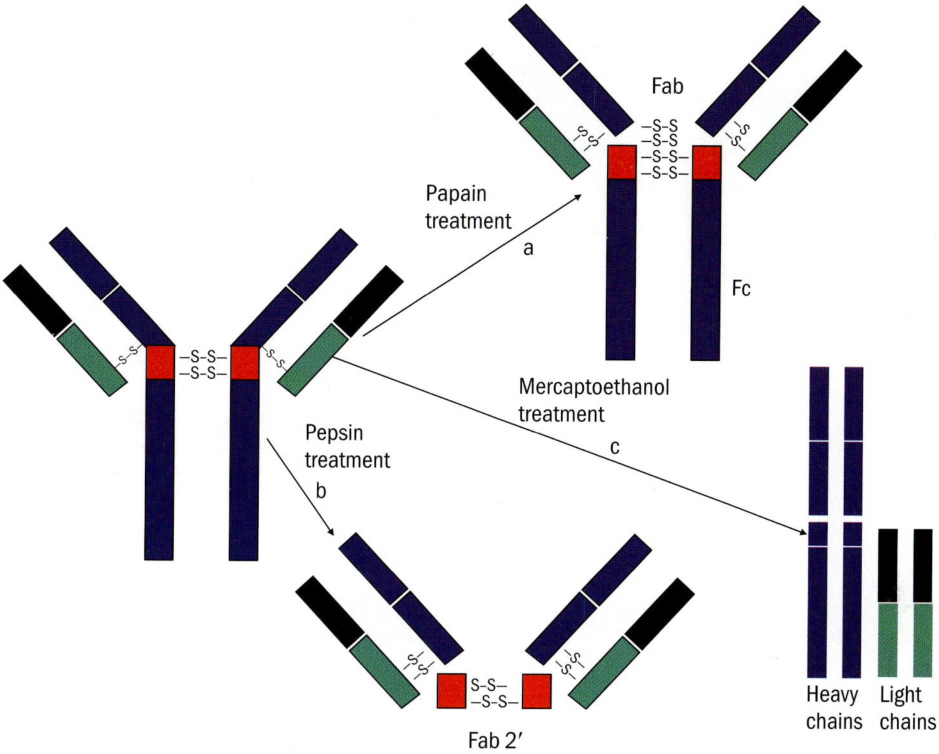

Plate 2 Digestion of immunoglobulin

Note Digestion with papain shows the presence of Fc and Fab regions; with mercaptoethanol shows the presence of disulphide bonds, two light chains, and two heavy chains; and with pepsin shows the Fab 2′ region, which contains the Fab and hinge regions.

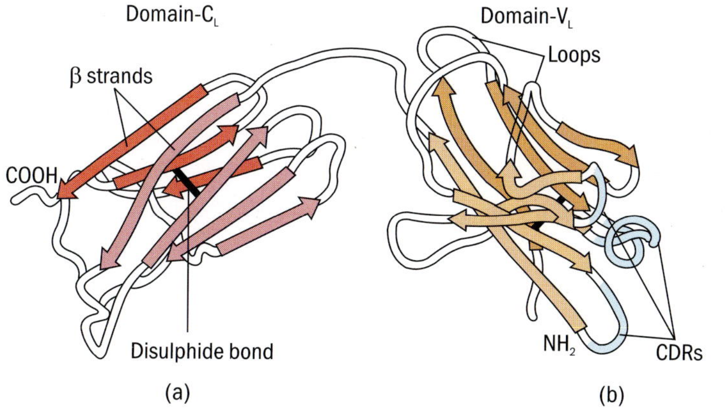

Plate 3 Immunoglobulin light chain with variable and constant domains

Note In each domain, β-pleated sheets are attached by hydrophobic interactions and conserved disulphide bonds. Complementarity-determining regions are hyper-variable regions.

Source Adopted from Schiffer, Girling, Ely, *et al.* (1973)

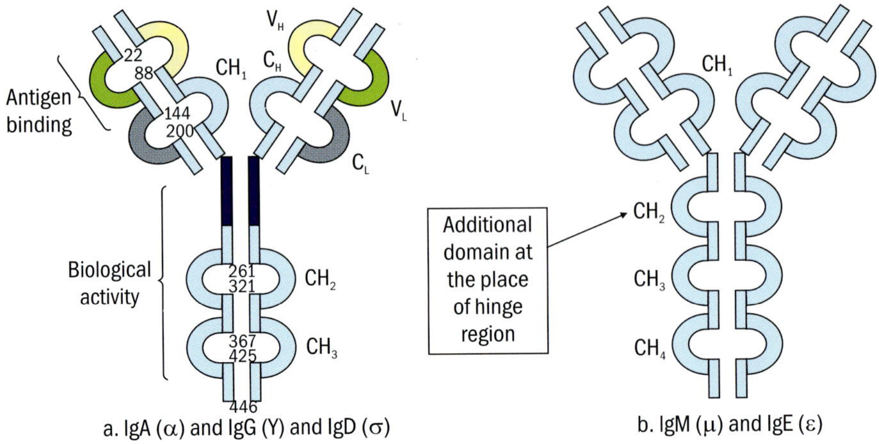

Plate 4 Division of antibodies on the number of domains.

Note (a) IgA with heavy chain-α IgG with heavy chain-γ and IgD with heavy chain-δ (b) Ig[V] with heavy chain-μ, IgE with heavy chain-ε.

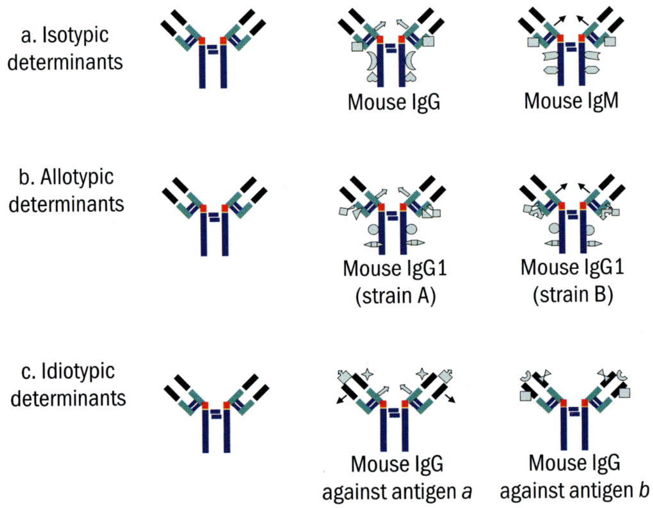

Plate 5 Types of antibody based on determinants

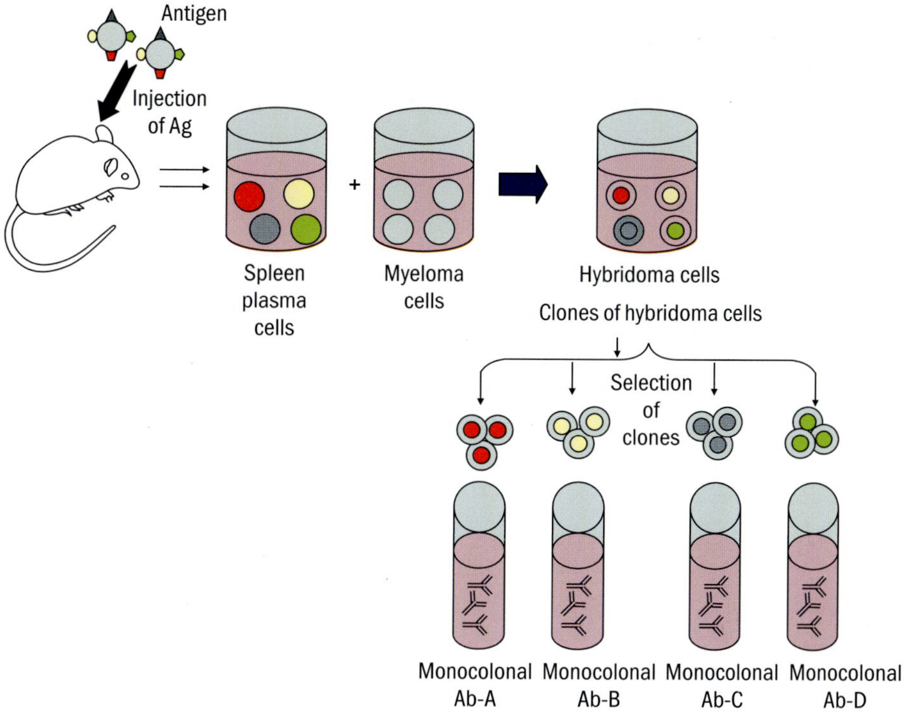

Plate 6 Monoclonal antibody formation using hybridoma technology

Chapter 5
Antigen and Antibody Interaction

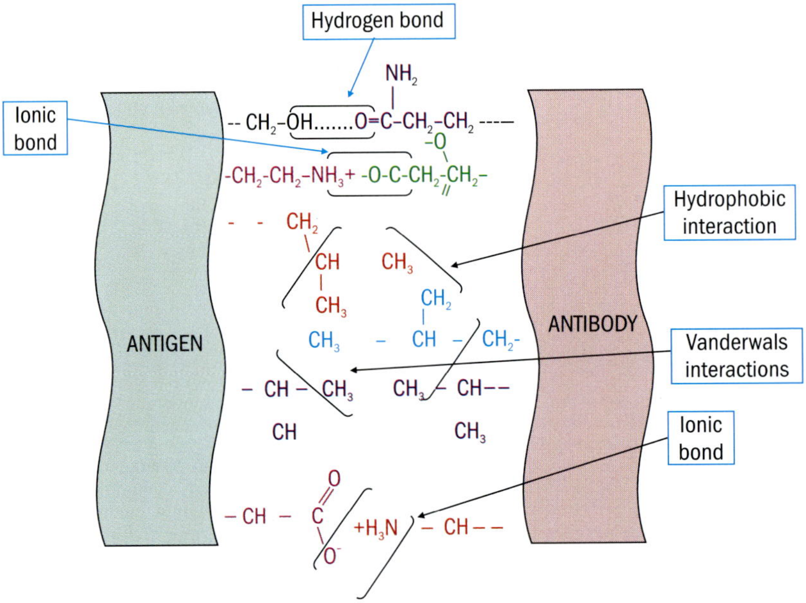

Plate 1 Different bonds and degree of complementarities during antigen–antibody interactions

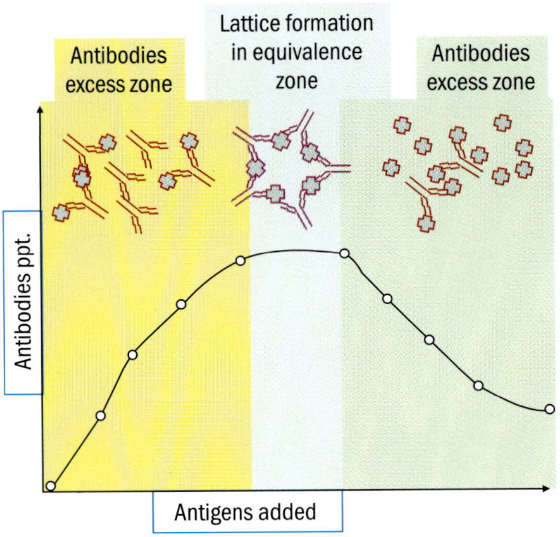

Plate 2 Lattice formation by polyclonal antibodies to form visible precipitation line in equivalence zone

Note The antigen concentration is low (antibody-excess zone) initially. The concentration of added antigens then becomes equal to the antibodies present forming an equivalence zone. Finally, the antigen concentration is more than the antibody concentration (excess antigen zone).

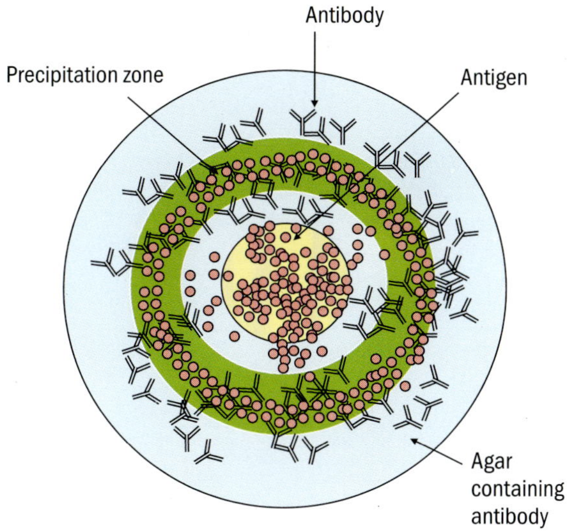

Plate 3 Radial immunodiffusion

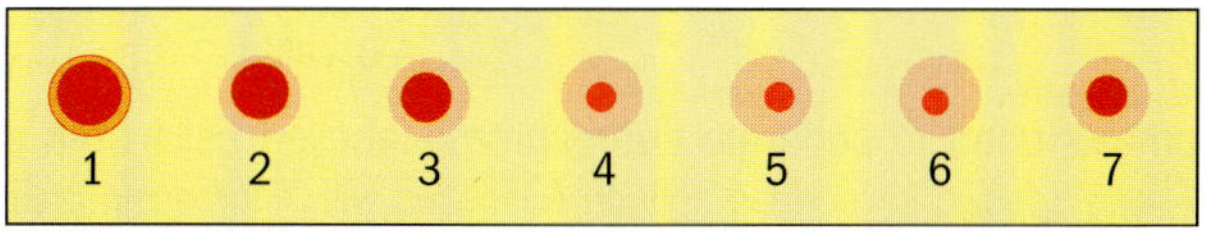

Plate 4 Pattern of haemagglutination reactions

Note Wells 1 to 6 contain a constant number of red blood cells and two-fold serially diluted antiserum. Well 7 is the control, which contains only blood cells. Wells 1 to 3 show positive haemagglutination.

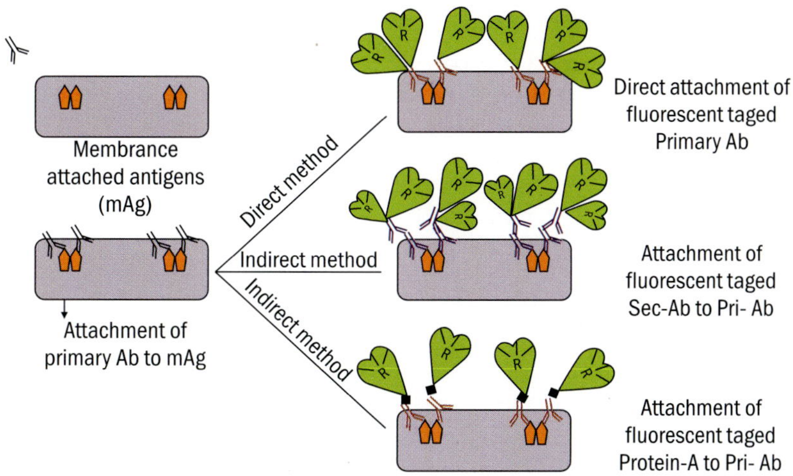

Plate 5 Immunofluorescence staining of membrane-tagged antigens

Note In direct staining, a labelled primary antibody is used, and in indirect staining, an unlabelled primary antibody and a labelled secondary antibody are used. A fluorescence microscope is used to view the stained membrane.

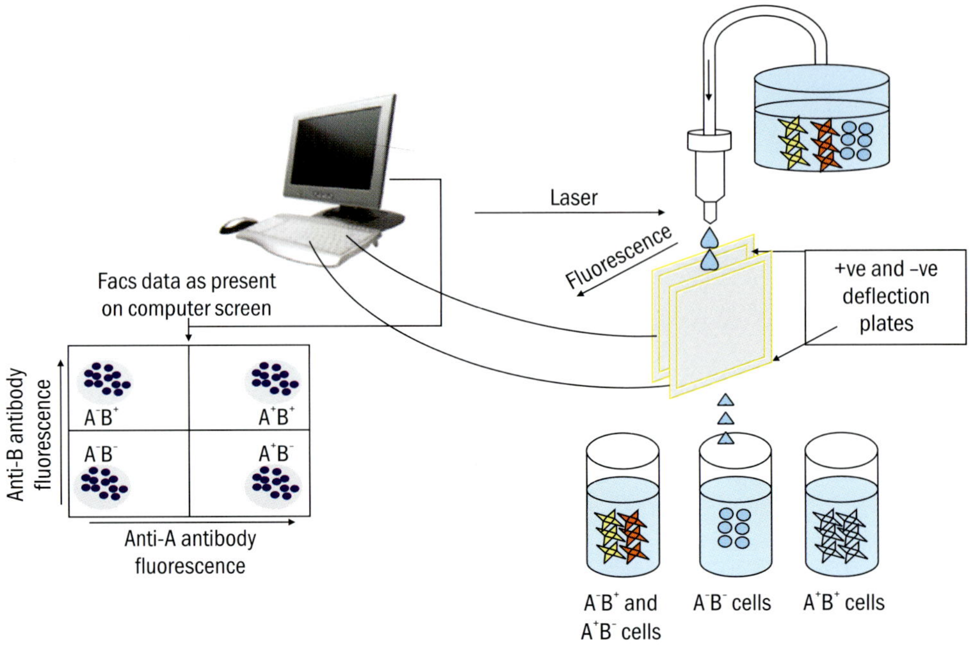

Plate 6 Flow cytometer used to separate fluorochrome-labelled cells from a mixed population

Note Two antibodies are used—one specific for surface antigen A labelled with fluorescein (yellow) and the other for antigen B labelled with fluorescein (red). The cells stained with both fluoresceins are loaded into the sample chamber of the cytometer. A single cell is expelled each time in microdroplets from vibrating nozzle. Each microdroplet contains a single cell, which receives a small electrical charge. The cell is detected by the computer controlling the flow cytometer when the drop passes through the beam of laser that excites the fluorochrome. The detector monitors the intensity of the fluorescence emitted by each droplet and displays the result on the computer screen. The passing time of each droplet through the deflection plate is fixed, and hence the droplets are collected in separate vessels. This makes possible the sorting of a population of cells into different profiles based on their surface marker.

The dots in the lower left corner represent cells not reacting with both antibodies (A^-B^-); the dots in the upper left side represent cells reacting with antibody B (A^-B^+); the dots in the upper right side represent cells reacting with both antibodies (A^+B^+); the dots in the lower right side represent cells reacting with only antibody A (A^+B^-).

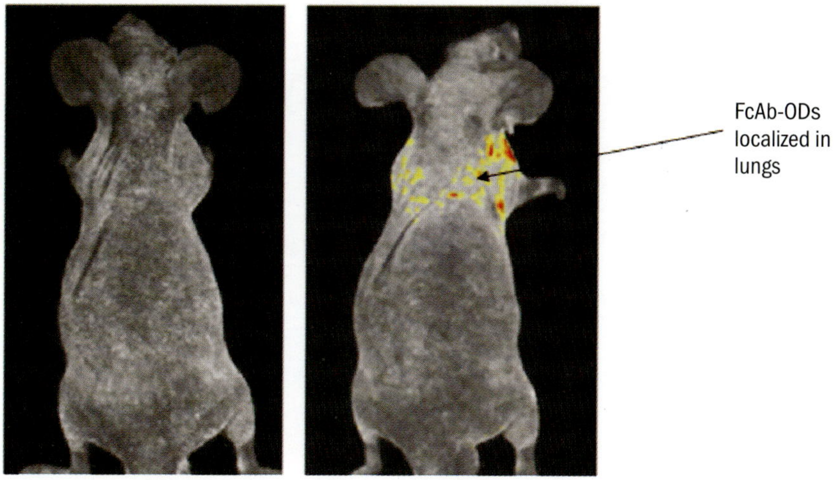

a. ODs without FcsAb associated

b. ODs with Ab associated FcsAb-ODs

Plate 7 (a) Mice injected with quantum dots without an antibody associated as control, showing distribution of nanoparticles all over the body. (b) Mice injected with quantum dots associated with Fc antibody (FcAb) to form complex FcsAb-QDs, which is localized only in lungs showing targeted delivery

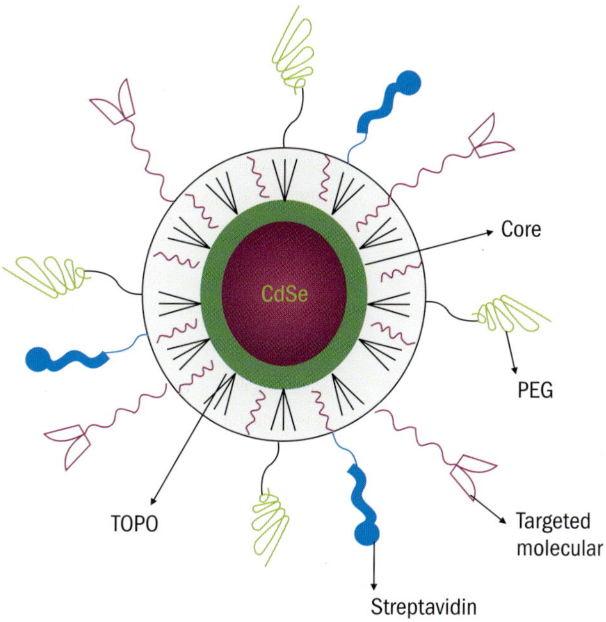

Plate 8 Structure of quantum dots
Source Peng and Yan (2010)

Chapter 6
Major Histocompatibility Complex

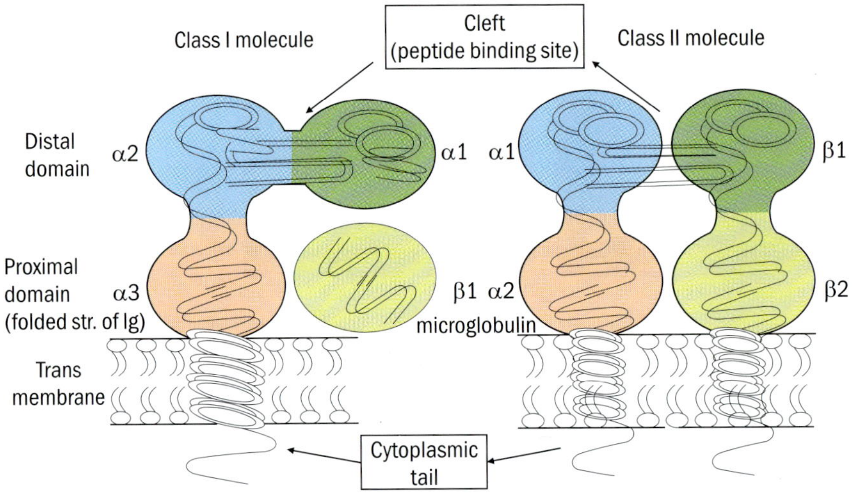

Plate 1 Structure of class I MHC and class II MHC molecule domain

Note Two spheres denote disulphide bonds: (a) cleft is formed by α_1- and α_2-domains and β_2-microglobulin interacts with α_3-domain; (b) cleft is formed of α_1 and β_1 for peptide binding. Cleft is composed of antiparallel β-strands and is able to accommodate 8–10 residues.

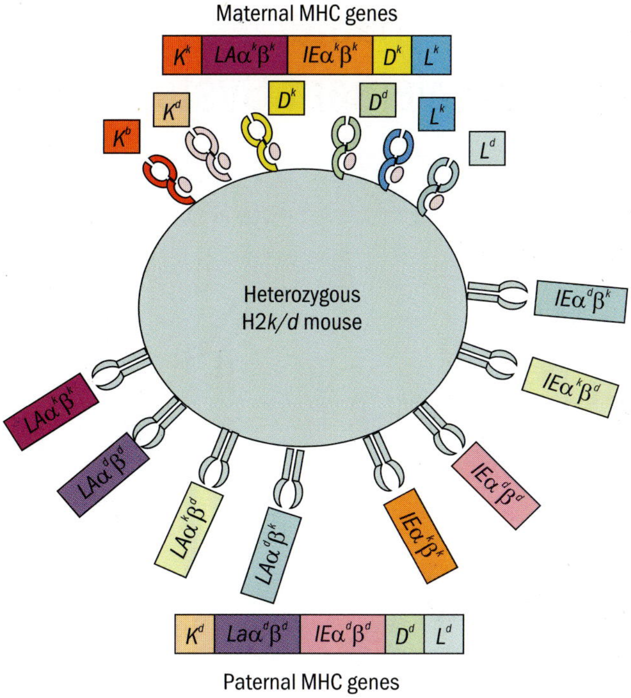

Plate 2 Expression of various MHC I and MHC II molecules on antigen-presenting cells of a heterozygous H-$2^{k/d}$ mouse

Note Both maternal and paternal genes are expressed. Pink colour molecule is of β2-microglobulin component of class I molecules. Since class II molecules are heterodimers, one maternal derived and one paternal derived chain are produced.

Chapter 7
Complement System

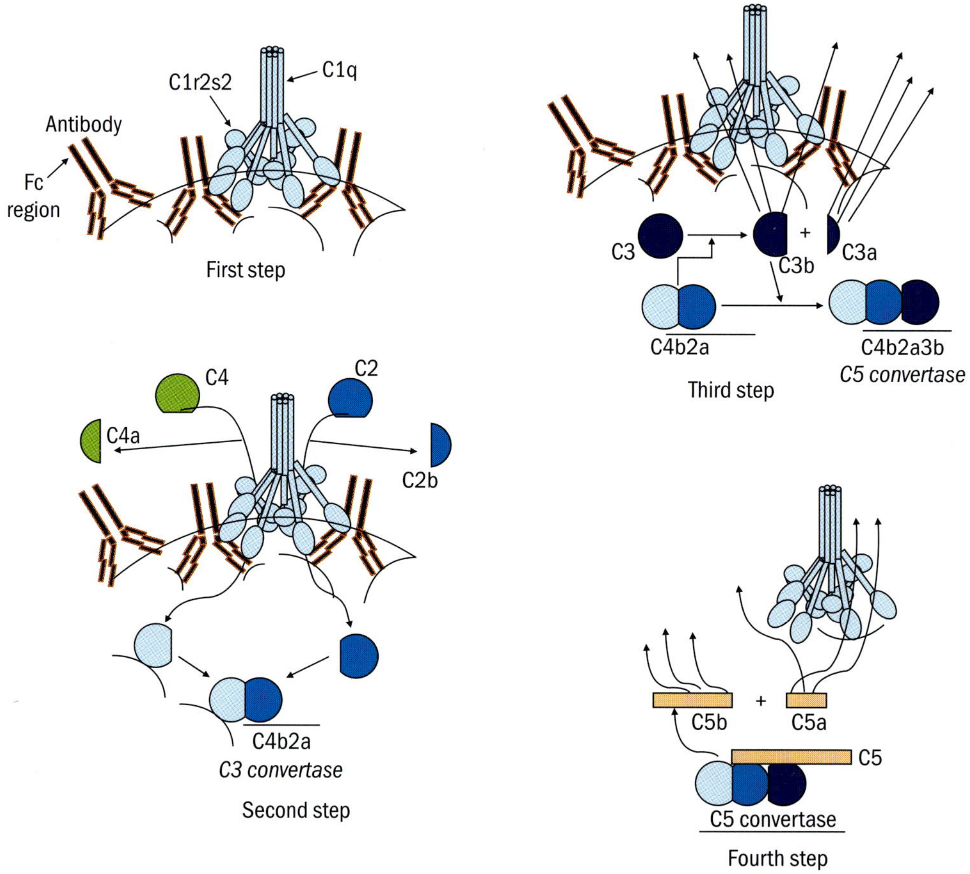

Plate 1a Steps of classical pathway of complement activation cascade leading to the formation of membrane attack complex (MAC) formation

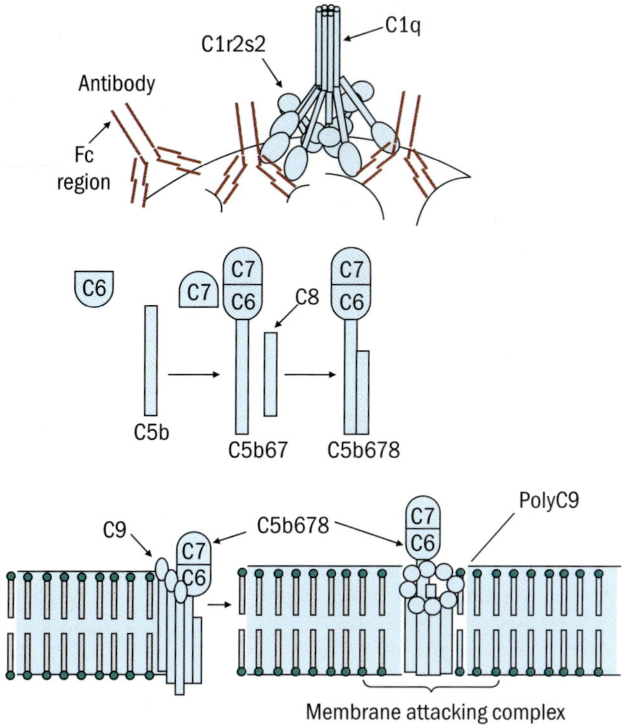

Plate 1b Steps of classical pathway of complement activation cascade

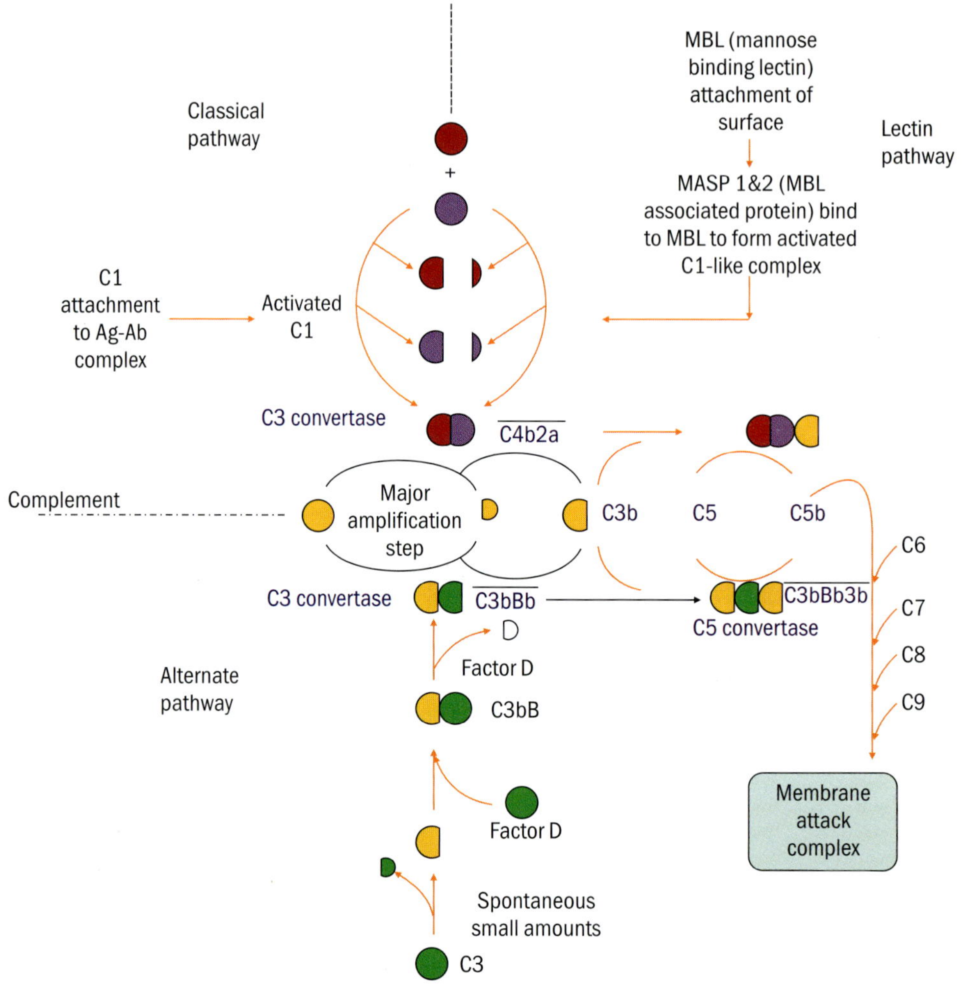

Plate 2 Complement activation pathways

Note C1-binding to antigen–antibody complex initiate classical pathway. Spontaneously generated C3b attach to cell wall of microbes to initiate alternative pathway. Serum protein MBL binds to pathogen surface to initiate lectin pathway. All the pathways produce C3 and C5 convertases and bound to C5b to form membrane attack complex (MAC). A major amplification pathway is C3b production and small amount of C5 convertase generation.

Chapter 8
Regulation of Adaptive Immune Response

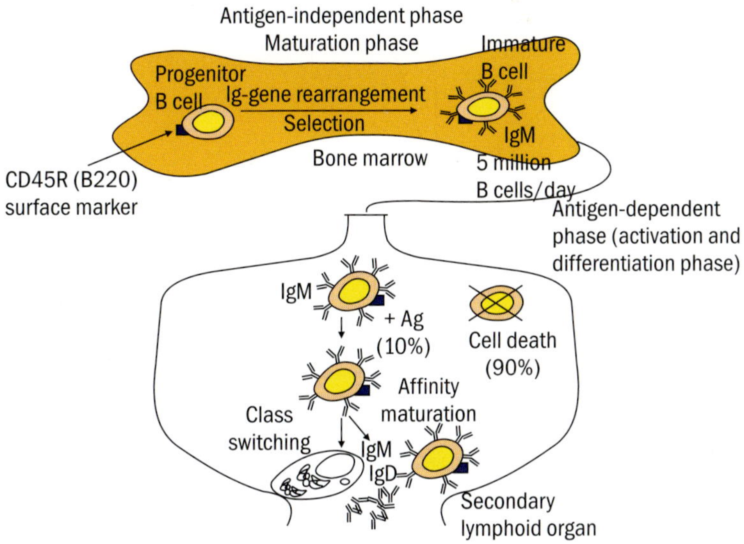

Plate 1 Characteristic stages in B-cell maturation in bone marrow and periphery in antigen-independent and antigen-dependent manners

Note Except 10%, rest of the B cells die before reaching maturity. The peripheral B cells only survive if they come in contact with soluble protein antigen and activated Th cells. After activation, the surviving B cells proliferate in secondary lymphoid organs. High-affinity mIg-bearing B cells differentiate into plasma cells and memory cells. After class switching, plasma cells express different isotopes Ig.

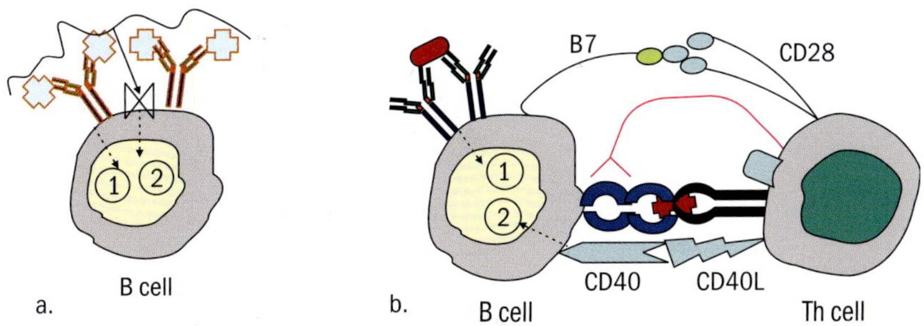

Plate 2 Mechanism involved in B-cell activation: (a) thymus-independent-1 (TI-1) antigen activation of B cells and (b) Th cell-mediated, thymus-dependent (TD) antigen activation of B cells

Notes TI-1 antigen provides both signals 1 and 2. Antigen binding provides signal 1, while CD40 and CD40L interaction provides signal 2.

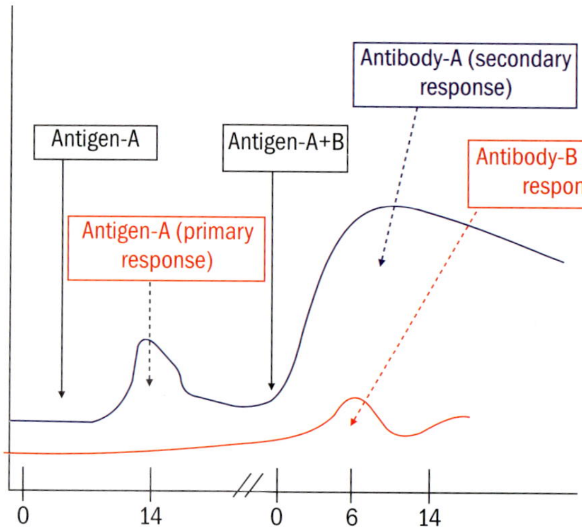

Plate 3 Primary and secondary humoral responses of B cells

Note Blue line represents primary and secondary response upon first-time interaction with antigen and second-time interaction with antigen-A and subsequent production of antibody. Red line represents only primary response upon first-time interaction with antigen and production of antibody in response.

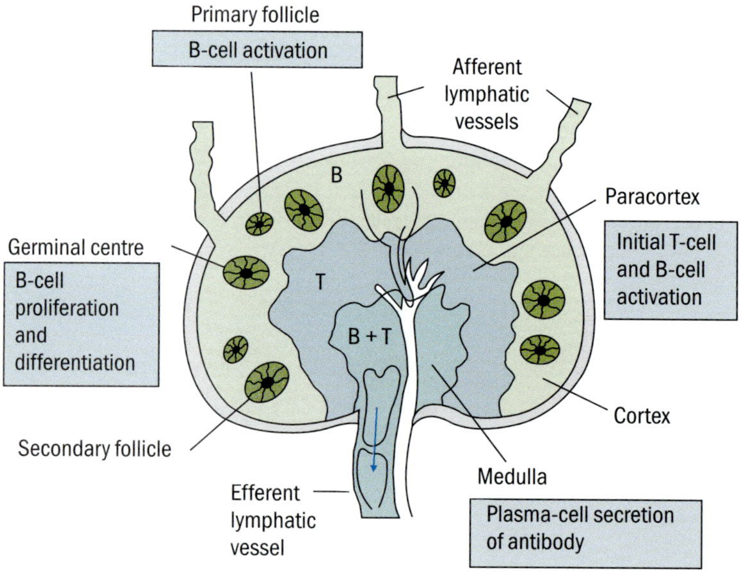

Plate 4 Anatomic sites of peripheral lymph node showing steps in B-cell activation, proliferation, and differentiation

Note Cortex contains B cells, paracortex contains T cells, and medulla contains both B cells and T cells. B-cell activation occurs in the primary follicle, while B-cell proliferation and differentiation occur in the germinal centre. The secondary follicle contains follicular mantle and germinal centre.

Chapter 9
Cell-mediated Immunity and Cytotoxicity

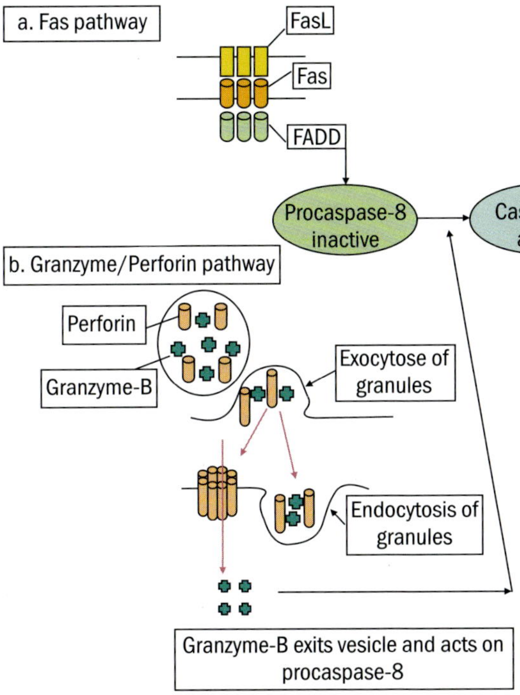

Plate 1 Apoptosis of cells mediated by Fas pathway and granzymes/perforin pathway

Chapter 10
Allergic Reactions or Hypersensitivity

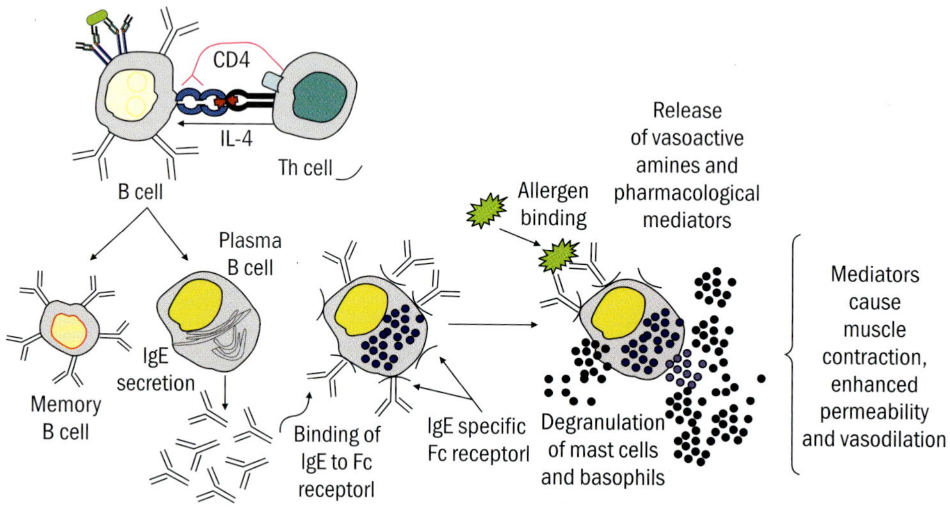

Plate 1 Type I hypersensitive mechanism

Note The hypersensitive mechanism includes secretion of allergen-specific IgE and degranulation of mast cells and basophils to produce vasoactive amines and pharmacological active mediators, which cause muscle contraction, enhanced permeability, and vasodilation.

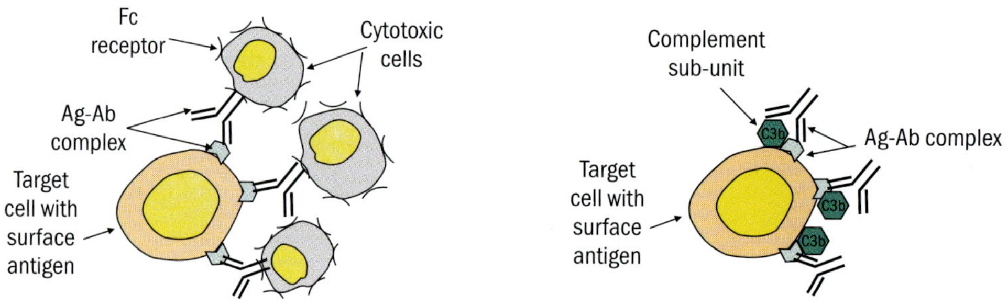

Plate 2 Type II hypersensitive mechanism mediated by cytotoxic cells and/or complement subunits

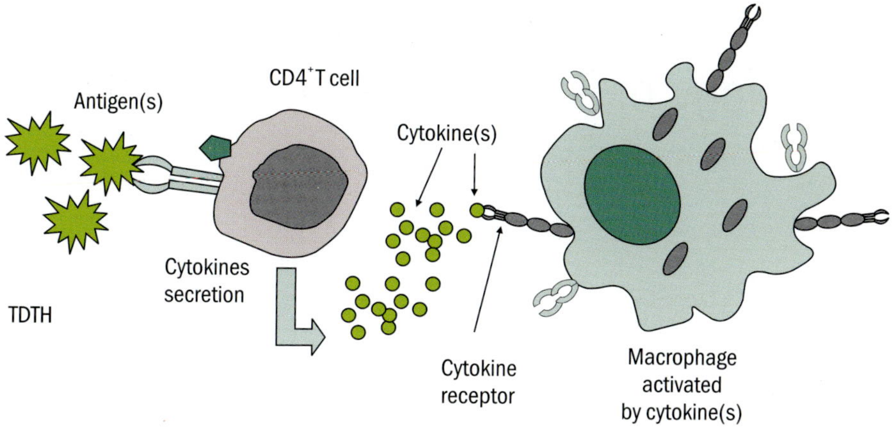

Plate 3 Mechanism of type IV hypersensitivity

Note Antigen binding activates CD_4^+ T cells called DTH to produce cytokines, which leads to the activation of macrophage.

Chapter 13
Immunology of Graft Rejection or Transplant Rejections

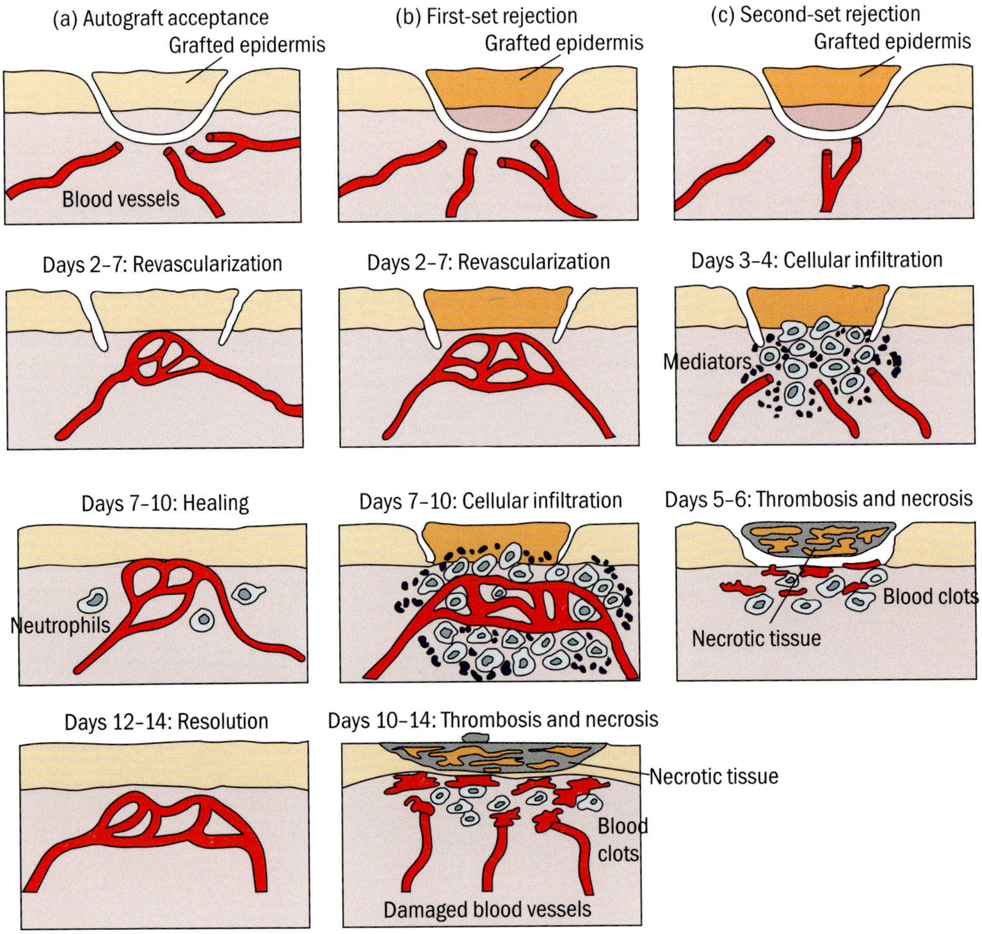

Plate 1 Step-wise representation of graft acceptance and rejection: (a) graft acceptance within 12–14 days; (b) first set rejection after 7–10 days and complete rejection by 10–14 days; (c) second set rejection within 3–4 days and complete rejection by 5–6 days

Note Graft-invading cells that cause rejection include lymphocytes, phagocytes, and inflammatory cells.

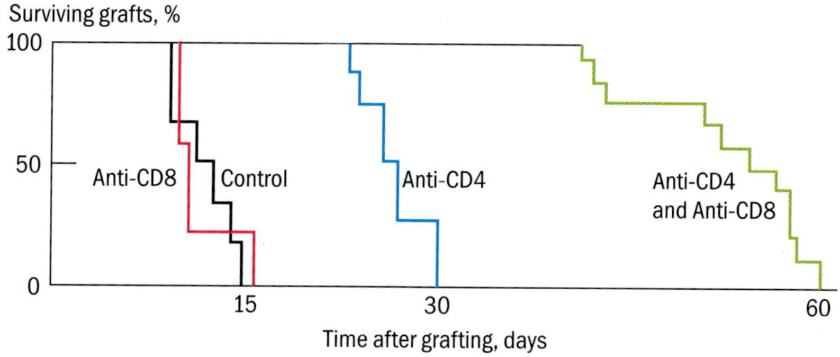

Plate 2 Role of CD_4^+ and CD_8^+ T cells on graft rejection

Note Anti-CD8 antibody is used to remove $CD8^+$ cells. Anti-CD4 antibody is used to remove $CD4^+$ cells. Anti-CD4 and anti-CD8 antibodies are used to remove both $CD4^+$ and $CD8^+$ T cells. Time after grafting (in days) represents the effect on graft survival.

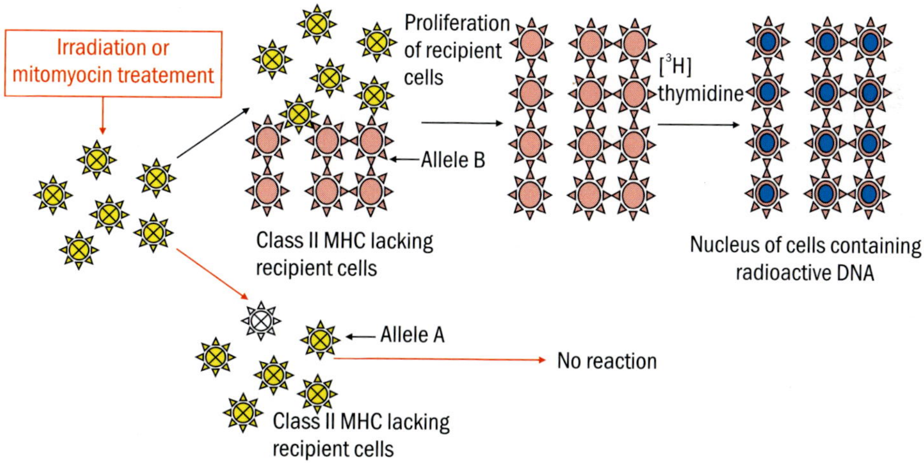

Plate 3 Identification of class II HLA antigens by mixed lymphocyte reaction between donor and recipient

Note Irradiated or mitomycin-C-treated lymphocytes of donor are added to the cells from recipient. If class II HLA antigens of donor and recipient differ from each other, the recipient cells divide very fast to incorporate radioactive nucleotides into newly synthesized DNA. The radioactive nucleotide uptake would be proportional to MHC class II differences between donor and recipient lymphocytes.

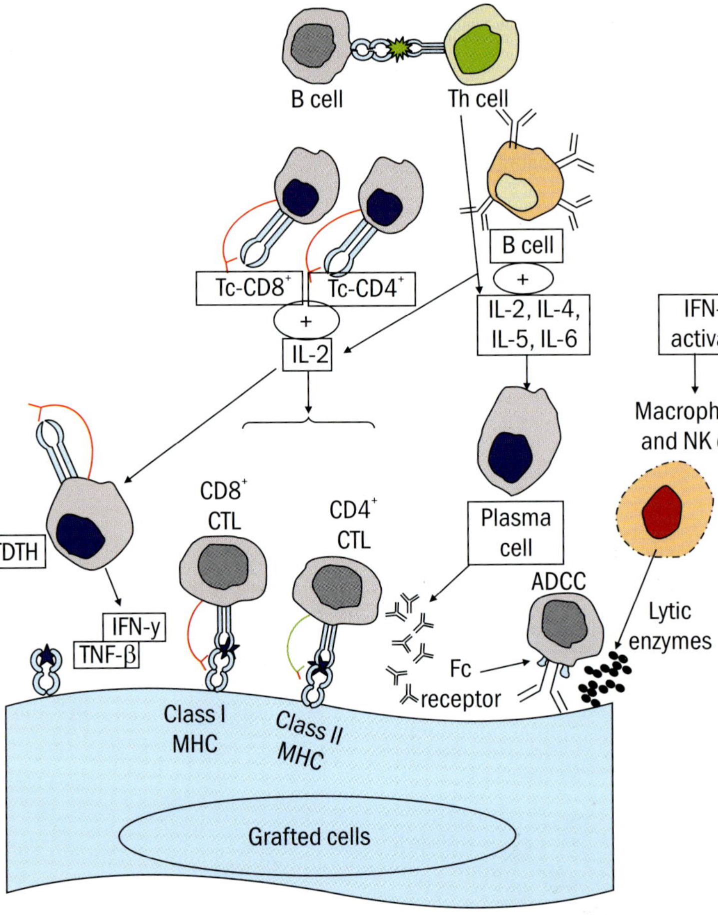

Plate 4 Allograft rejection based on effector mechanism

Note The activity of effector cells depends on the cytokines secreted by activated Th cells. Rejection is guided by antibody-dependent cell-mediated cytotoxicity and lytic enzymes produced by macrophages. T_{DTH} produces IFNγ and TNFβ, which cause damage to the grafted cells.

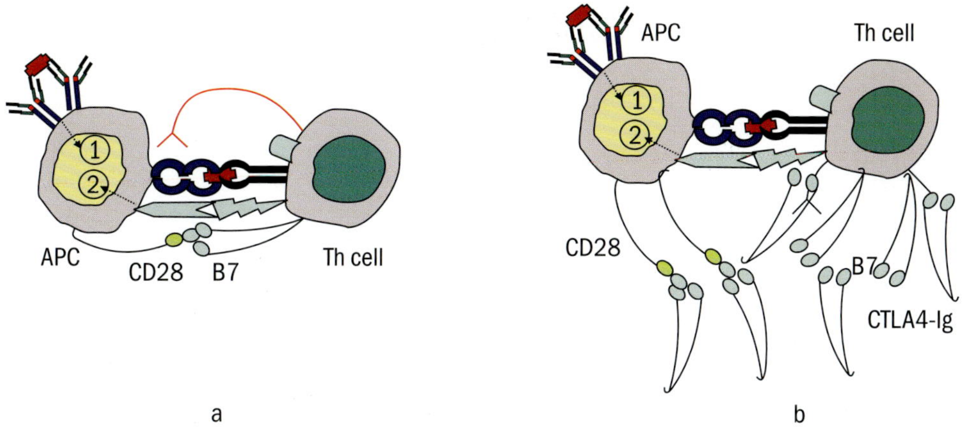

Plate 5 T-cell-mediated graft rejection procedure: (a) T-cells-interacting graft antigens become activated, causing graft rejection; (b) T-cells-interacting graft antigens lack costimulation and become anergic, causing graft survival.

Chapter 14
Cytokines

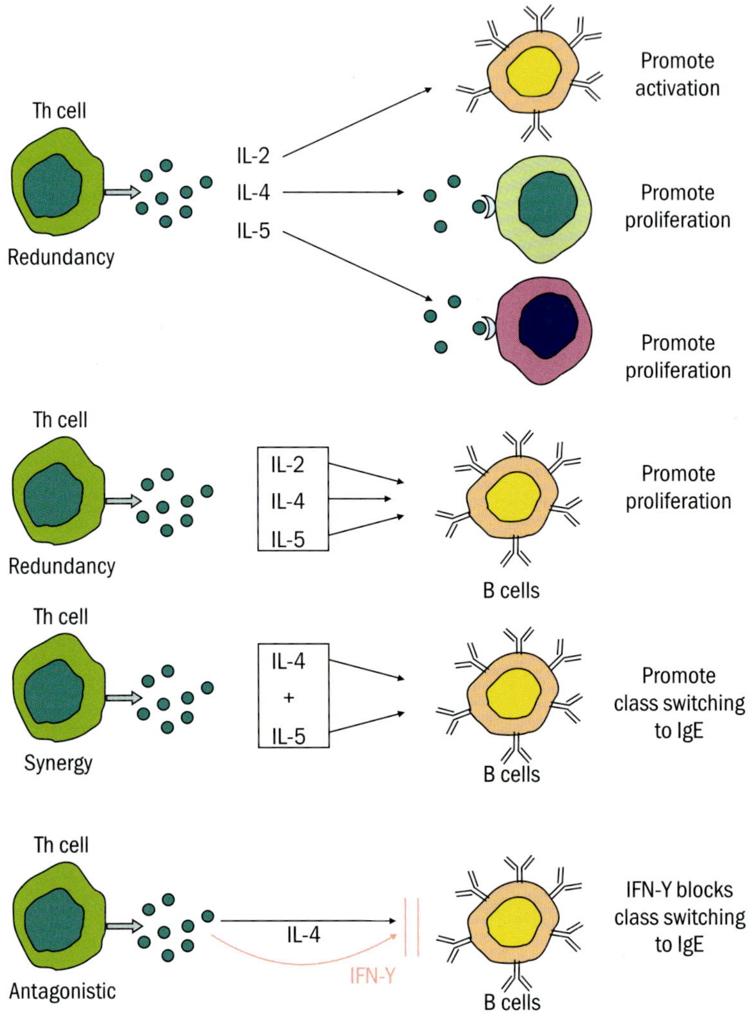

Plate 1 Effects of cytokines

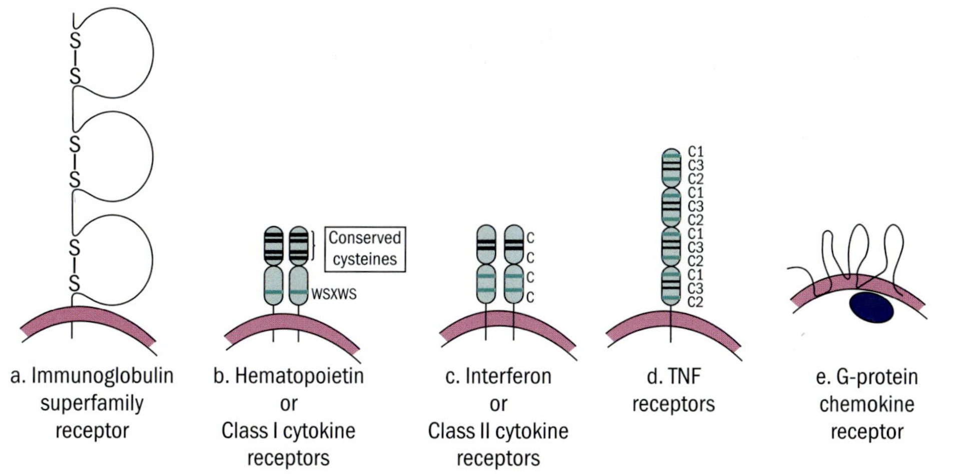

Plate 2 Cytokine-binding receptors

Note C represents conserved cysteine. Class I receptor contains most of the interleukin-binding receptors.

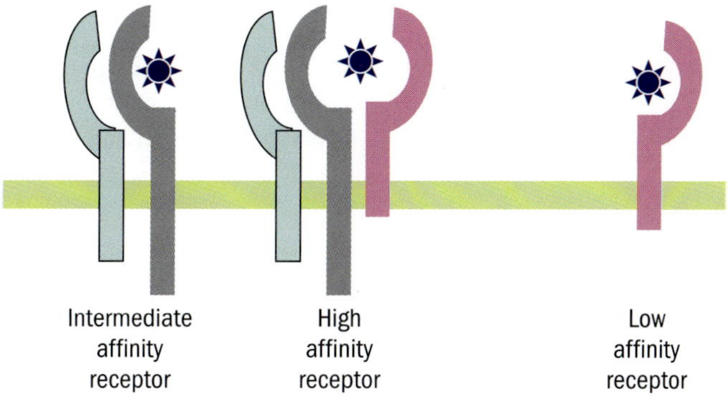

Plate 3 Representation of low-, intermediate-, and high-affinity receptors that bind to various cytokines

Note Low-affinity binding receptor shows poor signal transduction

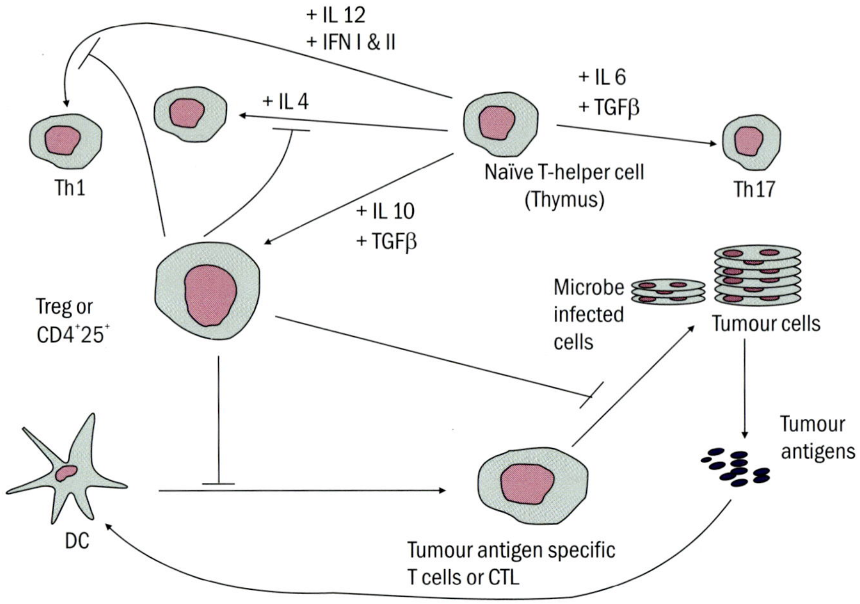

Plate 4 Tregs expression suppressing immune system in cancer patients by preventing differentiation of Th-1 cells, Th-2 cells, and conversion of normal dendritic cells to tumour antigen-specific T cells or CTLs to kill tumour cells ---------------| Suppression; ---------------→ Differentiation and activation

CHAPTER EIGHT 8

Regulation of Adaptive Immune Response

8.1 INTRODUCTION TO GENERATION AND MATURATION OF B CELLS

B-lymphocytes pass through three phases in their development: (1) B-cell generation (which occurs in the bone marrow), (2) maturation (generation of immunocompetent cells), and (3) activation/proliferation (of plasma cells and memory cells). The bone marrow of vertebrates generates B-lymphocyte cells by an orderly sequence of immunoglobulin gene rearrangements. These B cells are generated in the absence of antigens in the bone marrow. A mature B cell of the bone marrow expresses membrane-associated immunoglobulin mIgM and mIgD with specificity against one type of antigen. After maturation, B cells leave the bone marrow and are released into the blood and lymph to be transported to the secondary lymphoid organs (spleen or lymph nodes). The development of antigen-dependent B cells (Harriman, Volk, Defranoux, *et al*. 1993; Meffre, Casellas, and Nussenzweig 2000) includes affinity maturation, that is, continuous increase in the average affinity of the antibodies produced against a specific antigen, and class switching, that is, the change in the isotypes (μ to γ, α, or ε) of the antibody produced by the B cells.

A large number of naïve B cells are produced in the bone marrow of a vertebrate throughout its life. The cells circulate with a half-life of less than three days to approximately eight weeks. In this short lifespan, if they meet with antigens, they survive to reside in the lymphoid environment for further development; otherwise, they do not survive. In humans, more than 10^{12} different antibody molecules (10^9 in mice), called *pre-immune antibody repertoire*, are produced even in the absence of antigen stimulation. These antibodies have an ability to cross-react with other antigenic determinants. The large amount of pre-immune repertoire is essential to ensure the presence of an antigen-binding site fit to bind any antigenic determinant even with low affinity. After repeated and continuous stimulation of B cells by antigens, B cells can produce antibodies having an ability to bind antigens with higher affinity, a process called *affinity maturation.* The step-wise maturation of B cells into bone marrow and periphery lymphoid organs has been shown in Plate 1 and Figure 8.1.

8.1.1 Origin of B-lymphocyte Cells

At the embryo stage, B cells are generated from haematopoietic stem cells (HSCs) in the yolk sac, fetal liver, and fetal bone marrow. Generation continues even after birth in the bone marrow. The differentiation of HSC B cells or naïve B cells into mature B cells can occur in different stages depending on the phenotype and functional properties acquired by the B-lineage cell during the developmental process. The cells develop from progenitor-B cells (pro-B cells) to precursor-B cells (pre-B cells) to mature B cells (Berland and Wortis 2002).

8.1.2 Proliferation of Pro-B Cells

At first, lymphoid stem cells differentiate into B-lineage cells that give rise to pro-B cells (Hardy and Kayakawa 2001). Pro-B cells are identified by marker protein CD45R or B220, which is transmembrane tyrosine phosphatase. Pro-B cells also express CD43

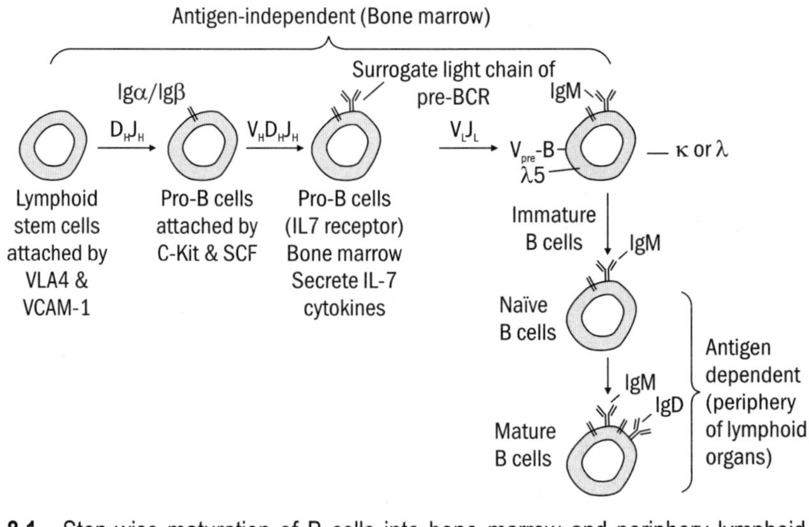

Figure 8.1 Step-wise maturation of B cells into bone marrow and periphery lymphoid organs

Note Precurson B cells (pre-B cells) are associated non-covalently with heavy (H) chain and Vpre-B and λ5 surrogate light chains. The immature B cells express κ or λ light chain and μ heavy chain. Pre-B cells have $V_HD_HJ_HC\mu$ along with κ or λ light chain. Mature B cells have membrane-bound IgM and IgD. Synthesis is done by recombination-activating enzymes encoded by two genes RAG-1 and RAG-2, which encode terminal deoxynucleotidyl transferase (TdT).

and c-kit. Pro-B cells proliferate in the bone marrow and fill the extra vascular space of the bone. The stromal cells of the bone marrow provide a proper environment for the differentiation of pro-B cells to Pre-B cells. Stromal B cells interact with pro-B cells or pre-B cells and secrete IL-7. pre-B cells that cannot grow on stromal cells are unable to produce IL-7. IL-7 receptors (IL-7Rs) present on pre-B cells are heterodimers composed of an α-chain and a γc-chain. The γc-chain is also shared by IL-2, IL-4, IL-9, and IL-15 for signalling. Interaction of the pro-B cells with the stromal cells is mediated by cell adhesion molecules such as VLA-4 of pro-B cells and VCAM-1 of stromal cells. After contact, the c-kit of pro-B cells interacts with the stem cell factor (SCF) of stromal cells. The c-kit has tyrosine kinase activity that activates pro-B cells to differentiate into a pre-B cell and acquire a receptor for IL-7. IL-7 initiates and supports maturation and the growth process. This enables pre-B cells to detach from stromal cells by the downregulating expression of adhesion molecules (Figure 8.1).

8.1.3 Rearrangement of Immunoglobulin Gene

The rearrangement of immunoglobulin genes starts in the bone marrow with pro-B cells. It first occurs in all pro-B cells, with an immunoglobulin heavy (H) chain D_H to J_H gene rearrangement followed by a V_H to D_HJ_H rearrangement. The successfully rearranged H-chain is expressed on the cell surface of the pro-B cells. After which, the cells are called pre-B cells. In the pre-B cells, immunoglobulin light (L) chains are rearranged and expressed on the surface of the cells to develop them into immature B cells. During gene rearrangement, the recombinase enzymes RAG-1 and RAG-2 are required and, hence, expressed at pro-B and pre-B cell stages. At pro-B cell stage, N-nucleotides are inserted into D_HJ_H and $V_HD_HJ_H$ coding joints with the help of the enzyme terminal deoxynucleotidyl transferase (TdT). At early pre-B cell stage, the expression of TdT is turned off, and N-nucleotides are not present because they are not required in the V_LJ_L joining (Blackwell and Alt 1989; Lewis 1994).

Immature B cells express both H- and L-chain on the cell surface. By allelic exclusion, the B cell expresses only one light chain isotype on its membrane. The antigenic specificity of immature B cells is determined by the V(D)J sequence of the heavy chain and the VJ sequence of the light chain. Immature B cells have membrane-IgM (mIgM) expressed on their cell surface at the end of the bone marrow developmental phase. During this phase, B cells are partially functional and death or anergy (unresponsiveness) is induced by antigens at the division and cell differentiation site. Coexpression of IgD along with IgM on the B-cell membrane is the sign of full maturation. IgD is a characteristic marker of mature naïve B cells but is not essential for development or response against antigen as proved by experiments on IgD knockout mice.

B-cell anergy is defined as the inability of an immune B cell to mount a complete response against its target antigens, that is, the *unresponsiveness* of the B cells against circulating antigens is known as anergy. B-cell anergy is often marked by the downregulation of surface-IgM expression and partial blockade of intracellular signalling pathways induced by exposure to soluble circulating antigens.

8.1.4 Development of Mature B cells

Two protein complexes Vpre-B (V-like sequence) and λ5 (C-like sequence) non-covalently attached to each other make a light chain-like structure called a *surrogate light chain* in the membrane of a pre-B cell. The surrogate light chain (Vpre-B and λ5 complexes) gets associated with the expressed μ-heavy chain (Cμ) and then comes in contact with membrane-associated Igα/Igβ heterodimer to form a pre-B-cell receptor. Only Cμ-expressing pre-B cells, which get associated with the surrogate light chain to form receptors after association with Igα/Igβ, proceed to follow the maturation pathway, recognizing the ligand on the stromal cell membrane. The pre-B cell receptor, which does not recognize ligand on stromal cells, receives a signal not to proceed V_H to D_HJ_H allelic rearrangement (Harriman, Volk, Defranoux, *et al.* 1993) thus causing allelic exclusion (Figure 8.1). The pre-B cells having effective receptors undergo multiple divisions to produce 32 to 64 progeny of pre-B cells. Rearrangement of light chain gene segments occurs in each of these pre-B cells to increase the diversity of antibody repertoire. The knockout mice experiment proved that λ5 is important to generate signals to proceed the pre-B cells to the immature B-cell stage.

The four important transcription factors involved in B-cell development are E2A, early B-cell factor (EBF), B-cell-specific activator protein (BSAP) (a product of *Pax-5 gene*), and Sox-4 (Henderson and Calame 1998). E2A and EBF are essential for RAG-1 (essential for D_HJ_H rearrangements), and λ5 is an important part of surrogate light chain expression. BSAP binds in the promoter region of specific genes of B cells, for example, Vpre-B and λ5, immunoglobulin switch regions and immunoglobulin heavy chain enhancer, important for early stages of B-cell development. Detailed information of the binding sites of Sox-4 is still unknown (Henderson and Calame 1998).

8.1.5 Marker Proteins of B cells

Different developmental stages of B cells have specific marker proteins on their cell surface. The pro-B cells express CD45R marker proteins (type of protein tyrosine phosphatase), Igα/Igβ (signal-transducing proteins), CD19 (B-cell coreceptor segment), CD43 (leukosialin), CD24 (heat stable antigen), and c-kit receptor (for stromal cells protein) on the membrane surface. At the pre-B-cell stage, c-kit and CD43 expression ceases and CD25, IL-2 receptor α-chain, and pre-BCR are expressed. Pre-BCR expression is a prominent feature of pre-B cells. Immature B cells are marked by the expression of immunoglobulin having both heavy and light chains, loss of CD25 and pre-BCR (Table 8.1).

Table 8.1 Status of different types of lymphoid B cells at different developmental stages

Developmental stage	Lymphoid progenitors	Pro-B cells		Pre-B cells		Immature B cells		Naïve mature B cells
Subcategories		Early Pre-B	Late Pre-B	Large Pre-B	Small Pre-B	Transitional I	Transitional II	—
Immunoglobulin gene status	Germ line	D → J_H	DJ_H → V_H DJ_H	V_HDJ_H $V_{L\to}J_L$		V_HDJ_H V_LJ_L		V_HDJ_H V_LJ_L
Rearrangement enzyme	N/A	RAG-1, RAG-2,TdT		RAG-1, RAG-2, TdT		RAG-1, RAG-2	RAG-1, RAG-2	Enzyme inactivated
Cell surface markers	CD43, IL-7R	B220, Igα, Igβ	c-kit, CD19, Pro-BCR, SL	c-kit, Pro-BCR, CD43, CD25, Pre-BCR	IL-7R	SL, Pre-BCR, CD25, $B220^{Low}$, IgM^{hi}, IgD^{rare}, $CD21^+$		$CD22^+$, $CD23^+$, L-selectin$^+$, Mostly IgM^{low}, IgD^+
Additional comments	30% of total pro-B cells, and IL7-responds	Differentiation but not proliferation controlled by IL-7, N-addition between V_H and D induces allelic exclusion		Proliferative	Non-proliferative	Gene edit when bound to antigen	Clonal deletion or anergization when bound to antigen	Activated when bound to antigen

8.1.6 Subsets of B cells

There are two subsets of B cells: B1 subsets and B2 subsets. B1 subsets consist of 5% of the total population of B cells. These cells are expressed during the fetal stage in the bone marrow. Their cell surface has IgM but little or no IgD. Human B1 subsets express CD5 marker protein, which is absent in rats and mice. The proliferation sites of B1 subsets in the post-natal stage are outside the bone marrow. B1 subsets show poor response to protein antigens but good response to carbohydrate antigens. They also show less somatic hypermutation (Papavasiliou and Schatz 2002) and class switching than the B2 subsets. Hence, they produce low affinity antibodies (Feinstein, Richardson, and Taussig 1986).

8.1.7 Negative Selection of B cells

Only 10% of the total B cells produced (in the bone marrow) is recruited to recirculate in the B-cell pool. The rest 90% of B cells is left to die in the bone marrow. This loss is due to negative selection and clonal deletion of immature B cells. Negative selection and clonal deletion is caused by the immature B cells in the bone marrow expressing autoantibodies against self-antigens. It has been experimentally proved in vitro that antibody against the μ-constant region causes death of immature B cells by apoptosis within the bone marrow. Similarly, immature B cells with self-reactive mIgM antibody bind to self-antigens present on the stromal cells leading to the death and elimination of immature B cells in the bone marrow in vivo.

8.1.7.1 Release from negative selection

The binding of self-antigens to the antibody of immature cells causes arrest of growth and maturation of immature B cells. Such cells edit the light chain gene of its receptor for further rearrangement by upregulating RAG-1 and RAG-2 genes. In some cells, the κ-light chain of self-antigen reactive antibody is replaced with λ-chain, which is encoded by endogenous λ-chain gene segments. The resulting antibody mIgM with replaced light chain loses its self-antigens reactivity, thus preventing negative selection. This allows the cells to leave the bone marrow to recirculate the immature B-cells pool.

8.1.7.2 Activation and proliferation of B cells

Activation and proliferation of cells occur in the periphery in the presence of antigens. Activation in the presence of antigen leads to the clonal selection of naïve B cells with the generation of plasma cells and memory B cells (McHeyzer-Williams and Ahmed 1999). The naïve B cells have a very short lifespan in the absence of antigen and die within a few weeks due to apoptosis. Based on the nature of antigens exposed to the B cell, the activation and proliferation adopt two routes: thymus-dependent antigens and thymus-independent antigens (Plate 2).

- *Thymus-dependent antigens:* There is direct contact of B cells with Th cells, T-cell-derived cytokines and antigens for activation and proliferation.
- *Thymus-independent antigens:* There is no direct contact of B cells with Th cells and antigens. Thymus-independent antigens are of two types: type 1 TI (TI-1) antigens and type 2 TI (TI-2) antigens.
- *TI-1 antigens:* TI-1 antigens are LPS (cell wall component of Gram-negative bacteria) and other bacterial cell wall components. Most of them are polyclonal B-cell activators called mitogens. They are able to activate B cells without depending on the specificity of antigens. At lower concentrations, TI-1 antigens activate only specific B cells that have a specific epitope for antigens. At higher concentrations of TI-1 antigens, one-third of the B cells gets activated and proliferates to secrete antibody by a mechanism, which is still unknown. For example, at lower concentrations of LPS, the antibodies produced are LPS specific, while at higher concentrations of LPS, the antibodies produced are of polyclonal type.
- *TI-2 antigens:* TI-2 antigens are polymeric proteins that are highly repetitious molecules, for example, bacterial flagellin and LPS with repeating polysaccharide units. They cross-link with mIg receptors of B cells to activate the respective B cells. TI-2 antigens differ from TI-1 antigens in the following three respects: (1) TI-1 antigens are not mitogens and are not polyclonal activators; (2) TI-1 antigens are able to activate both mature and immature B cells, while TI-2 antigens activate mature B cells and inactivate immature B cells; (3) TI-2 antigens require T-cell-derived cytokines, but not Th cells directly, for efficient B-cell proliferation and class switching from IgM to other isotypes (Feinstein, Richardson, and Taussig 1986; Snapper and Mond 1996; Stavnezer 1996).

  TI-2 antigens do not elicit B-cell response and production of antibody in T-cell-deficient mice called nude mice. This is because nude mice do not express either $\alpha\beta$T-cell receptors (TCRs) or $\gamma\delta$TCRs, the gene of which have been knocked out. If T cells are administered to these nude mice from outside, their B cells will be able to elicit B-cell responses against TI-2 antigens.

When compared to thymus-dependent antigens, thymus-independent antigens do not generate memory cells and have a comparatively weaker affinity maturation and a low level of class switching (IgM is predominant) to other isotypes. Thymus-dependent antigens, on other hand, are dependent on Th cells, which play an important role in generating memory cells, affinity maturation, and bringing about a high level of isotypes class switching (Feinstein, Richardson, and Taussig 1986).

8.1.8 Cell Cycle Stages of B cells

The progression of lymphocytes from G_0 to S phase leading to division occurs under the influence of two types of signals: *competence signals* and *progression signals*. First, the competence signals, which are of two types—*signal 1* and *signal 2*, drive the cells from G_0 to early G_1 phase, making the cell competent to receive progression signals. The progression signals then drive the cells from G_1 phase to S phase and finally division and differentiation of cells.

8.1.9 Intracellular Signalling Pathways of B cells

Intracellular signalling of B cells depends on Igα/Igβ heterodimers. The membrane-associated immunoglobulins mIgM and mIgD extend only three amino acids into the B-cell cytoplasm. The tails of mIgA, mIgG, and mIgE consist of 14 amino acids, 28 amino acids, and 28 amino acids, respectively. In each case, the length of tail is too short to generate intracellular signalling by interacting with tyrosine kinases and G-proteins (Hardy and Kayakawa 2001). It was later discovered that this problem is overcome by disulphide-linked single heterodimer Igα/Igβ interacting with mIg molecule to generate the receptor complex called B-cell receptor (BCR).

B-cell receptor is functionally made up of two parts: (1) immunoglobulin molecule and (2) Igα/Igβ heterodimer; the molecule is involved in ligand binding and the heterodimer in transducing the signal. The cytoplasmic tails of Igα and Igβ contain 61 amino acids and 48 amino acids, respectively. Both Igα and Igβ contain an 18-residue motif called immunoreceptor tyrosine-based activation motif (ITAM). BCR and other specific immunoglobulin receptors for the Fc regions, such as, FcεRI for IgE, FcγRIIA, FcγRIIC, FcγRIIIA for IgG, bind with ligands and transduce signals mediated through multimeric protein complex. These protein complexes are functionally compartmentalized. The ligand-binding protein complex part, such as mIg for BCR, resides on the cell surface and the signal-transducing part resides in the cytoplasm of the cells. BCR has no protein tyrosine kinase (PTK) activity of its own like TCR. The PTK activity for BCR is acquired by recruiting cytoplasmic kinases to the cytoplasm. Cross-linking of BCR cells surface immunoglobulin with antigen initiates a number of signalling cascades or signal transduction pathways resulting in the activation of respective B cells (Benschop and Cambier 1999). The activation and signalling pathways of B-lymphocytes are similar to T-lymphocytes in many ways. A failure or defect in the signal transduction pathway of B-cells development causes X-linked agammaglobulinemia. The roles of BCR and TCR activation are as follows.

- *Receptors and signalling unit:* Signalling pathways mediated by both B cells and T cells begin with the binding of antigens to a specific cell surface receptor, passing signals to the signalling unit. This unit has long cytoplasmic tails, which act as transducers of receptor signals.
- *Activation of PTKs:* The PTKs associated with BCRs or TCRs called Src-PTKs (such as Lck in T cells and Lyn, Blk, and Fyn in B cells) initiate phosphorylation to assemble a functional receptor signalling complex (Figure 8.2).

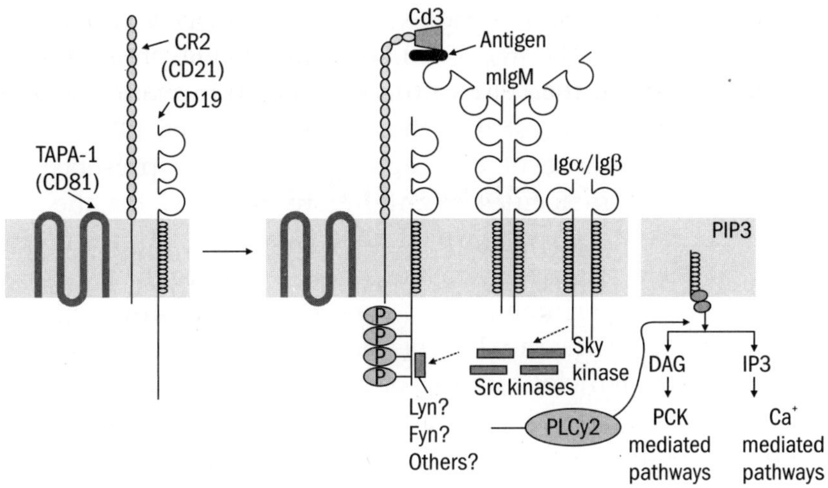

Figure 8.2 TAPA-1 (CD81), CR2 (CD21), and CD19 constituting B-cell coreceptor complex

Note Binding of CR2 to C3d a complement-derived component coated with antigen to mIg causes phosphorylation of CD19. Phosphorylated CD19 interacts with Src kinases Lyn, Fyn, and others to trigger signal transduction pathways.

- *Active complex formation with PTKs:* The ITAMs of BCRs and TCRs get phosphorylated at tyrosines to act as docking sites for molecules that provide PTK activity to these receptors, four example, ZAP-70 in T cells and Syk in B cells (Figure 8.2).
- *Activation of other signalling pathways:* BCR and TCR signalling produces second messengers, which are involved in other secondary signal transduction pathways. For example, IP3 is involved in the signalling of Ca^{2+} release, and diacylglycerol (DAG) causes protein kinase C (PKC) activation and initiates small G-proteins Ras and Rac signalling (Figure 8.2).
- *Affecting gene expression:* BCR and TCR signalling causes translocation of transcription factors from the cytoplasm to the nucleus. This regulates specific gene expression.

8.2 ROLE OF B-CELL CORECEPTOR

Three protein complexes—CD19, CD21 (CR2), and CD81 (TAPA-1)—are together known as the B-cell coreceptor. CD19 has a long cytoplasmic tail along with three extracellular domains and is a member of the immunoglobulin superfamily (Hunkapillar and Hood 1989). It contains six tyrosine residues in its tail and acts as a major substrate for tyrosine kinase. CR2 is a receptor for C3d and TAPA-1 or CD81. C3d is a breakdown product of the complement system, which acts as an effector molecule to destroy invaders. TAPA-1 is a transmembrane protein. The CR2 component of the complex binds with complement-coated antigen, which is captured by mIg of the B-cell membrane. Thus the coreceptor complex is cross-linked with BCR, providing a chance to the

CD19 component of the coreceptor complex to interact with Igα/Igβ. After tyrosine phosphorylation, CD19 allows binding of other signalling molecules such as PTK *Lyn*. In actuality, this coreceptor complex amplifies the BCR-transmitted activating signals. In vitro experimental results suggest that the antigen involves 10^4 molecules of mIgM to activate B cells in the absence of coreceptors. But in the presence of coreceptors CD19⁻CD2⁻TAPA-1, only 10^2 molecules of mIgM are required for the activation of B cells. In another experiment, it was proved that two–three copies of fusion protein of lysozyme and C3d have anti-lysozyme responses—1000 to 10,000 times more than that of lysozyme alone. Experimentally, it has also been observed that the CD19 component of the coreceptor has an independent role compared to that of the CR2 complement receptor. The artificial binding of anti-BCR antibodies to BCRs stimulates B-cell activation specific signal transduction pathways. Finally, CD19 knockout mice produces very low antibody response to most antigens (Figure 8.2).

8.3 ROLE OF Th CELLS IN B-CELL RESPONSES

The binding of soluble protein antigen to the mIg of B cell requires Th cells. Th cells assist in effective competence signalling for the activation of B cells. Cytokines are required for the proliferation of B cells. The activation mechanism of B cells is not simple. The antigen–mIg complex is internalized by B cells through receptor-mediated endocytosis so that it can be processed into peptides. It also upregulates cell-associated molecules such as class II major histocompatibility complex (MHC) molecules and costimulatory ligand B7 after receiving signal from BCR. The enhanced expression of these two protein molecules enables B cells to function as antigen-presenting cells (APCs) in the presence of Th cells. To function as APCs, the B-cell-internalized antigenic peptides are processed by endocytic processing pathway (within 30–60 min). The processed antigenic peptides get associated with class II MHC molecules to display on the B-cell membrane and be recognized by the Th cell. Since the internalization of antigen in B cells is through mIg, a very low concentration of antigens (100 to 10,000, lower than what is required for macrophage or dendritic cells) is effective. At a high concentration, antigens are processed and presented by macrophages and dendritic cells to Th cells; at a lower concentration, they are processed and presented by B cells. Th cells recognize antigens presented with class II MHC molecules on B cells or macrophage or dendritic cells to form T-B-cell conjugate, T-macrophage conjugate or T-dendritic cells conjugate, respectively. This reorganizes the Golgi apparatus and the microtubular organizing centre towards the junction of the B cell to secrete cytokines in the direction of antigen-specific B cells.

The released cytokines upregulate the Th cell membrane protein CD40L (CD154) to interact with the B-cell membrane protein CD40 to provide T-cell-dependent B-cell activation. CD40 is a cell surface protein of the tumour necrosis factor (TNF) family and is a soluble cytokine. It regulates cell proliferation and apoptosis (programmed cell death). CD40L is a TNF receptor (TNFR) family protein that interacts with CD40 on B cells to initiate the signal transduction pathway, which urges B cells into the G1 cell cycle. CD40 does not have kinase activity, but its interaction and binding with CD40L (Plate 2) lead to tyrosine kinase activation of Lyn and Syk. It also causes activation of phospholipase C and generation of secondary messengers IP3 and DAG along with the activation of transcription factors. The interaction and binding prompt the association of CD40 with TNFR-associated factor (TRAF) family, which leads to the activation of the NF-κB transcription factor. The role of the activated Th-cell membrane protein in B-cell activation was experimentally proved. Antigen and plasma membranes prepared from either activated or resting Th cell clones were used to incubate the naïve B cells. Only the naïve B cells incubated with antigen and plasma membranes of activated Th cells proliferated into B cells. This suggests that activated Th cells must contain one or more molecules expressed on their cell membrane to help B-cell proliferation by contacting BCRs. Antigen-stimulated B cells are able to proliferate when they are activated by anti-CD40 monoclonal antibodies in the absence of Th cells. This suggests the involvement of CD40 either as anti-CD40 antibody or as CD40L. These molecules are essential to pass signals to proliferate the B cells. The proliferated B cells require appropriate cytokines (added externally in the experimental system) to differentiate into plasma cells (Figure 8.1). Finally, it was found that B-cell activation terminates in the presence of anti-CD40L antibody, which blocks CD40–CD40L interaction.

8.4 CYTOKINES AND ITS ROLE IN B CELL

Cytokines are low-molecular weight regulatory proteins or glycoprotein molecules secreted by macrophages, white blood cells (WBCs), APCs, and other cells. These molecules pass messages from one cell to another. They also mediate the interactions between two cells. Some cytokines function as effector molecules or assist in the generation of immune effectors cells.

Cytokines bind to the membrane-associated receptors on the membrane of target cells leading to the initiation of the signal transduction pathway, which alters the expression of specific genes. The affinity of cytokine receptors to cytokines is very high, and hence a picomolar concentration of cytokines is sufficient to mediate biological effects. The action of cytokines might be autocrine (acting on the producing cells), paracrine (acting in close proximity), or endocrine (acting at a distance). Cytokines control the intensity and duration of activation, proliferation, and differentiation of various cells by controlling the secretion of antibodies or other cytokine molecules. They can cause the following effects.

- *Synergism:* Synergism is the combined effect of two or more cytokines to enhance the cellular activity more than each individual cytokines.
- *Antagonism:* Cytokines inhibit the effects of other cytokines.
- *Cascade induction:* Cytokines produced by one cell act on other target cells to produce cytokines that induce some other target cells. This is called cascade of induction reaction.
- *Pleiotropy:* Cytokines have different biological effects on different cells.
- *Redundancy:* Sometimes two or more cytokines have similar functions.

Cytokines secreted by lymphocytes are known as *lymphokines;* when secreted by monocytes or macrophages, they are known as *monokines;* and when secreted by leukocytes, they are known as *interleukins.* Cytokines having chemotaxis and other leukocyte behaviour are known as *chemokines.* Cytokines have many functions similar to that of growth factors and hormones.

Cytokines, growth factors, and hormones have the following similarities.

- They are all secreted as soluble factors.
- They are biologically effective even in picomolar concentrations.
- They all bind to membrane-associated receptor of targeted cells.

Cytokines, growth factors, and hormones have the following dissimilarities.

- Growth factors have constitutive expression for longer periods of time, while the other two have induced expression or are secreted due to discrete stimuli for a few hours or days.
- Hormones act in an endocrine fashion, while the other two act in an autocrine or paracrine fashion.
- Hormones are produced by specialized cells of a gland, while cytokines are produced by a variety of cells.

8.4.1 Specific Properties of Cytokines

Cytokines have a molecular mass lower than 30 kDa. On the basis of structure, cytokines are divided into four groups having four different types of receptors.

1. The haematopoietin group has haematopoietin receptors or the class I receptor family, such as IL-2 and IL-4.
2. The interferon group has interferon receptors or the class II cytokine receptor family.
3. The chemokine group has the chemokine receptor family.
4. The TNF factor group has the TNF receptor family.

The IL-2 and IL-4 groups have an α-helical structure and a very little or negligible β-sheet structure. Both groups also differ in amino-acid sequences.

8.4.2 Role of Th-cytokines

Apart from membrane-contact signals, Th-cytokine signals are also essential to induce B-cell proliferation and its differentiation into plasma cells. As mentioned earlier, membrane-contact signals are essential for proliferation, but cytokine signals are required for differentiation. It has also been revealed from electron micrographs that Th-cell membrane proteins and cytoskeleton elements are redistributed after interaction between B cells and Th cells. Interacting Th cells produce various cytokines such as IL-2, IL-4, and IL-5. Activated B cells express cytokine-specific receptors on the surface of their membrane. The interaction of cytokines with cytokine-specific BCRs helps the B cells to proliferate and differentiate into plasma cells and memory B cells. It also helps in affinity maturation and class switching.

8.5 NEGATIVE SELECTION OF MATURE SELF-REACTIVE B CELLS IN PERIPHERY

B cells expressing antibody (mIgM) specific to self-antigens, which are unable to cross and reach the bone marrow, are eliminated or made inactive in the periphery of the lymphoid tissue by a negative selection process.

The negative selection process of B cells in the periphery was proved after the experimental observation of C Goodnow and co-workers on transgenic mice. They made two transgenic mice—hen egg white lysozyme (HEL) mice and anti-HEL mice. Both were crossed to produce double transgenic mice carrying genes encoding HEL and anti-HEL (Figure 8.3). Flow cytometric analysis was used to observe the membrane IgM levels of peripheral B cells that were able to bind HEL. B cells were incubated with

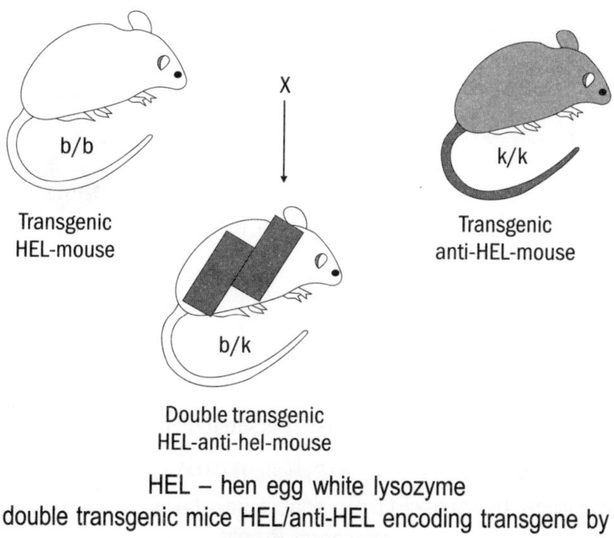

Figure 8.3 Generation of double transgenic mice HEL/anti-HEL encoding transgene by crossing HEL and anti-HEL transgenic mice

anti-mouse IgM antibody labelled with a fluorescent label. The IgM antibody fluorescent label is different from those used to label HEL, which measures the level of membrane IgM expressed on B cells. The non-transgenic mice produced B cells that expressed high levels of surface IgM. They normally express no cells that express HEL. Both anti-HEL transgenic mice and anti-HEL/HEL double transgenic mice produced large numbers of HEL-bound B cells (black coloured ≈ 10^{-9}M) (Figure 8.4). These cells were *anergic,* that is, functionally non-responsive. The level of membrane IgM was recorded to be 20 fold lower in the double transgenic mice. In these mice, B cells expressed both IgM and IgD on the surface indicating the introduction of anergy in mature B cells, not in immature B cells. The HEL transgene is linked with a metallothionein promoter controlled by zinc levels. Another group of transgenic mice had the light and heavy chain transgenes rearranged in the immunoglobulin that encoded the anti-HEL antibody.

Based on experiments on transgenic mice, Goodnow showed that double transgenic mice having ability to express high levels of HEL produce mature peripheral B cells. These B cells bear anti-HEL membrane antibody and were functionally non-responsive, that is, *anergic*. The double transgenic mice showed large numbers of anti-HEL B cells in which IgM expression was comparatively 20 fold less than the B cells of anti-HEL, single transgenic mice. Since the double transgenic mice B cells expressed both IgM and IgD on cell surface, it seems that anergy was not induced in immature cells but induced in mature cells. After giving a dose of HEL to mice, a few plasma cells that produce anti-HEL serum with low titer got induced.

Nemazee and Burki modified the transgenic mice system to understand the outcome of the expression of class I MHC only in the periphery rather than in the bone marrow. They made a construct of class I K^b gene with a liver-specific promoter to express this

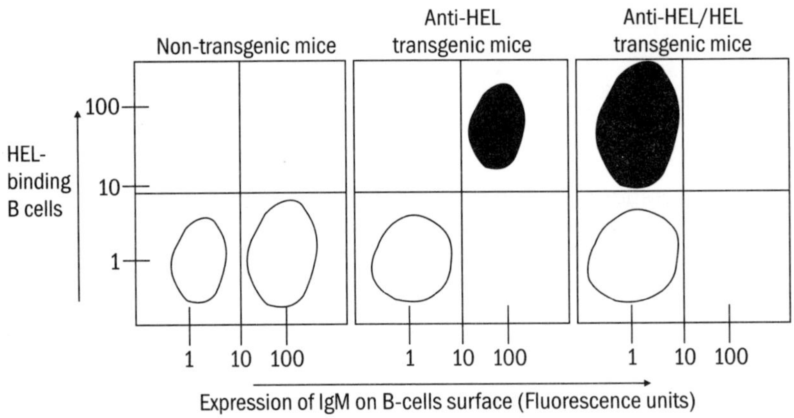

Figure 8.4 Determination of membrane-associated IgM levels by anti-mouse IgM antibody labelled with fluorescent in both anti-HEL transgenic mice and anti-HEL/HEL double transgenic mice using flow cytometric analysis of peripheral B cells

Note Both transgenic mice have large numbers of B cells associated with mIgM.

gene only in the liver. Other transgenic mice considered in the experiment were those with anti-K^b antibody producing on the B cells (Figure 8.5). Both types of mice were crossed to produce double transgenic mice in which the immature B cells with anti-K^b mIgM could not come in contact with class I K^b molecules present in the bone marrow. Immature B cells expressing the transgene-encoded anti-K^b cells (black) were present in the bone marrow but absent in the peripheral lymphoid organs (since K^b transgene expresses only in the liver and not in the bone marrow due to the presence of liver-specific promoter) as shown in flow cytometric analysis. From this, it could be concluded that mature self-reactive B cells were removed from the periphery (Figure 8.6). Flow cytometric analysis of both lymph nodes and bone marrow in single transgenic mice showed the presence of anti-K^b B cells.

8.6 PRIMARY AND SECONDARY HUMORAL RESPONSES OF B CELLS

Both primary and secondary humoral responses lead to the production of various isotypes of antibodies. These antibodies differ in the specificity of their effector functions.

Primary humoral response is generated by the very first contact of exogenous antigen (Plate 3), leading to the production of antibody secreting plasma cells and memory B cells. The level of production of serum antibody depends on the nature of antigen, presence and absence of adjuvant, the route of injection into the body, and the types of strain or species used for immunization.

The primary response to exposed antigen also depends on the lag phase. The nature of antigens determines the duration of the lag phase. During the lag phase, B cells

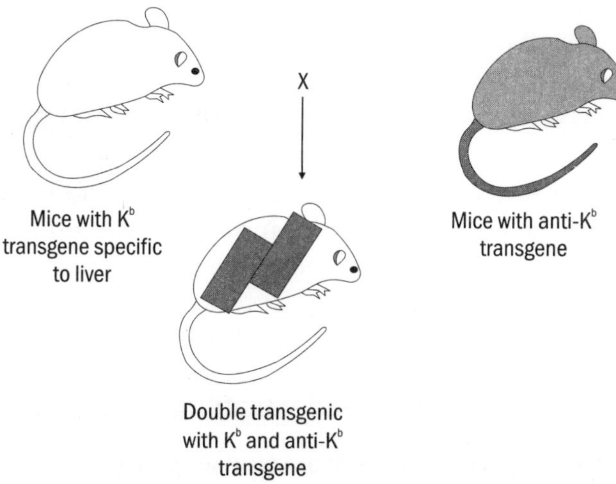

Figure 8.5 Production of double transgenic mice with transgenes K^b and anti-K^b containing liver-specific promoter

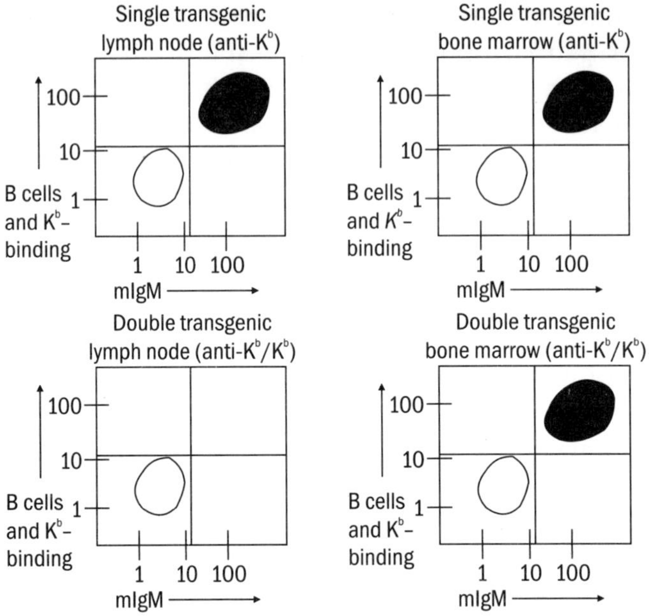

Figure 8.6 Analysis of bone marrow and peripheral lymph node B cells in double transgenic mice using flow cytometric analysis by mIgM-specific antibody

Note Black circle represents expression of Kb in single and double transgenic mice bone marrow and lymph node. Double transgenic mice lymph node does not express Kb, while double transgenic mice bone marrow and single transgenic mice lymph node and bone marrow express Kb.

proceed for clonal selection, clonal expansion, and differentiation into memory cells or plasma cells. For example, mice immunized with sheep red blood cells (SRBCs) have a lag phase of three to four days. In these mice, plasma and memory cells generate after 4 to 5 days and serum antibody is produced after 7 to 10 days. After 14 days, plasma serum titre is maximum. Initially, there is secretion of IgM, after which IgG is secreted. This triggers the primary response against that particular antigen, starting from days to several weeks (Plate 3).

Memory B cells enter into G_0 phase of the cell cycle and cease dividing. These cells remain alive throughout the whole life of the individual depending on the presence of and types of antigens. The secondary humoral response starts with the activation of memory cells. The efficiency and capacity of secondary humoral response depends on the total population of the memory cells of B cells and T cells.

8.6.1 Difference between Primary and Secondary Responses

Secondary response differs from primary response in many ways. The lag period of the secondary response is shorter than the primary response. This is due to the fact that memory cells for particular types of antigen are large in number and activated more easily than the naïve B cells. The magnitude and duration of secondary responses are more than the primary response. The secreted antibody is high in number and has higher affinity for the antigen in secondary response than in primary response. This gives effective defence for the host against further infection. Various isotypes of antibodies (except IgM) are predominant after secondary response to provide wide effector functions against pathogens. This is in contrast to primary response (Plate 3).

8.7 ORIGINAL ANTIGENIC SIN

The phenomenon of original antigenic sin was first described by Thomas Francis Jr in 1960. Richard Krause called it the *Hoskins effect.* According to this effect, the body's immune system preferentially utilizes the immunological memory of the first strain infection over a secondary infection of slightly different strain. This causes the immune system to be trapped by the antigen of the first infection and unable to mount defence over subsequent secondary infection (Francis 1960). The effect has been explained in relation to the infection of influenza virus, dengue fever, human immunodeficiency virus (HIV), and other viruses. Richard Krause cited the original idea of Thomas Francis as follows.

The antibody of childhood is largely a response to dominant antigen of the virus causing the first type-A influenza infection of the life time. The imprint established by the original virus infection governs the antibody response thereafter. This we have called the Doctrine of the Original Antigenic Sin.

8.8 ROLE OF T CELLS IN HUMORAL RESPONSES

Studies carried out on hapten–carrier protein conjugates demonstrated the generation of humoral antibody response and required recognition of different epitopes on antigens by both Th cells and B cells. Hapten is chemically coupled to a larger carrier molecule to induce humoral response. Very little or no antibody will be generated if only hapten is used for immunization without any carrier protein. Secondary antigen response is generated if an animal is immunized with the same hapten–carrier protein conjugate. A different unrelated carrier with the same hapten cannot trigger an anti-hapten response. This phenomenon is called the *carrier effect*. Experiments were also conducted to prove that carrier-primed cells are T cells and hapten-primed cells are B cells. Thus hapten-primed and carrier-primed cells are separate populations. Mice irradiated with X-rays were reconstituted with spleen cells from both DNP–BSA-primed mice and BGG-primed mice and challenged with DNP–BGG-generated secondary anti-DNP response. *Anti-Thy-1* antiserum was used to remove T cells from BGG-primed spleen cells; it did not show anti-DNP response. Similarly, *anti-Thy-1* antiserum was used to remove T cells from DNP–BSA-primed spleen cells; this also did not show anti-DNP response (Figure 8.7).

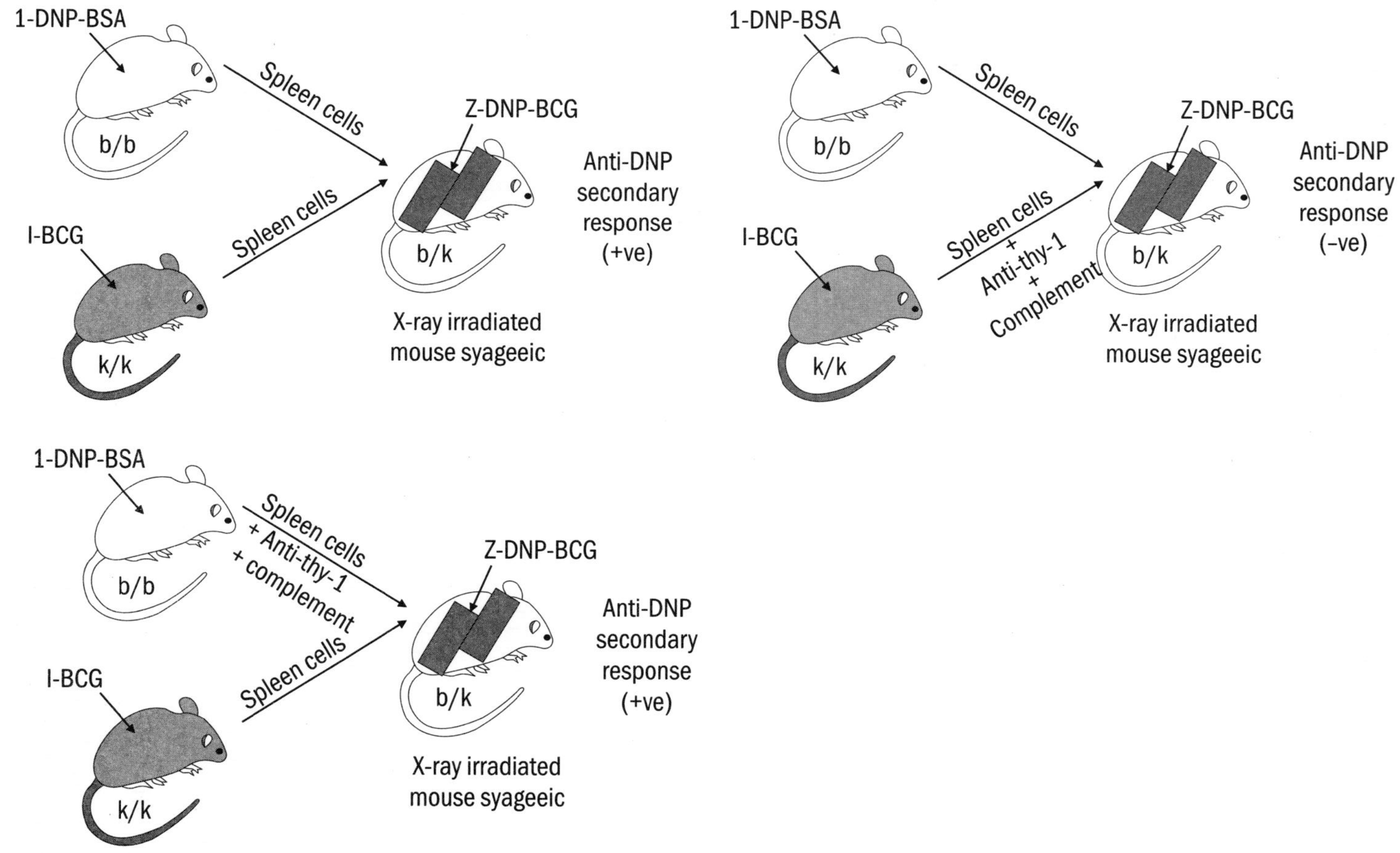

Figure 8.7 Experiment depicting hapten-primed and carrier-primed cells separating and showing different effects after transfer

Notes
1. X-ray-irradiated mouse reconstituted with DNP–BSA and BCG-primed spleen cells and later challenged with DNP–BSA generated a secondary anti-DNP response.
2. Anti-thy-1 antiserum-mediated removal of T cells from BCG-primed spleen cells lost secondary anti-DNP antibody response.
3. Anti-thy-1 antiserum-mediated removal of T cells from DNP–BSA-primed spleen cells had no effect on secondary response to DNP, that is, positive secondary antibody synthesis response.

Thus the above experiments suggest that both Th cells and B cells recognize antigenic determinants located on the same B-cell molecules. This kind of Th-cells and B-cells recognition feature is called *associative* or *linked recognition.*

Introduced antigen gets accumulated in various lymphoid organs. Antigen delivered into blood is filtered by the spleen, while the antigen delivered into the tissues is filtered by the lymph nodes located nearby or lymph nodules. Approximately 90% of antigen is filtered by the afferent lymphatic into the lymph nodes. Antigen is either filtered alone or filtered in the form of an antigen–antibody complex or by antigen-transporting cells, such as dendritic cells or Langerhans cells and macrophages. As it passes through the nodes, antigens encounter one of the following APCs: (1) dendritic cells located in paracortex; (2) node-situated macrophages, and (3) follicular dendritic cells (FDCs) or germinal centres dendritic cells. Humoral response against antigen occurs in a complex series of events taking place in different micro-environments within the lymph node. Primary and secondary pathways differ from each other because in secondary response, antigens form a complex with the circulating antibody. After antigen activation leads to the differentiation of plasma cells, small foci of proliferating B cells form the edges on the T-cells-rich zone. Similar events occur in the spleen plasma cells that secrete IgM and IgG isotypes. These foci contain most of the antibodies produced during the primary response. It is presumed that a few activated B cells, together with a few Th cells, migrate from the foci (of lymph nodes) to primary follicles. Later, these primary follicles develop into secondary follicles to provide a micro-environment for interaction between B cells, activated Th cells, and dendritic cells of follicles. FDCs differ from bone marrow dendritic cells (BMDCs) in the following ways: (1) FDCs do not express class II MHC molecules and do not present antigen to $CD4^{+}$ T cells; (2) FDCs have Fc receptors and complement receptors on its extended part to present antigen–antibody complexes for longer periods of time, such as for months. The thymus-dependent antigen-activated B cells and Th cells form a germinal centre at the centre of secondary follicles (Jacob, Kelsoe, Rajewsky, *et al.* 1991).

8.9 SECONDARY MODIFICATION OF B CELLS AT GERMINAL CENTRE

Proliferating and differentiating B cells pass through many secondary structural and functional modifications to play a pivotal role in immune systems. These secondary modifications happen in the cells at the germinal centre to counter infection and boost immunity (Smith and Fearon 2000).

8.9.1 B-cell Differentiation

Proliferating B cells in the dark zone of the germinal centre are called centroblasts. Centroblasts differ from B cells by their large size, diffused chromatin in an expanded cytoplasm, and the absence of surface immunoglobulin. Centrocytes, which are non-dividing B cells with no surface immunoglobulin, arise from centroblasts. They migrate from the dark zone to the light zone. In the light zone, centrocytes make contact with FDCs surface antigen–antibody complexes. B-cells differentiation occurs by the following set of events: affinity maturation, class switching, and the formation of plasma cells and memory cells (Lane 1996; Liu, de Bouteiller, and Fugier-Vivier 1997; Manis, Tian, and Alt 2002). Affinity maturation and memory cells formation occur in the germinal centre, while class switching and plasma cells formation occur outside (Smith and Fearon 2000).

8.9.2 Affinity Maturation

Affinity maturation occurs due to repeated somatic hypermutation followed by selection. It was first noticed by H N Eisen and G W Siskind. Due to affinity maturation, production of antibody is increased manifolds; within two to eight weeks after immunization with hapten–carrier complex (by DNP–BCG), it increases 140 fold.

8.9.3 Somatic Hypermutation

Germinal centre is the site for somatic hypermutation as reported by G Kelsoe and his colleagues. Such mutation occurs after infection in the immunoglobulin genes of B cells during immune response. Kelsoe and co-workers compared the B cells of germinal centres with B cells present outside the germinal centres to note the mutation frequencies in immunized animals. The animals were immunized with hapten NP (4-hydroxy-3-nitrophenylacetyl) conjugated with carrier protein chicken gamma globulin. The B cells from the germinal centre or outside the germinal centre were isolated, and immunoglobulin genes of each cell were amplified to study the mutation. Plenty of mutations were present in the immunoglobulin gene of germinal centre cells. A few mutations were noted in the activated non-germinal centre B-cells immunoglobulin genes. Mutated sequences in each immunoglobulin gene of various germinal centre B cells show similarity as they originate from common precursor cells. The prepared genealogic trees showed progressive somatic hypermutation.

The mutations that occurred were point mutations, deletions, and insertions. Most of the mutations were in the 5′-end (0.5 k^b) and 3′-end (1.5 k^b) of the V(D)J segments. Although the place or site of mutations is throughout the V-region, antigen-driven selection suggests that most of the mutations are in the three complementarity-determining regions (CDRs). The rate of mutations is 10^{-3}/base pair/division. The rate and target of such kind of mutations are unique to the immune system. In most cases, it is a random process. The mutations generate two kinds of B cells: (1) higher-affinity receptors containing B cells in which B cells are positively selected and (2) lower-affinity receptors containing B cells in which B cells die in the germinal centre or face negative selection.

8.9.4 Selection of B Cells

Selection of centroblasts occurs after somatic hypermutation (Papavasiliou and Schatz 2002) of heavy and light chain variable region genes in the germinal centre dark zone. On the other hand, the selection of non-dividing centrocytes occurs in the light zone. The selection is influenced by the ability of immunoglobulin molecules present on the membrane of centrocytes to recognize the antigen displayed on FDCs. The membrane of FDCs contains Fc receptors and complement receptors associated with antigen, which helps in binding to centrocytes immunoglobulin. Interaction generates a signal, which is essential for the survival of centrocytes; thus only higher-affinity centrocytes will survive. Centrocytes with lower affinity will not interact and, hence, will die. To survive, centrocytes must receive signals from $CD4^+$ Th cells. The CD40 molecules of centrocytes interact with the CD40L present on the helper T cell. It is also essential that class II MHC molecules on B cells or centrocytes interact with the TCR of Th cells. The centrocytes that do not receive signals from either Th cells or antigen-membrane immunoglobulin will die due to apoptosis in the germinal centre. The presence of condensed chromatin fragments is due to extensive apoptotic death of cells. The cell debris is removed by the phagocytic activity of macrophages (Plate 4).

8.9.5 Class Switching

Antibodies have two functions: (1) the effector function performed by the isotype of the heavy chain's constant domain (Fc region) and (2) the binding to specific antigen function performed by V_H and V_L domains (Fab region). Class switching is the association of a given V_H domain to any isotypes constant region. In this, the specificity of antigens does not change, but effector activities of the molecule will change. The class switching selection process is determined by a pool of cytokines for any IgM-bearing cells.

Thymus-dependent antibody production response against antigen is determined by molecules class switching to isotypes other than IgM. On the other hand, thymus-independent antibody production response against antigen is dominated by IgM. For class switching, it is essential for CD40 molecules of B cells to interact with CD40L molecules of Th cells. If Th cells fail to express CD40L, the immunodeficiency disorder called X-linked hyper-IgM syndrome occurs. Patient with this disorder will produce IgM but will not produce other isotypes. Memory cells will not be generated in such patients, and somatic hypermutation will not occur in the antibodies (Lane 1996).

8.9.6 Generation of Memory B Cells and Plasma Cells at Germinal Centres

High-affinity, mIg-bearing B cells are selected. They interact with antigen associated with FDCs in the germinal centre. Upon antigen interaction, some B-cells differentiate into plasma cells and others into memory cells. Plasma cells are also formed at other sites of the body. All plasma cells do not carry detectable amounts of membrane-attached immunoglobulin; they synthesize 1000 molecules of immunoglobulin per cell per second. RNA processing is altered in the plasma cells so that the immunoglobulin formed has a secreted type of heavy chain rather than a membrane-bound type. An increase in transcription rate has also been observed in the plasma cells by some scientists. This is due to the increase in synthesis of transcription factors and their binding to the enhancer region of immunoglobulin genes. Table 8.2 gives a comparison between naïve and memory B cells.

8.10 B CELLS DEVELOPMENT REGULATION

The transcription factors involved in the regulation of genes, which are important for the development of B cells, are NF-κB, BSAP, Ets-1, c-Jun, Ikaros, Oct-2, Pu.1, EBF, BCF, and E2A. These transcription factors either upregulate or downregulate the responsible genes by binding to the promoter region or enhancer region. Their importance was determined by knockout experiments. The knockout mice with transcription factor Ikaros had no lymphocyte development in the bone marrow.

The master regulator proteins of B cells are BSAPs. The BSAPs produced only by B-lineage cells regulate all stages of B-cell maturation. They are also involved in the differentiation of B cells leading to the development of memory B cells and plasma cells. The B cells that do not express BSAPs are not differentiated (Lane 1996). The genes regulated by BSAPs are λ5 and Vpre-B of the surrogate light chain, J-chain gene of IgM, and any one enhancer region of two 3′α heavy chains. Of the two α-chains, any one α-chain's 3′-end enhancer region helps in the binding of BSAP or other transcription factors such as NF-αP. BSAPs also bind to many other genes involved in B-cell activation and switch sites of various immunoglobulin heavy chains.

The enhancer of 3′α heavy chain ($E_3'\alpha$) contains binding sites for BSAPs and other transcription factors. The binding of BSAPs inhibits the binding of other transcription factors; for example, NF-αP binds to the enhancer site of 3′α heavy chain, thus inhibiting

Table 8.2 Types of cells and their characteristic properties

Type of cell	*Characteristic property*
Naïve B cells	Contain membrane markers such as IgM, IgD and low amount of complement receptors; are present in the spleen; have short lifespan; recirculate in the body; have low receptor affinity; and possess fewer ICAM-1 adhesion molecules
Memory B cells	Contain membrane markers such as IgM, IgD, IgG, IgA, and IgE and high amount of complement receptors; are present in the bone marrow, lymph nodes, and spleen; have long lifespan; recirculates in the body; have high receptor affinity due to affinity maturation; and possess high amount of ICAM-1 adhesion molecules

the transcription of 3′α heavy chain gene and triggering the formation of memory B cells. NF-αP binds at the enhancer site of 3′α heavy chain to initiate transcription of immunoglobulin gene heavy chain, which helps in the development of plasma cells.

8.10.1 Effector Responses and its Regulation

Antigens can provoke two kinds of responses in the body: (1) immunity to antigen and (2) unresponsiveness called *tolerance*. Both responses are guided and regulated by T-cells or B-cells lymphocytes and follow a series of consequences. Fresh regulatory decisions such as activation of immunity, response intensity, and duration are made each time when fresh antigen enters the body. Cytokines are key effector molecules taking part in this immune regulation. There might be other regulatory molecules that we are not aware of.

8.10.2 Antigens Compete with Each Other

The presence of competing antigens of previous encounters regulates the immune system against the antigen not related to the earlier antigen. Such an antigenic competition can be triggered in mice by immunizing them with competing antigens one or two days before application of the test antigen. For example, the two antigens could be horse red blood cells (HRBCs) and SRBCs. Prior immunization with one of these antigens reduces the effect of the other antigen. The molecular and cellular basis of such a phenomenon is still unknown.

8.10.3 Antibody Presence Suppresses Antigen Response

Synthesized antibody plays a role in feedback inhibition, that is, it inhibits its own synthesis. Due to this reason, infants are not immunized for measles and mumps before the age of one year. Till six months after birth, infants have a high amount of mother-acquired antibody IgG. Immunization against measles or mumps in the presence of maternal IgG will reduce the production of memory cells (inadequate number of memory cells), which helps in long period immunity. There are two reasons for antibody-mediated suppression. First, circulating antibody inhibits clonal propagation of B cells by competing with antigen-reactive B cells. Second, binding of the antigen–antibody complex to B-cell membrane-associated Fc receptors inhibits BCR signalling. Interaction between Fcγ receptors and BCR complex leads to dephosphorylation of BCR complex (Matsuuchi and Gold 2002). Dephosphorylation is carried out by phosphatases associated with Fcγ cytoplasmic tail and is essential for B-cell activation. More and more binding of IgG to antigen leads to downregulation of B-cell activity. In such a competition, only high-affinity antigen-reactive B cells are able to compete with the passive antibody to keep the B cells active.

8.11 T-CELL REGULATION

The maturation of non-self-reactive T cells from their progenitors (in thymus) and activation (in periphery) involve MHC molecules. Maturation follows two pathways—one for $CD4^+$ cells and the other for $CD8^+$ cells. $CD4^+$ and $CD8^+$ maturation involves class II MHC and class I MHC restriction, respectively. The activation of both $CD4^+$ and $CD8^+$ is guided by TCR interaction with peptides displayed on the surface along with MHC molecules, that is, TCR–antigen–MHC interaction. This interaction is well supported by other accessory molecules and coreceptors to promote activation, which in turn guides T cells to proliferate and differentiate into effector cells and memory T cells. T cells express two types of receptors—γδT-cell receptors and αβT-cell receptors. Among these two receptors, αβT-cell receptors are expressed by a vast majority of thymocytes.

8.11.1 MHC Restriction and Self-tolerance

Thymocytes follow two types of selection pathways—positive selection and negative selection. In positive selection, thymocytes have receptors that bind to self-MHC molecules to restrict them; this is known as MHC restriction. Cells that do not bind get eliminated by apoptosis in the thymus. In negative selection, thymocytes have receptors that bind to self-MHC to get eliminated, resulting in self-tolerance. The MHC-based positive and negative selections are explained by two hypotheses: avidity hypothesis and differential-signalling hypothesis.

Avidity hypothesis explains that the avidity, that is, the signal strength of the TCR–peptide–MHC interaction determines negative and positive selection. The differential-signalling hypothesis is a qualitative theory to explain the selection based on differential signals rather than the strength of a single signal. Here positive selection results for a weak signal and negative selection takes place for a strong and complete signal.

The first hypothesis based on avidity was experimentally proved by TAP-1 knockout mice for αβT-cell receptor, which was able to recognize LCM virus peptide presented with MHC. The TAP-1 knockout mice were unable to produce peptide–MHC complex in the absence of peptide (not supplied from outside). The concentration of exogenous peptide has an inversely proportional effect on the production of CD^-CD8^+ T cells. Low concentration of peptide means low avidity causing positive selection, which results in a normal level of $CD4^-CD8^+$. When no peptide was added, $CD4^-CD8^+$ T cells were not produced. An optimum high concentration of peptide means high avidity resulting in negative selection, which produces a high number of $CD4^-CD8^+$ T cells. Further increase after optimum concentration leads to sharp decline in $CD8^+$ cells number. A combination of different peptides added has more pronounced effects than a single peptide on the level of $CD4^-CD8^+$ T cells showed synergistic effects of two or more peptides over one peptide in their recognition ability in the selection process.

Two models have been proposed to explain the differentiation of double-positive thymocytes into single-positive thymocytes: (1) instructive model and (2) stochastic model. The instructive model explains that the interaction of one coreceptor with MHC molecules, that is, class I MHC or class II MHC, causes downregulation of the other coreceptor to produce either $CD4^-CD8^+$ or $CD4^+CD8^-$ T cells. The stochastic model suggests that the downregulation of either signal, that is, CD4 or CD8, in double-positive cells is a random process to produce either $CD4^-CD8^+$ or $CD4^+CD8^-$ T cells.

8.11.2 Gene Activation Mediated by TCR–Peptide–MHC Complex

The TCR-peptide–MHC complex interaction leads to the activation of many genes important for the activation of T cells.

- *Immediate gene* expression synthesizes many transcription factors such as c-fos, c-myc, c-jun, NFAT, and NF-κB. It occurs within half an hour after antigen interaction.
- *Early gene* expression leads to synthesis of cytokine proteins such as IL-2, IL-2R, IL-3, IL-6, and IFNγ. The activation occurs within 1–2 h after antigen interaction.
- *Late gene* expression leads to synthesis of adhesion protein molecules. The activation occurs after two days of antigen interaction.

8.11.3 TCR–Peptide–MHC Complex Signal Transduction

The TCR–peptide–MHC complex interaction causes the generation of a signal transduction pathway for the synthesis of the required proteins.

8.11.3.1 G-protein signal transduction pathway

Two types of G-proteins are involved in this kind of signal transduction mechanism: (1) Monomeric G-Ras protein, which is of 21 kDa molecular weight and normally takes part in receptor tyrosine kinase (RTK) activation and (2) heteromeric large G-protein composed of α, β and γ subunits taking part in glucose metabolism, olfaction, vision, and leukocyte chemotactic movement.

The second messengers generated by such a pathway are cyclic nucleotides (cAMP, cGMP), calcium ion (Ca^{2+}), DAG, and inositol triphosphate (IP3).

Kinases get activated to phosphorylate the tyrosine, serine, or threonine site of target proteins to activate them. These proteins get inactivated by activated phosphatases after removal of the phosphate group, that is, by de-phosphorylation of the target protein. This means phosphatases reverse the action of kinases. Adaptor proteins and enzyme cascades are involved to enhance the signal.

8.11.3.2 Other signal transduction pathways initiated by the interaction

- *Phospholipase Cγ (PLCγ) pathway* It is involved in the hydrolysis of phospholipids component to generate IP3 (inositol-1,4,5-trisphosphate) and DAG, which regulate the Ca^{2+} channel. This can activate the NFAT transcription factor, which regulates the expression of cytokines IL-2 and IL-3 to promote T-cell growth.
- *PKC pathway* The PKC pathway is important to activate NF-κB, which in turn helps to activate T cells and also protect them from apoptotic death.
- *Mitogen-activated protein* (MAP) *kinase pathway* The MAP pathway is initiated by Ras-protein, which is a small monomeric G-protein. The MAP kinase pathway is also known as the extracellular-signal-regulated kinase (ERK) pathway, which activates a transcription factor called Elk responsible for the expression of transcription factor fos. Fos regulates Jun to form AP-1, which has an important role in T-cell activation.

8.11.3.3 Role of costimulatory signal in T-cell activation

The activation and proliferation of T cells to become effector T cells require more than one signal, which is facilitated by the interaction of costimulatory molecules. The TCR–CD3–antigenic peptide interaction generates the *first signal*, while the CD28 molecule and CTLA-4 (CD152) molecule of T-cell membrane interact with the membrane molecule B7 of APCs (dendritic cells, activated macrophages, and activated B cells) to generate the *second signal*. Thus B7 molecules of APCs have two ligands expressed by activated Th cells as homodimers—CD28 and CD152 (CTLA-4). Both CD28 and CTLA-4 are structurally similar but functionally antagonistic to each other. Since both ligands contain immunoglobulin-like domain, they are placed in the immunoglobulin superfamily.

CD28 is expressed by resting T cells as well as activated T cells, while CTLA-4 is expressed only by activated T cells. CD28 delivers positive costimulatory signals, while CTLA-4 produces inhibitory signals. CTLA-4 is able to block the positive signal of CD28 by competitive binding with B7; therefore, it is used for clinical purposes in autoimmune diseases and transplantation. Anti-CD28 Fab portion of the antibody is able to block interaction between CD28 and B7, thus preventing the interaction between normal APCs and Th cells to cause anergy.

8.11.3.4 Clonal anergy or absence of costimulatory signal

The TCR–peptide–MHC complex interaction sometimes leads to non-responsiveness of immunity called *clonal anergy*. Non-responsiveness causes non-proliferation of cells after interaction due to the absence of costimulatory signal 2, which is generated by the interaction of CD28 present on Th cells and B7 present on APCs. This means that signal 1 generated due to the interaction of MHC-associated peptide with TCRs becomes anergic in the absence of signal 2 generated by costimulatory molecules. In vitro experiments have been performed to prove these effects, but there is still no animal model to study the anergy effect in vivo. APCs fixed with glutaraldehyde cannot express B7 molecules; hence, they are unable to generate signal 2, although they can generate signal 1 by the interaction of Th cells with MHC-associated peptide. The resting Th cells interact with fixed APCs due to the absence of costimulatory signal Th cells to produce a minimum amount of cytokines IL-2.

8.11.3.5 Role of superantigens in T-cell activation

Antigens binding to the V_β domain of TCR and the α-chain of class II MHC molecules at the same time are known as superantigens. Examples of these antigens are bacterial and viral proteins. Superantigens are of two types—exogenous superantigens and endogenous superantigens. Exogenous superantigens are bacteria-produced soluble proteins such as exotoxins, enterotoxins, toxic shock syndrome toxin, and exfoliate dermatitis toxin. Endogenous superantigens are viral membrane proteins able to infect the cells or proteins produced by the host DNA-integrated nucleic acid of retrovirus. The host DNA-integrated retrovirus nucleic acid expressed protein is translocated on the host cell membrane called minor lymphocyte-stimulating (MLs) superantigens. MLs-superantigens bind in the same way as other superantigens do. An example of an MLs-superantigen is mouse mammary tumour virus (MMTV)-produced protein. Superantigen-mediated activation is polyclonal because it affects a significant number of the Th cell population, causing overproduction of Th-cytokines. Overproduction of cytokines causes systemic toxicity, such as food poisoning by staphylococcal toxin and toxic shock syndrome.

Superantigens help in negative selection of T cells in the thymus leading to a large amount of deletion of T cells containing V_β domain.

8.11.4 Differentiation of T cells

Naïve T cells, both $CD4^+$ and $CD8^+$, circulate between the blood and lymph systems passing through secondary lymphoid organs till they encounter antigens in the lymph nodes. The rate of encounter of antigen and T cells is 1 in 10^5 naïve T cells. Those naïve T cells that do not encounter antigens pass again into the blood through the thoracic duct. This process continues every 12–24 h.

Those naïve T cells that get the chance to encounter antigens get activated to enlarge into blast cell, start repeated division, and then differentiate into either effector T cells or memory T cells. Enlargement of T cells is known as *primary response* of activation, and differentiation into memory cells and effector cells is known as *secondary response*. Before activation, the naïve T cells remain in the G_0 phase of the cell cycle. After activation, T cells enter into the G_1 phase of the cell cycle. In the G_1 phase, T cells produce a large amount of IL-2 cytokines and express IL-2 receptors on the surface. The produced IL-2 cytokines bind to TCRs to initiate the proliferation and differentiation into memory cells and effector cells.

The following differences can be noted between effector T cells and memory T cells.

- The half-life of effector T cells is shorter (a few days to a few weeks) while that of memory T cells is comparatively longer.
- Memory T cells are able to generate secondary response, while effector T cells cannot.
- Effector cells are generated to conduct specialized functions such as cytokines secretions, cytotoxic killing activity (CTLs or $CD8^+$ cells), and assistance in B-cell activation ($CD4^+$ Th cells).
- $CD4^+$ effector cells have two subpopulations on the basis of the secretion of cytokines Th-1 and Th-2 subset (for details, refer to Chapter 14).
- Memory T cells can be derived from both naïve T cells and effector T cells, while effector T cells can be derived only from naïve T cells after antigen encounter.
- Memory cells, like naïve T cells, reside in the G_0 phase of the cell cycle.
- Naïve T cells get activated only by dendritic cells, while memory T cells get activated by macrophages, dendritic cells, and B cells.

AT A GLANCE

- After development in the bone marrow, B cells interact with antigens in the periphery to activate and differentiate into plasma cells and memory cells.
- Immunoglobulin gene arrangement takes place in sequence to transform a progenitor B cell (pro-B cell) into an immature B cell, which express mIgM having specificity against a single antigen. The immature B cell further develops into a naïve mature B cell, which expresses both mIgM and mIgD.
- Negative selection of self-reactive immature B cells occurs in the bone marrow after the expression of self-reactive B-cell receptors (BCRs). The self-reactive immature B cells either die by apoptosis or proceed to BCR editing to produce non-self-reacting mIg for survival. Self-antigens reacting B cells become anergic in periphery.
- The B cells in periphery interact with protein antigens or other antigens to produce antibody as a thymus-dependent response, that is, with the help of T-helper cells (Th cells). Thymus-independent response, that is, response without the help of Th cells, occurs against bacteria LPS or repeating epitopes.
- B-cell activation is defined as a BCR-mediated signal transduction process leading to changes in the expression of certain specific genes.
- B-cell and T-cell activation have the following common points: (1) both include receptor subunit based compartmentalization of functions; (2) protein tyrosine kinases activation (TKA); (3) recruitment of signalling complexes to tyrosine kinases to generate many signal transduction pathways.

- The activating signal during primary responses is intensified by B-cell coreceptors to overcome the low concentration of antigens.
- Thymus-dependent antigens-mediated activation and survival of B cells require B cell's CD40 coreceptor to interact with Th cell's CD40L coreceptor. This leads to germinal centre formation, generation of memory cells, and somatic hypermutation.
- Primary and secondary antibody responses are not the same in an organism. Primary antibody response has a longer lag time period, a slow logarithmic phase, a short time period plateau, and a rapid decline with IgM as the main antibody produced. On the other hand, secondary antibody response has a shorter lag time period, a comparatively rapid logarithmic phase, a longer time period plateau, and a comparatively slow decline. The predominant antibody produced during secondary response is IgG. Other antibodies are also produced, including IgM. The antibody produced during this response has a higher average affinity than the average affinity of antibody produced during primary response.
- The germinal centre, which is a site for the somatic hypermutation of immunoglobulin genes (rearranged), is produced within a week after exposure to thymus-dependent antigens.
- The germinal centre, which is a site for the somatic hypermutation of immunoglobulin genes (rearranged), is produced within a week after exposure to thymus-dependent antigens.
- T cells originate in the bone marrow and mature in the thymus by the rearrangement of T-cell receptor (TCR) genes. Most T cells become $\alpha\beta$T-cells, and some become $\gamma\delta$T-cells.
- In the beginning, T cells are $CD4^-CD8^-$. Due to negative selection at the development stage, they either become $CD4^+CD8^-$abT cells or $CD4^-CD8^+\alpha$bT cells.
- Those T cells that are unable to detect self-major histocompatibility complex (MHC) are eliminated by positive selection, whereas T cells with high affinity receptors that bind to self-MHC molecules are eliminated by negative selection. The self-antigen and self-MHC interaction generates self-tolerance.
- The activation of Th cells requires interaction of the TCR–CD3 complex with the peptide–MHC complex of antigen-presenting cells (APCs) supported by costimulatory molecules along with the coreceptors CD4 and CD8.
- $CD4^+$ Th cells recognize class II MHC molecule to act as helper cells (Th cells), while CD8+ Th cells recognize class I MHC molecule to act as T cytotoxic cells (Tc cells).
- Clonal anergy or immune ineffectiveness is caused by the absence of the costimulatory signal 2.
- The cell cycle stage of naïve T cells is in the G0 phase, while activated T cells that differentiate into effector cells and memory cells are in the G1 phase. Effector T cells functioning as cytotoxic cells, or as helper cells, are short lived, while memory T cells are long-lived cells. Memory T cells have secondary functions.

QUESTIONS

1. Where does development of B cells take place? Discuss the rearrangement of immunoglobulin genes in developing B cells.
2. What are B-cell maker proteins? Discuss the negative selection and its importance.

3. What is negative selection of B cells? Discuss the thymus-dependent and thymus-independent procedures of selection of B cells.
4. What types of coreceptors are present in B cells? Discuss the role of intracellular pathway in the development of B cells.
5. What are cytokines? Discus the role of cytokines in the development and maturation of B cells. How do cytokines differ from growth factors and hormones?
6. What do you mean by primary and secondary humoral responses of B cells? Differentiate between primary and secondary humoral responses.
7. What is antigenic sin? How does it originate? Discuss.
8. Write short notes on the following.
   (i) B-cell differentiation
   (ii) Affinity maturation
   (iii) Somatic hypermutation
   (iv) Selection of B cells
9. Generation of memory B cells and plasma cells
10. What do you understand by MHC restriction and self-tolerance? Discuss TCR, peptide, and MHC-complex-mediated activation and signal transduction pathway.
    (i) What do you understand by the following?
    (ii) Role of costimulatory signal in T-cell activation
    (iii) Clonal anergy or absence of costimulatory signal
    (iv) Role of superantigens in T-cell activation
    (v) Differentiation of T cells

OBJECTIVE QUESTIONS

A. Fill in the blanks.

a. Development of B cells occurs in ________ .
b. B-cell activation occurs by ________ to trigger differentiation in the ________ region.
c. B cells differentiate into ________ cells and ________ cells.
d. Immature B cells express antibody subtypes ________.
e. Mature naïve B cells express antibody subtypes ________ and ________.
f. Centroblasts that reside in ________ zone divide into centrocytes to move into ________ zone where it interacts with ________ residing on the ________ as antigen–antibody complex.

B. Match the following.

i. CD40	a. Anergic
ii. CD40L	b. B cells
iii. Thymus-dependent antigens	c. Th cells
iv. Thymus-independent antigens	d. Proteins
v. Germinal centre	e. Lipids
vi. Naïve B cells	f. Dark zone
vii. Centroblasts	g. Spleen
viii. Self-antigen reactive B cells in periphery	h. Somatic hypermutation site
ix. Centrocytes	i. Apoptosis or receptor editing of self-reactive immature B cells
x. Negative selection	j. Light zone

CHAPTER NINE 9

Cell-mediated Immunity and Cytotoxicity

9.1 INTRODUCTION

Cell-mediated cytotoxicity is the process by which pathogen-infected cells are identified and eliminated. It includes the identification and elimination of specific antigen-producing altered cells, such as cancerous cells. Antigen-specific cells or antigen-non-specific cytotoxic cells eliminate altered cells. Antigen-specific cells include $CD8^+$ Tc cells, also known as cytotoxic T-lymphocytes (CTLs), and $CD4^+$ Th cells, which have a role in delayed-type hypersensitivity (DTH). Antigen-non-specific cells include natural killer (NK) cells, macrophages, neutrophils, and eosinophils. The functionality of these cells depends on the types of cytokines. These cells recognize/kill target cells by using antibody receptors present on the cell surface. The antigen–antibody complex activates a complement cascade to generate chemotactic peptides, which pull the phagocytic cells to act on affected cells.

9.2 MECHANISMS OF CELL TARGETING AND LYSIS MEDIATED BY CYTOTOXIC T CELLS AND NK CELLS

There are two mechanisms involved in the elimination of altered cells—antigen dependent and antigen independent. The mechanisms are based on the cytokines that form the complex with the antibody to target the cells. Two types of cells take part in the whole mechanism—effector T cells or CTLs and NK cells (Berke 1994).

9.2.1 Effector T cells or CTLs

T cells normally eliminate microbial infections using the antigen-dependent mechanism. Children lacking thymus T cells are unable to tackle intracellular infection of bacteria, virus, and fungi.

There are two types of cell-mediated responses. The first type is caused by cells having specificity (such as CTLs) or non-specificity towards antigens (such as NK cells and macrophages), which are directly involved in the killing of malignant cells, virus-infected cells, chemically conjugated cells, and allogenic cells. The second type is caused by cells (such as $CD4^+$ T cells) that mediate DTH reaction to eliminate the infected cells.

Effector cells include $CD4^+$, Th1, Th2, and $CD8^+$ CTL cells. Their characteristic features include non-strict activation, expression of a large amount of adhesion molecules and other membrane-associated molecules or soluble molecules.

9.2.1.1 Activation and association of effector T cells with target cells

For activation at different stages, T cells require costimulatory signals apart from T-cell receptors (TCRs). Naïve TCRs require primary costimulatory signals during the interaction of the TCR complex and CD4 or CD8 coreceptor with the major histocompatibility complex (MHC)–foreign peptide molecule complex. Costimulatory signals may be from naïve T cells or antigen-presenting cell (APC) membrane molecules. But effector T cells that have encountered antigens and memory cells can respond to TCR-mediated signals without or with a little help of costimulation.

Preliminary research suggests the rationale behind different activation of naïve and activated T cells. Both naïve and activated effector T cells express different isoforms, either CD45RA or CD45RO, which originate from C45 gene RNA transcript splicing. CD45RA are expressed by naïve T cells, while CD45RO are expressed by activated effector T cells. Although both CD45RA and CD45RO are expressed by memory T cells, CD45RO are expressed more. CD45RO gets associated with TCR complex, CD4 and CD8 coreceptors better than CD45RA. The association of CD45RO isoforms of CD45 makes activated and memory T cells more sensitive towards the peptide–MHC complex than naïve T cells.

Apart from CD45 isoforms, the predominant presence of CD2 and integrin LFA-1 cell adhesion molecules on effector T cells enables them to bind to target cells, including APCs, better than naïve T cells. Both CD2 and integrin LFA-1 bind to LFA-3 and intercellular adhesion molecule (ICAM), respectively, on target cells or APCs. Initially, the TCRs of effector T cells scan APCs and target cells for membrane-specific peptide–MHC complexes. In the absence of peptide–MHC complexes, effector T cells disassociate from the target cells or APCs. In the presence of peptide–MHC complexes, TCRs generate signals to increase the affinity between LFA-1 and ICAMs and prolong the interaction; for example, Th-1 cells and Th-2 cells remain associated with peptide–class II MHC complexes on macrophages and B cells, respectively, while CTL effector cells remain associated with the peptide–class I MHC complex displayed on the membrane of virus-infected target cells.

Other soluble or membrane-bound effector molecules secreted by effector T cells include Fas ligand by $CD8^+$ or CTLs, TNFβ by Th-1 cells, and CD40 ligand by Th-2 cells. CTLs also secrete soluble effector molecules such as cytotoxins such as perforins and granzymes and cytokines such as IFNγ and TNFβ. The other two effector cells, Th-1 and Th-2, secrete large numbers of sets of overlapping cytokines.

Fas ligand, perforins, and granzymes enable destruction of target cells by CTLs; TNFβ, IFNγ, and GM–CSF cause activation of macrophages mediated by Th-1 cells; and CD40 ligand, IL-4, IL-5, and IL-6 cause activation of B cells mediated by Th-2 cells (Kägi, Vignaux, Ledermann, *et al.* 1994; Wallach, Varfolomeev, Malinin, *et al.* 1999).

9.2.1.2 Function of CTLs

Cytotoxic T-lymphocytes are generated by the activation of T-cytotoxic cells (Tc cells). CTLs work by the recognition and lysis of altered self-cells, such as cells infected with virus, grafted cells or tumour cells. Both CTLs and $CD8^+$ cells are restricted to the recognition of class I MHC–peptide complexes. Even $CD4^+$, which is restricted to class II MHC–peptide complexes, functions rarely as CTLs. In general, all nucleated cells express class I MHC–peptide complexes, hence they are recognized and destroyed by CTLs.

9.2.1.2.1 Stages in CTLs' functions

Cytotoxic T-lymphocytes function in three steps to eliminate altered cells: (1) activation and differentiation of naïve Tc cells into active effector CTLs; (2) recognition of class I MHC–peptide complexes on specific target cells; and (3) destruction of target cells.

- *CTL-mediated apoptotic cell death:* The initiation of two mechanisms activates the signalling pathway leading to the CLT-mediated apoptotic death of target cells.

  CTL-mediated killing of target cells follows the following two pathways.

  (i) Perforin/granzyme pathway—entry of cytotoxic proteins, perforin, and granzymes into the target cells

  (ii) Fas pathway—interaction of CTLs Fas ligand with the Fas receptor of target cells (Kägi, Vignaux, Ledermann, *et al.* 1994; Wallach, Varfolomeev, Malinin, *et al.* 1999) (Figure 9.1).

  Apoptotic death is due to caspase activity. Caspase is a cysteine protease that cleaves protein after aspartic acid residue. The term "caspase" is derived from the "C" of cysteine, "AS" of aspartate, and "PASE" of protease. Initiator caspase is derived

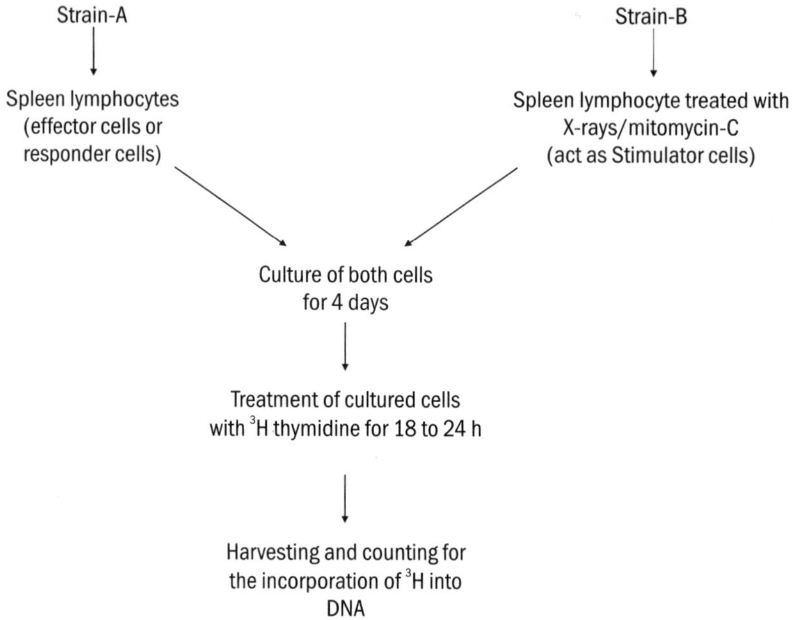

Figure 9.1 Experimental proof of lymphocyte proliferation

from inactive "procaspases" to produce more active caspases after they are cleaved. There are more than dozen caspases depending on their specificity (Bleackley, Lobe, Duggan, *et al.* 1988).

Granzymes of CTLs enter into target cells either by pore formation by perforin or by endocytosis to produce initiator caspase from inactive procaspase; this initiates the caspase activity. Fas ligand, which is associated with the Fas-associated protein with death domain (FADD), gets associated with procaspase-8 to generate active caspase-8. Active caspase-8 is also generated by granzyme-B from procaspase-8 in a perforin-dependent process. Activated caspase-8 acts on procaspase-3 and converts it into active caspase-3 to initiate an apoptotic substrate. This substrate is converted into active apoptotic effectors to initiate apoptosis (Figure 9.1). Thus it seems that CTLs do not kill target cells; rather, they initiate suicidal death cascade pathways present inside each target cells (Rathmell and Thompson 1999).

- *Mitochondria-mediated pathways:* Granzyme and activated caspase-8 act on mitochondria to release cytochrome-c. Cytochrome-c forms a complex with procaspase-9 and Apaf-1 to form activated caspase-9. Caspase-9 acts on procaspase-3 (inactive) to convert it into caspase-3 (active) that undergo apoptosis by converting apoptosis substrate into active apoptotic effectors.

9.2.2 Natural Killer Cells

Natural killer cells are lymphoid cells derived from naïve progenitor T cells in bone marrow. The lineage of NK cells is not very clear since its membrane contains markers similar to the markers present on monocytes, granulocytes, and typical T cells. Two NK cells also differ from each other in the sets of membrane molecules expressed. NK cells express membrane molecules CD2 (75 kDa β-subunit of IL-2 receptor), CD16 (FcγRIII), and IgG Fc region specific receptors. Peripheral blood cells deficient in CD16 molecules lack all activity of NK cells. Like T-lymphocytes, the development of NK cells does not occur exclusively in the thymus. There is evidence of development of NK cells in transgenic mice lacking thymus. Such NK cells developing in thymus-deficient nude mice are functionally active. There is also no rearrangement of receptor genes in NK cells, while rearrangements of receptor genes occur in T cells and B cells. Development of NK cells also occurs in mice deficient in RAG-1 and RAG-2 recombinase genes. RAG-1 knockout mice, which lack T cells and B cells, thus lacking in adaptive immunity, are still able to survive and fight against infection due to the presence of NK cells and innate immunity mechanisms (Lopez-Botet and Bellon 1999; Natarajan, Dimasi, Wang, *et al.* 2002; Yokoyama 1998).

9.2.2.1 Killing mechanisms of NK cells

Both NK cells and CTLs have the same killing mechanism for tumour cells and cells infected with virus. In this mechanism, Fas-associated target cells are traced by Fas ligand containing NK cells to eliminate the altered target cells (Kägi, Vignaux, Ledermann, *et al.* 1994). Granzymes and perforin granules are always present in NK cells. Thus they are constitutively cytotoxic and do not need activation to appear in cells as in case of CTLs. Degranulation occurs at the site of interaction of NK cells. It alters target cells resulting in the release of granzymes and perforin. Both granzymes and perforin help in the initiation of apoptosis in target cells by a mechanism somewhat similar to that of CTLs (Figure 9.1). Although killing mechanisms in both cells are similar, they do have some differences. In NK cells, TCRs or CD3 are absent, and hence the recognition of target cells is not restricted to MHC. The second injection or prior priming has no effect on NK cells, while it increases CTLs activity. This is due to the absence of memory cells in NK cells and its presence in CTLs (Kägi, Vignaux, Ledermann, *et al.* 1994; Kägi, Ledermann, Bürki, *et al.* 1994). The differences and similarities in the mechanism of killing target cells by NK cells and CTLs are given in Table 9.1.

9.2.2.2 NK cell receptors and opposing two-signal model

The two-receptor model explains how NK cells differentiate between normal self-cells and altered self-cells. The activation receptors of NK cells interact with counter ligands present on normal self-cells and altered self-cells to kill them. At the same time, two

Table 9.1 Differences and similarities in the mechanism of killing target cells by NK cells and CTLs

Features	*NK cells*	*CTLs*
Granzymes and perforin	Present	Present
Fas ligand	Present	Present
Memory cells	Absent	Present
Activation receptors	Present	Present
Inhibition receptors	Present	Absent

groups of inhibitory receptors are present: (1) c-type lectin-like inhibitory receptors (CLIRs) and (2) immunoglobulin superfamily inhibitory receptors (ISIRs) or killer cell inhibitory receptors (KIR). Although differing from each other chemically, the two groups are together known as inhibitory receptor superfamily (IRS) (Lanier and Phillips 1996; Long 1999; Lopez-Botet and Bellon 1999). Two glycoproteins CD94 and NKG2 form a heterodimer with disulphide bonds to form CLIRs in humans. CLIRs recognize HLA-E on target cells. HLA-E is transported to the cell surface only in combination with HLA-A, HLA-B, or HLA-C. Thus the amount of HLA-E is an indicator of the overall synthesis and level of class I MHC molecules in human cells. The CD94/NKG2 inhibitory receptor does not have HLA allele-specific expression to kill target cells. On the other hand, KIRs contain more than 50 family members and have HLA loci-specific expression. NK cells express many KIR molecules, each specific to different sets of MHC molecules. The inhibitory signals (CD94/NKG2 or KIR) dominate the activator signal and thus block the killing of target cells mediated by NK cells.

Interaction of KIRs and CD94/NKG2 receptors, which are present and expressed on NK cells with class I MHC molecules, on normal self-cells passes inhibitory signals to counteract the killing signals. Thus the expression of class I MHC molecules on normal cells protects them from being killed, as they pass inhibitory signals; the absence of class I MHC molecules on virus-infected altered cells prevents inhibitory signals, leading to NK cells killing the altered cells. Since there is a decrease in the expression of class I MHC molecules, the signals to kill the altered cells predominate (Natarajan, Dimasi, Wang, *et al.* 2002). The inhibitory signals are due to the presence of ISRs that detect normal cells and override the killing signals. Signals activating NK cells are well supported by cytokines such as IFNα and IFNβ, TNFα, IL-12, and IL-15 present as soluble molecules. Inhibitory signals prevent secretion of IFNγ and TNFα, including NK cell proliferation. C-type lectin, a member of the calcium-dependent carbohydrate-binding proteins such as NKR-P1, helps in activation signals. LFA-3 (adhesion molecules) receptor CD2 and FcγIII receptor CD16 are also involved in the activation signal. CD16 works in an antibody-dependent manner, killing using NK cells. Apart from these, NKp30, NKp44, and NKp46 play a significant role in NK cells activation.

9.3 ANTIBODY-DEPENDENT CELL-MEDIATED CYTOTOXICITY

Apart from NK cells, the other cells that work in an antibody-dependent cell-mediated cytotoxicity (ADCC) against virus-infected altered cells include macrophages, monocytes, neutrophils, and eosinophils. After the antibody binds to the surface antigens of the target cell, these killing cells bind to the free end of Fc receptors to become metabolically active. This results in an increase in the amount of lytic enzymes in lysosomes and granules. These lytic enzymes released at the site of contact cause perforin-mediated damage of membrane and killing of the target cells by granzymes. Additionally, TNFs secreted by the killer cells have cytotoxic effect on the attached target cells.

9.3.1 Experimental Proof of Cell-mediated Cytotoxicity

Mixed-lymphocyte reaction (MLR) assay occurs in in vitro condition. It proves that CTLs are generated by coculturing lymphocytes on mouse fibroblast or by coculturing allogeneic spleen cells. Radioactive [^{3}H] thymidine that gets incorporated into the DNA of proliferating cells is used to measure the degree of proliferation (Plate 1).

Mixed-lymphocyte reaction assay can be generated either by inactivating a population of cells by mitomycin-C treatment or by lethal X-ray radiation treatment. The inactivated cells act as stimulator cells to help the proliferation and differentiation of other cocultured cells that function as CTLs. MLR assay is also used to study the role of Th cells in the generation of CTLs. By removing $CD4^+$ Th cells from coculture using anti-$CD4^+$ proves that CTLs cannot be generated in the absence of $CD4^+$ Th cells. This assay experiment also proves that accessory cells such as macrophages are also essential for the generation of functional CTLs. This is because macrophages activate class II MHC-restricted Th cells (Russell and Ley 2002).

The effector cytotoxic function of cells is assayed in vitro by cell-mediated lympholysis. This assay is based on the incorporation of labelled chromium-51 (^{51}Cr) in target cells by incubating the cells in Na_2CrO_4. The amount of chromium released is directly proportional to the number of cells lysed due to incubation of CTLs. In general, CTL activities are restricted to class I MHC restriction. They kill only those cells that present class I MHC-associated antigens.

Cell-mediated cytotoxicity is assayed in vivo in graft-versus-host (GVH) reaction. This reaction occurs in immunocompromised animals receiving immune component lymphocytes. Due to the non-identical genetic nature of donors and recipients, grafted lymphocyte cells attack host cells. The host becomes non-immune responsive to the graft. This happens in humans after transplantation of bone marrow in leukaemia and in immunodeficiency anemic patients (Haddad, Le Deist, Blanche, *et al.* 1995; Lekstorm-Himes and Gallin 2000). The symptoms of GVH reaction include diarrhoea, skin lesions, jaundice, and spleen enlargement, finally leading to death. The intensity of GVH reaction is measured as follows.

$$\text{Spleen index} = \frac{\text{Experimental spleen weight/total weight of body}}{\text{Control spleen weight/total weight of body}}$$

Enlargement of spleen by 1.3 or greater due to proliferation of $CD4^+$ and $CD8^+$ indicates positive GVH reaction.

9.4 MACROPHAGE-MEDIATED CYTOTOXICITY

Macrophages are present throughout the human body, for example, stromal macrophage in bone marrow, sinus-lining macrophage in liver (Kupffer macrophage), alveolar macrophage, thymic macrophage or endothelial or epithelial lining in Langerhans cells, and interstitial fixed macrophage or serosal cavity-free macrophage. The markers of fixed or circulating mature macrophages are F4/80 antigen. These macrophages are recruited at the site of injury in response to immune stimuli generated by cytokines (IFNγ, TNFα, IL-4, IL-13, and IL-10) or chemokines (macrophage chemotactic protein-1 or MCP-1), which provide them chemotactic movement. Cytokines also modulate the phenotypes of macrophages. Macrophages are involved in the following types of functions.

9.4.1 Activation and Phagocytosis

Macrophages get activated in response to inflammatory cytokines to express class II MHC and complement receptor CR3. They promote phagocytosis in microbes and altered cells. The accumulation of macrophages in local tissues is called diapedesis.

9.4.2 Opsonization-mediated Phagocytosis

Macrophages express two types of receptors—opsonic receptors and non-opsonic receptors. Opsonic receptors include Fc receptors and carbohydrate receptors present on the macrophage surface. They bind with bacterium to engulf them. After phagocytosis, phagosomes are formed, which secrete acid, cytotoxic molecules (reactive oxygen, nitrogen intermediate), and proteolysis enzyme.

In non-receptor-mediated opsonization, endotoxins are produced by bacteria-activated macrophages and endothelial cells to produce TNFα, IL-1, and IL-6 cytokines. These cytokines and endotoxins-activated macrophages engulf or phagocyte in non-receptor-mediated manner. Also surfactant proteins called collectins kill bacteria by disrupting their lipid membranes to make an aggregate of bacteria to be phagocytosed easily by macrophages. Under this mechanism, macrophages can directly expand its pseudopodia-like structure around the microbes to make vesicle-like structures called phagosomes. These phagosomes fuse with lytic enzymes containing lysosome to kill and digest the engulfed bacteria.

AT A GLANCE

- Cell-mediated immunity depends on two types of effector cells based on the type of immunity: $CD8^+$ T cells or cytotoxic T-lymphocytes (CTLs) and $CD4^+$ T cells that perform delayed-type hypersensitivity (DTH) reactions. Effector cells are activated in a more easy way than naïve T-helper cells (Th cells) and T-cytotoxic cells (Tc cells). The activated effector cells express comparatively more amount of cell adhesion molecules along with soluble and membrane-bound effector molecules. They have different trafficking pathways than naïve cells.
- CTL-dependent immune response includes activation and differentiation of Tc cells known as CTL-precursor (CTL-P) cells. This is the first phase of CTL-mediated immune response. Identification of $CD8^+$ cells is based on the labelling of major histocompatibility complex (MHC) tetramers.
- The second phase of CTL-dependent immune responses include (1) recognition of target cells by CTL–MHC; (2) CTL–target cells conjugate formation; (3) accumulation and orientation of cytoplasmic granules of CTL at the contact point of target cells; (4) release of granules and formation of pores in the target cells at the contact point; and (5) CTL dissociation leading to the death of target cells.
- Two pathways are involved in the CTL-mediated killing of target cells: (1) perforin–granzyme mediated and (2) Fas/Fas ligand mediated.
- Non-MHC-dependent cytotoxic cells having no specificity, such as natural killer (NK) cells, neutrophils, eosinophils, and macrophages, can kill target cells in an antibody-dependent manner. The Fc region of immunoglobulin binds to target cells, initiating the release of lytic enzymes, perforin, and tumour necrosis factor (TNF), which damage the membrane of target cells and kill them. This is called antibody-dependent cell-mediated cytotoxicity (ADCC).
- Like CTLs, NK cells follow the perforin-mediated pathway to kill target cells, such as tumour cells and virus-infected cells.
- Normal cells express high numbers of class I MHC molecules to protect them from being killed by NK cells. NK cells are regulated by two types of signals: (1) positive activating signals (such as NKR-P1) and (2) negative inhibitory signals transmitted by inhibitory receptors (CD94/NKG2 and the KIR family) to kill or leave the target cells.

QUESTIONS

1. Why are NK cells in any host able to kill only virus-infected cells but not normal cells? Discuss with a suitable diagram.
2. Discuss the Fas ligand and perforin-mediated killing of infected cells with a suitable diagram.
3. How does cell-mediated immunity work? Discuss the important properties of Th cells, Tc cells, and CTLs.
4. What is GVH reaction and spleen index? Discuss the importance of spleen index.
5. What are effector T cells or CTLs? Discuss the various steps of CTL functions with a suitable flow chart.

OBJECTIVE QUESTIONS

A. Fill in the blanks.

a. The cell-mediated immune system involves two types of effector cells specific to particular antigen. They are _________ and _________.
b. Effector cells express _________ the level of adhesion molecules compared to naïve cells.
c. CTL precursors (CTL-Ps) are CTL mediated _________ and _________ of Tc cells.
d. TCR–MHC-mediated recognition of target cells is _________ phase CTL immune response.
e. CTL-mediated cell death involves two pathways: _________ and _________.
f. Non-MHC-dependent cytotoxic cells are _______, _______, _______, and _______.
g. Non-specific CTLs bind to the ______ region of antibody to bind with target cells.
h. Non-specific cells release ______, ______ or ______ to kill target cells.
i. High expression of _____ by normal cells protects them from _____ mediated killing.
j. Spleen index is the ratio of _____ divided by total body weight and ______ divided by total body weight.

B. Match the following.

i. First-phase CTL immune response	a. Mitochondria
ii. Second-phase CTL immune response	b. Active
iii. ADCC cytotoxicity	c. Inactive
iv. NKR-P1	d. Apaf-I
v. CD94/NKG2	e. Th-1 cells
vi. Cytochrome-c	f. Negative signals of NK cells
vii. Procaspase-3	g. Positive signals of NK cells
viii. Caspase-3	h. Non-specific cytotoxic cells
ix. Procaspase-9	i. CTL-target cells conjugate
x. IL-2	j. CTL-Ps

CHAPTER TEN 10

Allergic Reactions or Hypersensitivity

10.1 INTRODUCTION

Allergy or *hypersensitivity* is an inappropriate immune response caused by the activation of effector molecules to eliminate antigen. The mechanism involved is responsible for inflammation and sometimes causes extensive damage to the host tissues. Sometimes inflammatory responses are so severe that it causes large-scale tissue damage leading to deleterious effects, even death. It occurs in either humoral or cell-mediated immune responses. Allergy is due to allergens, which are non-parasitic antigens. Some common allergens are given in Table 10.1.

Paul Portier and Charles Richet were the first to report an allergic reaction in a person taking bath in the Mediterranean Sea. Both French scientists concluded that the localized reaction that took place was due to the presence of jellyfish toxins. They also hypothesized that successful immunization may lead to phylaxis (protection). The scientists performed experiments by injecting purified jellyfish toxin into a dog as vaccine. After the final booster dose, instead of producing antibodies in response to the toxins, the dog overreacted with symptoms like diarrhoea, vomiting, and asphyxia

Table 10.1 Common allergens present in the environment

Allergen/biochemical nature	*Example of allergen*
Proteins	Used in vaccines
Pollen grains of plant/grasses	Birch, rye, ragweed, Timothy grass
Food products	Eggs, peas, beans, milk, nuts, sea foods
Common drugs	Penicillin, sulphonamides, local anesthetics, salicylates
Products of insects/animals	Venom of bee, wasp, and ant; mites, calyx of cockroach; dandruffs and hair of animals

and finally died. This overreaction was termed anaphylaxis by Portier and Richet. Anaphylaxis is opposite in meaning to the Greek word prophylaxis. In 1913, Richet was awarded the Nobel Prize for physiology or medicine for discovering anaphylaxis.

10.2 CLASSIFICATION BASED ON GELL AND COOMBS REACTION

Based on the generation of effector molecules, hypersensitive reactions are divided into various categories. Hypersensitivity mediated by antibody IgE induces degranulation and the release of many biologically active molecules, including histamine. Antibodies IgG and IgM activate complement molecules. The product of splitting complement molecules, such as C3a, C4a, and C5a, acts as effector molecules, which attack the cell membrane. Cytokines secreted by Th cells or Tc cells act as effector molecules in delayed-type hypersensitivity (DTH). Based on antibody and antigen–antibody complexes, Gell and Coombs classified hypersensitivity into type I (antibody IgE mediated), type II (antibody IgG or IgM mediated), type III (immune complex mediated), and type IV (cell-mediated called delayed type).

10.3 ANTIBODY-DEPENDENT ALLERGIC REACTIONS

Allergic reaction or hypersensitivity is triggered by an antibody. IgE is an important antibody generated by B cells after interaction with allergens.

10.3.1 Type I Hypersensitivity

10.3.1.1 Components of type I hypersensitivity

The main components of type I hypersensitivity are allergens, reaginic antibody (IgE) and its receptors, mast cells and basophils, and primary and secondary mediators. Type I hypersensitivity differs from normal humoral response by the production of antibody IgE. In type I hypersensitivity, IgE is produced, while it is not produced by plasma cells. Allergens or antigens bind to mast cells and basophil-associated IgE to release primary and secondary mediators (Plate 1).

Primary and secondary mediators: Histamine, proteases, eosinophil chemotactic factor (ECF), neutrophil chemotactic factor (NCF), and heparin are primary mediators. Platelet-activating factors (PAFs), leukotrienes, prostaglandins, bradykinins, and cytokines are secondary mediators.

Allergens: In normal conditions, if an individual has parasitic infections, the IgE level in the serum goes up till the parasites are removed completely from the body. But in some individuals, due to hereditary predisposition, the body stimulates an uncontrolled production of IgE against non-parasitic antigens causing damage to nearby tissues, for example, type I hypersensitivity. Such abnormality is called *atopy*. These non-parasitic antigens having the ability to trigger type I hypersensitivity are called *allergens*, and the individuals are called *allergic individuals* or *atopic individuals*. Atopic individuals have a genetic abnormality, which is hereditary, to produce an abnormally high amount of IgE. Thus atopic individuals have high levels of circulating IgE and hence are more susceptible to allergies such as hay fever, eczema, and asthma. Atopy is multigenic and has several loci such as 5q and 11q. The locus 5q encodes cytokines IL-3, IL-4, IL-5, IL-9, IL-13, and GM-CSF, while 11q encodes the β-chain of IgE high affinity receptor. Common allergens are listed in Table 10.1. The two major allergens of ragweed fractions are E and K fractions. The minor allergens of ragweed fractions are Ra3, Ra4, and Ra5. Allergens are proteins or protein-associated molecules of molecular weight 15,000–40,000. The allergenicity depends on the dose, route of entry, adjuvant, and genetics of the individual (Hoyne, Kristensen, Yssel, *et al.* 1995; Lympany, Welsh, Cochrane, *et al.* 1992).

Reaginic antibody or IgE antibody: Upon entry of allergens into the body by inhalation or ingestion, IgE is produced on the mucous membrane in large quantities. These allergens reacting with antibodies were first identified by Carl Otto Willy Prausnitz and Heinz Küstner in 1921 by a test known as the Prausnitz–Küstner (P–K) reaction. The reaction is a local wheal-and-flare response at the site of injection on the skin. The serum developed against these allergens was called P–K antibodies or *reagins* or *reaginic antibodies*. The biological activity of this reaginic antibody cannot be neutralized by rabbit IgG, IgA, IgM, and IgD isotype antiserum. It is rather neutralized by rabbit antiserum against atopic sera of humans. The serum from ragweed-sensitive individuals, when injected into rabbit, caused the rabbit antiserum to prevent or neutralize the P–K tests. This new antiserum isotype was called IgE (molecular weight 190,000) due to E-antigen of ragweed discovered in 1996 by Kimishige Ishizaka and Teruko Ishizaka. Normal individuals contain 0.1–0.4 μg/ml of IgE. Even in highly allergic individuals, the amount does not go up more than 1 μg/ml. It was thus impossible to study the physiochemical behaviour of IgE due to its low concentration till the discovery of IgE myeloma by S G O Johansson and H Bennich in 1967. The molecular weight of IgE is more than IgG (150,000) due to the presence of an additional domain present in the constant region (C_H4). C_H4 causes conformational change in the Fc region to enable it to bind to blood basophils and tissue mast cells. In the free state, the half-life of IgE is two to three days, but when it gets attached to receptors present on mast cells or basophils, it shows stability up to weeks.

Basophils are blood-circulating (0.5%–1.0%) white blood cells with granulated cytoplasm containing glycogen granules or membrane-bound granules, multi-lobed nucleus, and a few mitochondria.

Mast cells: Precursors of mature mast cells are formed in the bone marrow. These precursor mast cells mature in vascularized peripheral tissues into mature mast cells. The mature mast cells are particularly localized in blood and lymphatic vessels and are also present throughout connective tissues. The concentration of these cells is highest in skin (10,000 mast cells/mm^3) and the mucous membrane of the respiratory and gastrointestinal tracts.

IgE-binding receptors: IgE binds to receptors specific to the Fc region of ε heavy chain, called FcεR, present on mast cells. Based on its affinity, receptors are of two types (Beaven and Metzger 1993; Novak, Kraft, and Bieber 2001; Razin, Pecht, and Rivera 1995): (1) a high-affinity IgE receptor FcεRI (Figure 10.1) and (2) a low-affinity IgE receptor FcεRII, also called CD23.

- *FcεRI:* FcεRI contains four polypeptide chains—α-chain, β-chain, and two identical disulphide-linked γ-chains having an external part, a transmembrane (Tm) part, and a cytoplasmic part. The cytoplasmic part is carboxy terminal (–COOH), while the external part is the amino terminal (–NH_3). The external region of the α-chain has two 90-amino-acid domains (homologous to immunoglobulin), due to which this molecule is considered part of the immunoglobulin superfamily. The two α-chain domains of FcεRI interact with the two domains of IgE C_H3/C_H3 and C_H4/C_H4, respectively. The β-chains stretch four times into the plasma membrane space and connect the homodimer γ-chain to the α-chain (Figure 10.1). Both the amino and carboxy terminals of the β-chain reside in the cytoplasm. The two homodimer γ-chains are attached to each other by disulphide bonds. Its amino terminal resides in the extracellular space, while the carboxy terminal resides in the cytoplasmic space. This carboxy terminal contains a conserved sequence called

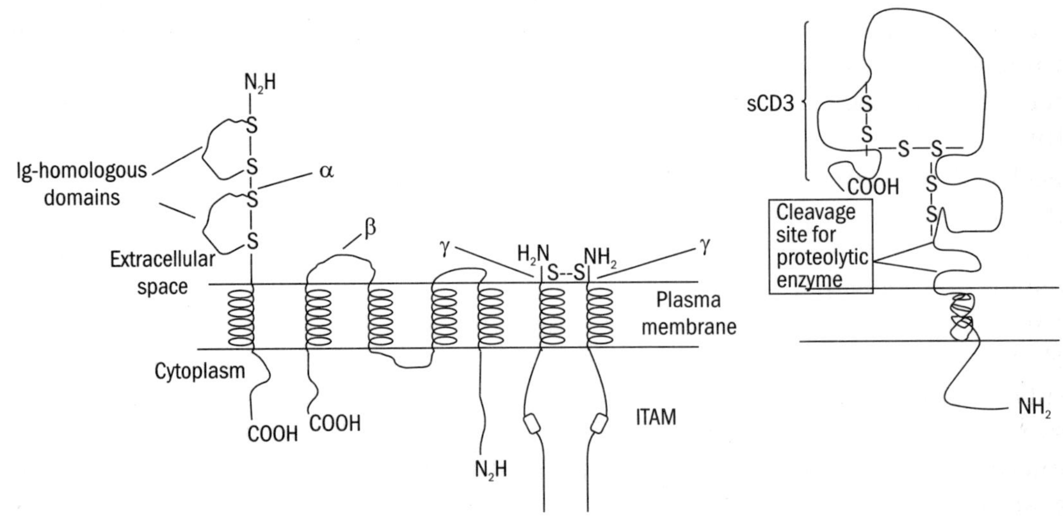

Figure 10.1 FcεRI receptor

the immunoreceptor tyrosine-based activation motif (ITAM). CD3 and the ζ-chain of the T-cell receptor complex and the Igα/Igβ-chains of membrane immunoglobulin on B cells contain the same motif. By blocking these motifs with peptide, allergic responses can be modulated (Daser, Meissner, Herz, *et al.* 1995). The ITAM motif transduces activating signals to the cells through protein tyrosine kinases (PTKs). Quick tyrosine phosphorylation is caused by allergen-driven cross-linking of FcεRI-associated IgE, leading to degranulation in mast cells. By using anti-IgE, allergic reaction could be stopped; thus the respective diseases can be cured (Busse and Neaville 2001). The affinity of this receptor is high (K_D = 1–2 × 10^{-9} M), and it can bind even in low-level concentrations up to 1 × 10^{-7} IgE.

- *FcεRII (CD23):* The soluble form of FcεRII is called sCD23. It is generated from membrane receptors by autoproteolysis. FcεRII is a low-affinity receptor (K_D = 1 × 10^{-6} M), which interacts with the C_H3/C_H3 domain of IgE. Binding of allergens activates B cells and other cells such as alveolar macrophages and eosinophils. Secretion of IgE by activated B cells decreases when the FcεRII receptor is blocked by monoclonal antibody but is enhanced in the presence of sCD23. The level of CD23 on lymphocytes/macrophages and sCD23 in serum is found to be higher in atopic individuals. IL-4 interacts with β2-adrenoceptor to regulate CD23 and thus the production of IgE in humans (Aubry, Pochon, Graber, *et al.* 1992; Holt 1994; Paul-Eugene, Kolbe, Calenda, *et al.* 1993).

10.3.1.2 Degranulation of mast cells or basophils

Extracellular events: Degranulation in mast cells or basophils (Teixeira, Williams, and Hellewell 1995) is triggered essentially by cross-linkage of IgE, which is bound to IgE receptor, or cross-linkage of IgE receptor by anti-receptor antibody, or non-receptor mechanism using ionophore. Ionophore increases Ca^{2+} influx/permeability. Cross-linkage is done by allergen or anti-isotype/idiotype antibody or by anti-receptor antibody or chemicals that bind with the receptor (Razin, Pecht, and Rivera 1995).

Intracellular events: Cross-linkage of FcεRI leads to the activation of PTKs, which then leads to the phosphorylation of the tyrosine of γ-subunit-associated ITAM, β-subunit residues, and phospholipase-C. At the same time, there is methylation of membrane phospholipids within 15 s causing the formation of Ca^{2+} channels and enhancing the fluidity of the membrane. The increase in Ca^{2+} level in the cell is observed by the uptake and release of Ca^{2+} from the endoplasmic reticulum and subsequent formation of arachidonic acid. Later, arachidonic acid is converted into prostaglandins and leukotrienes, two potent mediators. The increase in Ca^{2+} helps microtubule assembly and microfilaments contraction to guide the trafficking of granules to the plasma membrane. The drug disodium cromoglycate (cromolyn sodium) blocks Ca^{2+} influx to prevent allergies. This shows the importance of Ca^{2+} ions in allergies. Other events that occur include increase in adenylate cyclase activity leading to transient increase in the level of cAMP. The produced cAMP causes activation of cAMP-dependent protein kinases to phosphorylate membrane-bound granules, thereby increasing permeability. Later, the level of cAMP falls to facilitate degranulation. Drugs causing increase in the level of cAMP causes blocking of degranulation (Bonnefoy, Pochon, Aubry, *et al.* 1993; Chang 2000; Hoyne, Kristensen, Yssel, *et al.* 1995).

Mediators produced in degranulation

- *Histamine:* Histamine is the primary type of mediator for a type I hypersensitive reaction produced by the decarboxylation of amino-acid histidine in mast cells. After release, histamine binds to specific receptors H1, H2 or H3 located in different tissues. Binding to H1 receptor induces contraction of the smooth muscles of intestine and bronchi. This increases permeability of venules and mucus secretion by goblet cells. Binding of histamine with H2 receptor increases dilation, vasopermeability, and also stimulates the exocrine glands. When histamine is attached to H2 receptors located on mast cells and basophils, it causes suppression of degranulation.
- *Leukotrienes and prostaglandins:* Leukotrienes are secondary mediators. They are generated by the enzymatic cascade breakdown of phospholipids occurring after degranulation. In comparison to histamine, the effects are long lasting and more pronounced. Leukotrienes cause increase in vascular permeability, mucus production, and bronchoconstriction. The prostaglandin D2 causes only bronchoconstriction. Initially, the bronchial and tracheal smooth muscle constriction is due to histamine, but after a person crosses 30, it is mediated by leukotrienes and prostaglandins. The constriction effect of leukotrienes is 1000 times more than histamine and causes prolonged bronchospasm, mucus secretion, and vascular permeability.
- *Cytokines:* Cytokines come under the category of secondary mediators and are secreted by mast cells. The ratio of Th-1 to Th-2 cytokines is the key regulator of type I hypersensitive responses. Th-1 cytokines (IFNγ) reduce allergy, while Th-2 increases the allergic response. Th-2 cytokines are IL-3, IL-4, IL-5, IL-6, TNFα, and IL-10 (Akdis and Blaser 1999). IL-4 causes increase in the production and class switching of IgE along with clonal expansion of those B cells regulated by IgE. IL-3 and IL-5 are important in recruiting, maturing, and activating eosinophils and neutrophils. TNFα contributes to systemic anaphylaxis. The ratio of IL-4 (Th-1) to IFNγ (Th-2) determines IgE production. Systemic anaphylaxis occurs within 60 s of a type I hypersensitive reaction and is often fatal. Downregulation

of IL-4 cytokine could be one of the possible treatments for allergy (Borris and Rosenwasser 1997; Daser, Meissner, Herz, *et al.* 1995; Finkelman, Katona, Urban, *et al.* 1988; Kühn, Rajewsky, and Müller 1991).

10.3.1.3 Clinical symptoms of type I hypersensitivity

Degranulation, receptor cross-linkage, systemic anaphylaxis, and localized anaphylaxis such as asthma, hay fever, and eczema are the clinical symptoms of type I hypersensitivity.

Systemic anaphylaxis: Systemic anaphylaxis is a fatal shock that occurs within minutes after a type I hypersensitive reaction starts. Characteristic symptoms of systemic anaphylaxis differ in different animals due to differences in the distribution of mast cells and their granules content. Guinea pigs are used in experiments for anaphylaxis study because anaphylaxis can be induced easily in guinea pigs and the symptoms match closely with that of humans. A single injection of egg albumin and other foreign proteins produces active sensitization in guinea pigs. After injection, the animal is kept for approximately two weeks. Then the same protein is injected intravenously. Immediately, the animal's blood pressure drops. It becomes restless, urinates, and the smooth muscles of the gastrointestinal tract and bladder contract. Within 2–4 min, the animal dies of anaphylaxis. Death occurs due to massive oedema, shock, and bronchoconstriction caused by anaphylaxis. A list of antigens that triggers anaphylaxis is given in Table 10.1.

Localized anaphylaxis or atopy: Localized anaphylaxis is limited to the tissues and organs at the entry site of the allergen. It is inherited and hence called atopy. It causes allergic rhinitis (hay fever), asthma, eczema (atopic dermatitis), and allergies due to food (CSGA 1997). Approximately, 20% of the population in developed countries is affected by atopic allergies. Examples of atopic allergies are as follows.

- *Allergic rhinitis or hay fever:* Hay fever is an atopic disorder caused by airborne allergens that sensitize mast cells present in the nasal mucosa and conjunctivae. Sensitized mast cells release active mediators that cause localized vasodilation and enhance permeability of local capillaries. The symptoms of hay fever include sneezing, coughing, watery exudation of the conjunctivae, and mucosa of the nasal and upper respiratory tract.
- *Asthma:* Asthmatic attack is caused by pollens, dusts, fumes, insect antigens and their products, or viral antigens. It can also occur during exercise or due to cold. These allergens cause degranulation of mast cells releasing mediators in the lower side of the respiratory tract. This causes constriction of bronchial muscles and inflammation due to the secretion of cytokines by eosinophils and neutrophils IL-4, IL-5, IL-6, TNFα, ECF, and PAF. Infiltration of eosinophils and neutrophils to respiratory tract and bronchial muscles secretes oxygen radicals and lytic enzymes

causing tissue damage. At early response, bronchoconstriction and vasodilation occur due to primary mediators such as histamine, LTC4 (leukotrienes), and PGD2 (prostaglandins). Histamine causes secretion of mucus that incorporates inflammatory cells and epithelial cell fragments. Cockroach allergen largely affects inner-city children, causing asthma and leading to high child morbidity (Barnes and Marsh 1998; CSGA 1997; Romagnani 2001; Rosenstreich, Eggleston, Kattan, *et al.* 1997).

- *Allergies due to food:* Food allergen causes localized smooth muscle contraction and vasodilation leading to vomiting and diarrhoea. Other symptoms include urticaria (hives), oedema (swelling), erythema (red eruptions), and P–K reactions.
- *Eczema or atopic dermatitis:* Due to Th-2 cells increase in the number of eosinophils, atopic dermatitis or eczema occurs.

10.3.1.4 Clinical tests of type I allergies and drugs for treatment

Radioimmunosorbent test (RIST) and radioallergosorbent test (RAST) can detect allergens in nanograms.

RIST

- Anti-IgE-coated agarose beads or paper disc + patient IgE $\rightarrow$ IgE–anti-IgE complex.
- Washing to remove non-specific IgE.
- IgE–anti-IgE complex + radiolabelled anti-anti-IgE (I^{125}) $\rightarrow$ IgE–anti-IgE–anti-anti-IgE complex.
- Washing to remove excess of unbound radiolabelled anti-anti-IgE (I^{125}).
- Counting of radiolabelled anti-anti-IgE.

RAST

- Allergen-coupled agarose beads + patient IgE $\rightarrow$ allergen–IgE complex.
- Washing to remove non-specific IgE.
- Allergen–IgE complex + radiolabelled anti-IgE (I^{125}) $\rightarrow$ allergen–IgE–anti-IgE complex.
- Washing to remove excess of unbound radiolabelled (I^{125}) anti-IgE.
- Counting of radiolabelled anti-IgE (I^{125}).

Drugs used for treatment

- Anti-histamine drugs are used to block H_1 and H_2 receptors present on target cells.
- Cromolyn sodium is used to block Ca^{2+} influx into mast cells.
- Theophylline is used to inhibit enzyme phosphodiesterase responsible for cleavage of cAMP into 5′-AMP, thus prolonging high cAMP levels in mast cells.

- Adrenalin or epinephrine binds with β-adrenergic receptors of mast cells to stimulate the production of cAMP.
- Cortisone prevents conversion of histidine to histamine, which reduces histamine levels and stimulates cAMP production in mast cells.

10.3.2 Type II Hypersensitivity

Antibodies activate complement factors to develop pores in non-self cells or the Fc region of antibody binds to the Fc receptors of cytotoxic cells and antibody binds to the antigen of target cells to activate killing. Antibody can coat non-self cells by binding to antigens to kill by opsonization or by complement activation. Typical examples of type II hypersensitivity (Plate 2) include blood transfusion, erythroblastosis fetalis, and autoimmune haemolytic anaemia.

10.3.2.1 Blood transfusion

Red blood cell (RBC) membrane contains proteins and glycoproteins encoded by different genes or many alleles. The product of one allelic form may be recognized as foreign; this triggers antibody response. In other cases, normal flora of microorganisms of the gut induces antibody formation against various antigenic determinants to form ABO blood group. Antibodies developed against antigens A, B, and O are called *isohaemagglutinins* normally of the IgM group. An individual can develop isohaemagglutinins against only one type of antigen determinants (such as A) and show self-tolerance against other type of antigens (such as B). Even minor allelic differences in these antigens trigger different antibody development causing blood transfusion with other blood types. The antibody developed is of the IgG group. The immediate reaction of blood transfusion is blood group ABO incompatibilities and IgM isohaemagglutinins-triggered complement-mediated lysis. This is evident by the presence of free haemoglobin in the plasma within hours causing haemoglobinuria. Bilirubin resulting from haemoglobin lysis is toxic in high amounts and causes fever, chills, nausea, lower back pain, urine with haemoglobulin, and blood clotting within vessels. If haemoglobin is allowed to accumulate in the kidney, it will cause acute tubular necrosis. Late response occurs due to repeated blood transfusion that causes clonal selection and IgG production against RBC membrane antigens called Rh, Kidd, Kell, and Duffy. IgG is less effective than IgM to activate complement system, and hence transfused RBCs are destroyed mostly by agglutination, opsonization, and phagocytosis. Transfusion reaction occurs when blood type A is injected into individuals having blood type B due to the presence of anti-A isohaemagglutinins that bind to type A cells causing complement-mediated lysis.

10.3.2.2 Erythroblastosis fetalis

Erythroblastosis fetalis is a haemolytic disease of the newborn that develops due to the presence of Rh^+ factor of foetus RBCs and Rh^- factor of the mother triggering development of rhogam (Rh antibody) against the foetus. In first pregnancy, very few Rh-specific B cells are produced due to the separation of foetus blood RBCs from the mother RBCs by trophoblast (placenta). At the time of birth, foetal umbilical cord RBCs enter the mother's blood circulation to activate a large number of Rh-specific B cells. These B cells differentiate into plasma cells to produce IgM antibody, which removes Rh^+ foetal RBCs from the mother's blood. During second pregnancy, the remaining memory B cells form IgG anti-Rh antibodies. These new antibodies cross the placenta to destroy foetal RBCs.

Treatment

- Anti-histamine drugs are used to block H1 and H2 receptors of target cells to prevent degranulation.
- Cromolyn sodium or disodium cromoglycate is used to stop Ca^{2+} influx into mast cells to prevent ionophore-mediated degranulation.
- Theophylline, normally an oral inhaler, is used to increase the level of cAMP in mast cells by blocking phosphodiesterase. Phosphodiesterase reduces cAMP level by converting it into 5'-AMP.
- Epinephrine or adrenalin is used to stimulate β-adrenergic receptors of mast cells to enhance cAMP production.
- Cortisone is used as an anti-inflammatory drug to reduce histamine levels by preventing conversion of histidine to histamine. It also increases cAMP production.

10.3.3 Type III Hypersensitivity

In type III hypersensitivity, an antigen–antibody complex is formed and deposited in tissues, inducing complement activation. Complement activation leads to inflammatory response followed by infiltration of neutrophils at the site of deposition. This reaction is denoted by localized Arthus reaction within 4–8 h. The symptoms of Arthur reaction include serum sickness, necrosis of vasculitis, glomerulonephritis, rheumatoid arthritis, systemic lupus erythematosus, meningitis, and hepatitis. Arthur reaction can result in oedema (accumulation of fluid), erythema (RBCs), mild swelling and redness in tissue at the site. Examples of reactions are those caused by insect bites, entry of bacterial spores and fungi, inhalation of actinomycetes, which cause farmers lung, and the occurrence of pigeon fancier disease due to the inhalation of dried pigeon faeces serum protein. Antigens that form complexes could be bacterial, viral, and parasitic causing skin rashes and arthritic symptoms. In systemic lupus erythematosus and synovial arthritic symptoms, the complex is formed with anti-DNA antibody raised against DNA.

The magnitude of type III hypersensitive reaction (Figure 10.2) depends on the formation and presence of immune complex in the body. Localized reaction results when the complex is formed and deposited at the site of entry of the antigens. If the complexes are formed in the blood, the reaction can be visible at its deposition sites, for example, at the wall of the blood vessel, at the joints in the synovial membrane, in the kidney at the membrane of glomerular basement, and in the brain at choroids plexus. Deposition is followed by the recruitment and release of granules from neutrophils.

These immune complexes activate a complement system cascade, such as C3a, C4a, C5a, and anaphylotoxins, to cause degranulation of mast cells and enhanced permeability of the vascular membrane. The chemotactic properties of C3a, C5a, and C5b67 attract neutrophils at the deposition site of the immune complex. Deposition of immune complex causes lesions in the membrane due to secretion of lytic enzymes by neutrophils as they attempt to phagocytose the complex. C3b opsonizes immune complex and can also bind to the CR1 membrane receptor of neutrophils. C3a, C4a, and C5a can also bind to the receptor present on the mast cells surface to degranulate them. These subsets of the complement system induce aggregation of platelets causing microthrombi due to the release of clotting factors (Bonnefoy, Pochon, Aubry, *et al.* 1993).

10.4 ANTIBODY-INDEPENDENT OR CELL-MEDIATED HYPERSENSITIVITY

Antibody-independent hypersensitivity does not depend either on antibody or the complement system; rather, it depends on phagocytic cells. These phagocytic cells are directly responsible for the damage of altered cells and causing type IV hypersensitive

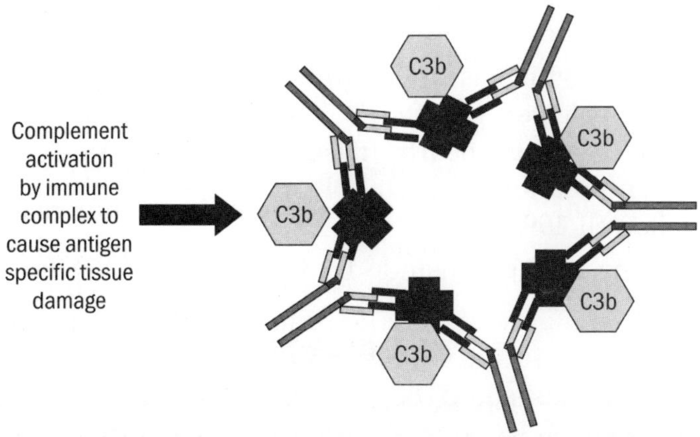

Figure 10.2 Mechanism of type III hypersensitivity

Note In type III hypersensitivity, the antigen–antibody complex induces C3b complement activation to induce neutrophils filtration and inflammation.

reactions (Plate 3). Such reactions occur after a certain period of time because they depend on secreted molecules, such as cytokines. Thus they are called DTH reactions (Romagnani 2001).

10.4.1 Type IV Hypersensitivity

Activated Th-1 cells secrete cytokines to induce localized inflammatory reactions called type IV hypersensitivity or DTH. Secreted cytokines recruit and activate macrophages and Tc cells. After activation, secretion of cytokines does not take place immediately; rather, it takes an hour and hence this type of reaction is called DTH. The activated macrophages and Tc cells are directly involved in killing cells, including microbes and other pathogens. Examples of such reactions are contact dermatitis, tubercular lesions, and rejection of grafts. DTH occurs in the following steps.

- Antigens activate and expand $CD4^+$ Th cells and, in a few cases, $CD8^+$ Th cells. Earlier, these T cells were denoted as T_{DTH} cells to show their functionality in type IV hypersensitivity.
- Antigen-presenting cells (APCs), such as Langerhans cells and macrophages, present antigen in association with class II major histocompatibility complex (MHC).
- DTH reactions become apparent after 24 h. Maximum reaction peak is observed at 48–72 h when the second contact of antigens occurs. This delay is due to the time taken by cytokines to activate the macrophages.
- Activated macrophages show enhanced phagocytosis, increased killing ability, increased expression of class II MHC molecules, and increased level of cell adhesion molecules.
- Lytic enzymes secreted by macrophages cause non-specific killing of nearby cells apart from killing microbes and infected cells.
- It forms a granuloma, which is an accumulation of activated macrophages in a tissue, to form multinucleated giant cells. The granuloma secretes large amount of lytic enzymes, which damage nearby tissues and blood vessels causing necrosis. There might be accumulation of eosinophils by pharmacological drugs (Teixeira, Williams, and Hellewell 1995).
- Cytokines participating in DTH include IL-3, IFNγ, TNFβ, TNFα, and IL-1.

AT A GLANCE

- Hypersensitivity is an inflammatory reaction mediated by humoral cells. It leads to large-scale tissue damage culminating in death. Hypersensitivity is of four types, each having specific clinical symptoms and generating characteristic effector molecules.
- Type I hypersensitive reaction is mediated by IgE. The Fc region of IgE binds to mast cells or basophils. Allergens cross-link mast/basophile cells by two IgE molecules to trigger degranulation

with the release of mediators. Mediators are pharmacologically active molecules that cause smooth muscle contraction and vasodilation. The clinical symptoms of type I hypersensitivity include systemic anaphylaxis having life-threatening effects and localized responses having asthma and hay fever effects.

- Type II hypersensitive reaction is due to the interaction of antibody to antigenic determinant on the cell surface causing cell damage, death or cell lysis. It is mediated by a complement called antibody-dependent cell-mediated cytotoxicity (ADCC). Examples of this reaction include haemolytic disease in newborns and transfusion reactions.
- Type III hypersensitive reaction is due to the activation of complement mediated by immune complexes, which splits the complement into subunits. These subunits serve as immune effector molecules to cause vasodilation and accumulation of neutrophils at the site by chemotactic attraction. Immune complexes get deposited at the site of antigens entry to induce an Arthus reaction. An Arthus reaction is a release of lytic enzymes by neutrophils and complement subunits causing local tissue damage.
- Type IV hypersensitive reaction involves cell-mediated (specifically Th-1 cells) release of cytokines that cause accumulation and activation of macrophages at the site to release lytic enzymes, which damage local tissues.

QUESTIONS

1. What is hypersensitivity? Discuss different types of hypersensitivity reactions with suitable examples.
2. Discuss the cause of type I hypersensitivity reactions with suitable examples.
3. What is wheal-and-flare reaction? In which type of hypersensitivity reactions, is it important? Discuss.
4. What is delayed-type hypersensitivity reaction? Discuss the role of cytokines in different kinds of hypersensitivity reactions.
5. Discuss the difference between type I hypersensitivity reaction and type II hypersensitivity reaction with suitable diagrams. Which of them is known as the antibody-dependent hypersensitivity reaction and why?

OBJECTIVE QUESTIONS

A. Fill in the blanks.

a. Antibodies involved in hypersensitivity reactions are _____; _____; and _____.
b. Hypersensitivity reactions classification is called _____ classification.
c. Antigens inducing type I hypersensitivity are called _____.
d. Antibody-mediated hypersensitivity is called _____.

e. Type III hypersensitivity is due to the formation of _____ complex.
f. Membrane-bound IgE of mast cells and basophils is cross-linked to cause _____.
g. Molecular weight of allergens stands between _____.
h. Reaginic antibody is _____.
i. Local _____ and _____ reaction occurring when allergen is injected into a sensitive person is called _____ reaction.

B. Match the following.

i. IgE-mediated allergy	a. Delayed-type hypersensitivity (DTH)
ii. Complement-mediated lysis of opsonized RBC	b. Type I hypersensitivity
iii. Damage of tissues due to poison oak	c. Type II hypersensitivity
iv. Immune complex-mediated allergy	d. High-affinity receptor
v. Enhanced Ca^{2+} influx	e. Type III hypersensitivity
vi. FcεRI	f. Ionophone
vii. Fatal shock	g. Rh antigen
viii. FcεRII	h. Localized anaphylaxis
ix. Atopy	i. Systemic anaphylaxis
x. Rhogam	j. Low-affinity receptor

CHAPTER ELEVEN 11

Autoimmunity

11.1 INTRODUCTION TO AUTOIMMUNE DISEASES

Paul Ehrlich termed the action of the immune system against self-antigens instead of foreign antigens as *horror autotoxicus*. Self-tolerance is a term used when individuals are protected from self-reactive lymphocytes. Earlier in the 1960s, it was believed that during the development of lymphocytes in bone marrow and thymus, all self-reactive lymphocytes get eliminated. But in the 1970s, experimental results suggested that some self-reactive lymphocytes are left alive during lymphocyte development. This is the reason why even healthy individuals possess self-reactive mature lymphocytes in circulating condition. These peripheral circulating self-reactive lymphocytes do not cause autoimmune reaction, and their activity is controlled or regulated by clonal suppression or *clonal anergy*. Any disruption in regulation causes activation of self-reactive clones of B cells and T cells (Hausmann and Wucherpfennig 1997; O'Garra, Steinman, and Gijbels 1997). Such self-reactive clones cause serious damage to self-cells and organs leading to fatal consequences. List of organs, names of diseases, and affecting immune molecules have been given in Table 11.1. Different types of autoimmunity are shown in Figure 11.1.

11.2 ORGAN-SPECIFIC AUTOIMMUNE DISEASES

Organ-specific autoimmune diseases are limited only to specific organs. Damage in organs is caused by cell-mediated mechanisms. The resulting autoantibodies either block the function or overstimulate the affected organ causing cell lysis or inflammation. This leads to decline in the function of the organ. Damaged cells/tissues in an organ are replaced by scar tissues or connective tissues. Some autoimmune diseases are discussed as follows (King and Sarvetnick 1997; Liblau, Singer, and McDevitt 1995; Lockshin 1998; Silverstein and Rose 1997).

Table 11.1 Organs, diseases, and affecting immune molecules

Name of cell/organ	*Name of disease*	*Immune cell*
Adrenal cells	Addison's disease	Self-antibodies
Red blood cell (RBC) membrane	Haemolytic anaemia	Self-antibodies
Lung and renal membrane	Goodpasture's syndrome	Self-antibodies
Receptor of thyroid-stimulating hormone	Graves' disease	Self-antibodies
Cells and proteins of thyroid	Hashimoto's thyroiditis	T_{DTH} cells, self-antibodies
Proteins of platelet membrane	Idiopathic thrombocytopenic purpura	Self-antibodies
Pancreatic β cells	Insulin-dependent diabetes mellitus	Self-antibodies, T_{DTH} cells
Acetylcholine receptors	Myasthenia gravis	Self-antibodies
Heart	Myocardial infarction	Self-antibodies
Gastric parietal cells; intrinsic factor	Pernicious anaemia	Self-antibodies
Kidney	Post-streptococcal glomerulonephritis	Self-antibody complexes
Sperm	Spontaneous infertility	Self-antibodies
Premature ovarian failure (POF); hormone-producing organs	Stops or intermittently produces eggs, estrogen, and progesterone	Self-antibodies
Systemic autoimmune diseases		
Vertebrae	Ankylosing spondylitis	Immune complexes
Brain or white matter	Multiple sclerosis	Self-antibodies, Th-1 and Tc cells
Connective tissue, IgG	Rheumatoid arthritis	Self-antibodies and immune complexes
Heart, lung, kidney, gastrointestinal tract, nuclei	Scleroderma	Self-antibodies
Salivary gland, liver, kidney, thyroid	Sjögren's syndrome	Self-antibodies
DNA, RBC, nuclear protein, platelet membranes	Systemic lupus erythematosus	Self-antibodies, immune complexes

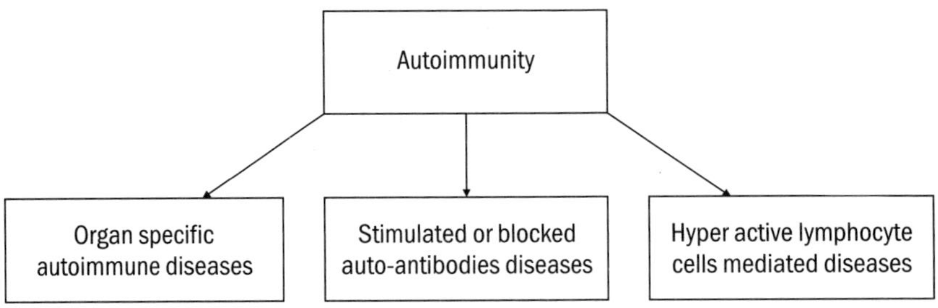

Figure 11.1 Proposed mechanism for inducing autoimmune responses

11.2.1 Hashimoto's Thyroiditis or Goitre or Enlargement of Thyroid Gland

- *Antigens* Thyroid proteins, including thyroglobulin and thyroid peroxidase, act as antigens.
- *Effects* Autoantibody production causes activation of Th-1 cells and T_{DTH}. Damage of peroxidase causes iodine uptake inhibition. Infiltration of lymphocytes, macrophages, and plasma cells leads to the formation of lymphocytic follicles and germinal centres.
- *Final effect* No iodine uptake or decrease in the production of thyroid hormones, called *hypothyroidism,* is observed.

11.2.2 Autoimmune Anaemia

1. **Pernicious anaemia**

- *Antigens* Proteins of small intestine or gastric cells act as antigens. Autoantibody developed destroys intrinsic factors that help in the uptake or absorption of vitamin B12. Vitamin B12 is essential for haematopoiesis.
- *Effects* Destruction of red blood cells (RBCs) causes decrease in their number.
- *Treatment* Injection of vitamin B12 restores absorption by cells.

2. **Haemolytic anaemia or Rhesus incompatibility**

- *Antigens* RBCs are altered to become antigenic due to the effect of drugs such as penicillin, anti-hypersensitive drugs, or methyldopa.
- *Test* Coombs' test is used in which isolated fatal RBCs agglutinate to goat IgG developed against maternal IgG.
- *Effects* Phagocytosis or opsonization of RBCs by autoantibody and complement-mediated lysis are observed.

11.2.3 Goodpasture's Syndrome

- *Antigens* Renal and lung basement membrane cells act as antigens. Autoantibody formed causes damage of glomerular and alveolar basement membrane.
- *Effects* Pulmonary haemorrhage and kidney damage as well as linear deposition of IgG and C3b are observed.

11.2.4 Insulin-dependent Diabetes Mellitus

- *Antigens* Pancreatic β cells, which develop autoantibodies, act as antigens.
- *Effects* Effects include insulitis due to cell-mediated delayed-type hypersensitivity (DTH) or antibody-dependent cell-mediated cytotoxicity (ADCC) or antibody-complement fragment-mediated lysis of β cells. Inhibition of insulin production

leads to increase in blood sugars. Infiltration of CTLs and activation of macrophages occur. Activated macrophages produce lytic enzymes and cytokines such as IFNγ, TNFα, and IL-1.

- *Metabolic problem* Metabolic problems include ketoacidosis (increased urination), atherosclerotic vascular lesions causing gangrene formation, renal failure and blindness, finally leading to death (Streilein, Dana, and Ksander 1997).

11.3 DISEASES DUE TO STIMULATED OR BLOCKED AUTOANTIBODIES

11.3.1 Graves' Disease

- Developed autoantibody (IgG) mimics thyroid-stimulating hormone (TSH) and binds to the TSH receptor to block TSH binding.
- *Effects* Blocking of TSH receptor prevents adenylate cyclase activation causing inhibition in the production of thyroxine and triiodothyronine.

11.3.2 Myasthenia Gravis

- *Antigens* Acetylcholine receptors present on motor end plates or muscles act as antigens.
- *Effects* Autoantibody developed causes complement-mediated degradation of receptors resulting in weakening of skeletal muscles.
- *Symptoms* Drooping eyelids and inability to retract the corners of mouth are observed.

11.4 DISEASES MEDIATED BY HYPER-ACTIVE LYMPHOCYTE CELLS

11.4.1 Systemic Lupus Erythematosus

- The ratio of female to male humans affected with this disease in the 20–40 years age group is 10:1 (Lockshin 1998; Vyse and Kotzin 1998). It is an increase in the expression of type III complement receptor (CR3) on neutrophils mediated by the split complement C5a.
- *Effects* Neutropenia and vasculitis are observed. Neutropenia is the aggregation and attachment of activated neutrophils with CR3 receptor to vascular endothelium blocking neutrophils circulation. Vasculitis is the development of new blood vessels.
- *Symptoms* Symptoms include fever, weakness, arthritis, skin rashes, neurosis, kidney dysfunction, and type III hypersensitivity due to four-fold higher concentration of split complements C3a and C5a than normal.

- Autoantibody is developed against autoantigens such as DNA, histones, RBCs, platelets, leukocytes, and blood-clotting factors.

11.4.2 Multiple Sclerosis

- Autoreactive T cells infiltrate into cerebrospinal fluid causing inflammatory lesions and destroying the myelin sheath (Steinman, Oksenberg, and Bernard 1992; Weiner, Mackin, Matsui, *et al.* 1993).
- *Effects* An abnormality in the central nervous system is observed.

11.4.3 Rheumatoid Arthritis

- Autoreactive T cells target and damage cells and tissues at joints.
- *Effects* Joint pain and inhibition of free movement at joints are observed.

11.4.4 Scleroderma

- Overproduction of collagen in the connective tissues.
- *Effects* Hardening of skin.

11.4.5 CREST Syndrome

- *Calcinosis*—calcium deposit on skin
- *Reynaud's phenomenon*—abnormal blood flow in cold or response to cold
- *Esophageal dysfunction*—problems in swallowing
- *Sclerodactyly*—tightening of the skin
- *Telangiectasia*—formation of red spots on hands, palms, forearms, face, and lips.

11.5 ROLE OF CD4$^+$ T CELL, MHC, AND TCR IN AUTOIMMUNITY

Autoimmune diseases occur due to inappropriate humoral or cell-mediated response of immune system against self-antigens as shown in Figure 11.2. The primary mediator in an autoimmune disease was identified in various animal models as CD4$^+$ Th-1 cells. It is possible to transfer autoimmune diseases from one animal to another with the help of T cells developed against myelin basic protein (MBP) or proteolipid protein (PLP) after vaccination with Freund's adjuvant or T-cell lines (Weiner, Mackin, Matsui, *et al.* 1993). Activated T cells were developed by culturing experimental autoimmune encephalomyelitis (EAE) rat's lymph node cells with MBP. When it was injected intravenously into normal syngeneic animals, it induced demyelination, which developed into EAE disease within four to five days. A similar result was obtained when thyroglobulin from experimental autoimmune thyroiditis (EAT) and *Mycobacterium tuberculosis* from amino-acid animal-specific T cells were injected into normal animals.

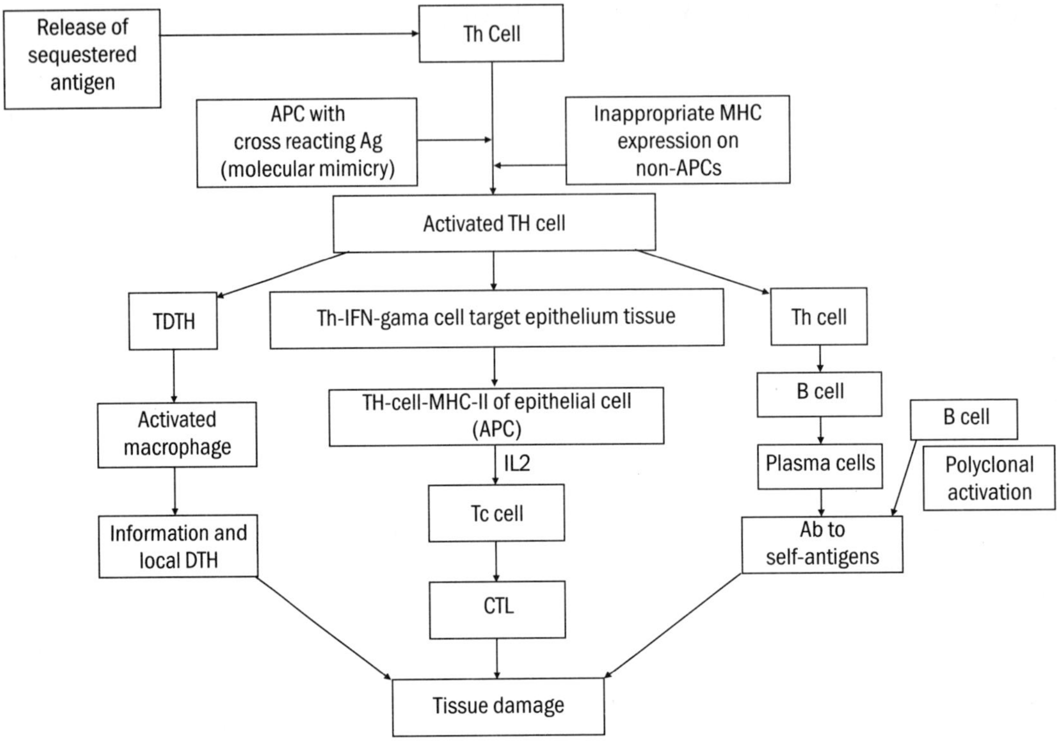

Figure 11.2 Mechanism of autoimmunity

This proved that T cells bear CD4 membrane marker in diseased animals. The developed disease can be treated by the administration of anti-$CD4^+$ monoclonal antibody. This proves the role and involvement of $CD4^+$ T cells in autoimmune diseases.

T cells are of two types—Th-1 cells and Th-2 cells—classified on the basis of secretion of two different kinds of cytokines. Th-1 cells (which secrete IL-2, TNFα, and IFNγ) have been found to develop autoimmune diseases, whereas Th-2 cells (which secrete IL-4 and IL-12) protect induction and progression of such disease. In other words, IL-2, TNFα, and IFNγ promote, while IL-4 and IL-12 inhibit the occurrence of autoimmune diseases (such as EAE). The ratio of Th-1 cells to Th-2 cells determines whether autoimmune diseases will occur or not (Charlton and Lafferty 1995; Theofilopoulos 1995a; Theofilopoulos 1995b).

The involvement of $CD4^+$ T cells occurs after the interaction of MHC-associated peptide with its T-cell receptor (TCR). Thus the trimolecular complex of T cells, TCR, and MHC molecules triggers autoimmune diseases. For example, ankylosing spondylitis (inflammation at the vertebral joints, up to 90% of the cases are found in males) occurs due to the presence of HLA-B27 allele. HLA-B27 peptide residue 70 (KAQ TDRED L) showed homology to klebsiella pneumoniae nitrogenase peptide residue 186 (SRQ TDRED E). The TCR contains mostly Vα and Vβ domains.

11.6 AUTOIMMUNE DEVELOPMENT MECHANISM

There are more than one mechanisms involved in the development of autoimmune diseases. Gender differentiation also affects the occurrence of autoimmune diseases. Most probably, the first step is the release of sequestered antigens, which are not self-reactive and are not exposed to immature thymocytes. These antigens are normally eliminated, and they do not develop self-tolerance. If these sequestered antigens reappear, they trigger activation of mature T cells to cause autoimmune diseases. These antigens reappear in circulation due to trauma to tissues or due to viral or bacterial infection. For example, sperm antigens released after a vasectomy, lens protein after eye damage, or antigens from heart muscle after myocardial infarction. These antigens develop antibody to cause autoimmune diseases (Theofilopoulos 1995a,b).

Viral or microbial antigens also possess antigenic determinants identical to normal host cell components. Such molecular mimicry led Michael Oldstone to propose the mimicry-based development of autoimmune diseases. It has been found that monoclonal antibodies developed against viral proteins are able to bind to normal tissues (up to 3%) to cause autoimmune diseases. For example, post rabies, encephalitis occurs in those individuals who were vaccinated with rabies vaccine developed from rabbit brain cell antigens. The antibodies and T cells developed against these antigens react with the recipient's normal brain cells due to the presence of similar antigenic determinant to cause encephalitis and other diseases (Benoist and Matis 2001; Horwitz, Bradley, Harbertson, *et al.* 1998; Levin, Lee, Kalume, *et al.* 2002; Wilder 1995; Zhao, Granucci, Yeh, *et al.* 1998). Another example is that antibodies develop against antigens of *Streptococcus* after infection damages the heart muscle in rheumatic fever.

Proteins of infectious organisms mimic human host proteins due to similarity in their peptides. For example, residue 79 of human cytomegalovirus IE2 (PDP LGRP DED) showed sequence homology with HLA-DR molecules (VTE LGRP DAE). Poliovirus VP2 residue 70 (STT KESRGT T) showed sequence homology with acetylcholine receptor residue 176 (TVI KESRGT K). MBP peptides mimic viral peptides. Amino acid residues 61–69 (EIS FK LGQE) have high percentage of homology with a measles virus P3 peptide amino acid residue 31 (EIS DN LGQE). Insulin receptor residue 66 (VYG LESLKD S) and residue 764 (N KESLVI SE) show homology with papilloma virus E2 residue 76 (SLH LESLKD L) and rabies virus glycoprotein residue 147 (T KESLVI IS), respectively. Hepatitis B virus polymerase enzyme peptide shows 60% homology with encephalitogenic MBP peptide sequence. The peptide showing homology is able to induce formation of antibodies and proliferation of T cells, which in turn induce autoimmune diseases damaging host cells and respective organs. Most strikingly, the human IgG constant region residue 466 (GVETTTPS) shows homology with human immunodeficiency virus (HIV) p24 residue 160 (GVETTTPS). Any antibody synthesized against the residue 160 of HIV p24 can damage IgG to weaken the immune system.

11.6.1 Class II MHC Mediated Autoimmune Diseases

Very little class I MHC genes, and no class II MHC genes, are expressed in normal healthy β cells of pancreas, while in insulin-dependent diabetes mellitus (IDDM), both class I MHC and class II MHC genes are expressed in high levels. In Graves' disease, the thyroid acinar cells express class II MHC on their membrane. Class II MHC expresses only on antigen-presenting cells (APCs). However, their inappropriate expression leads to the sensitization of Th-1 cells, including activation of B cells or Tc cells against self-antigens causing autoimmune diseases. It has also been experimentally proved that increased concentration of IFNγ upon trauma or viral infection in organs due to inflammatory response can induce class II MHC expression on non-APCs. This in turn triggers activation of Th cells against self-antigens to cause autoimmune consequences. For example, higher IFNγ titre was found in systemic lupus erythematosus (SLE) patients having active disease.

Many AIDS patients, due to infection with Gram-negative bacteria or cytomegalovirus or Epstein–Barr virus (EBV), produce high levels of non-specific autoantibodies against RBCs and platelets (Rose 1998; Wucherpfennig 2001). The IDDM autoimmune disease transgenic mice model (Erikson, Mandik, Bui, *et al.* 1998) was developed by the expression of insulin promoter-mediated expression of IFNγ transgene. Insulin promoter was used to express the transgene only in pancreatic islets of Langerhans to develop IDDM.

11.7 USE OF IMMUNOSUPPRESSIVE DRUGS

The immunosuppressive drugs used include corticosteroids, azathioprine, and cyclophosphamide to slow down lymphocytes proliferation. Cyclosporin A or FK506 blocks T-lymphocytes-mediated signal transduction pathway.

- *Disadvantages* Use of drugs suppressing autoimmune diseases makes patients more infection prone and promotes or causes development of cancerous cells. FK506 inhibits only antigen-activated T-lymphocyte cells.
- *Use of T-lymphocyte cells vaccination* T cells are cross-linked with cell membrane components along with formaldehyde or glutaraldehyde for the vaccination.
- *Mechanism* Vaccination elicits regulatory T-lymphocyte cells that are specific to the TCR variable region determinants of the autoimmune clones.
- *Use of synthetic peptides* Synthetic peptides are used to block the antigen-binding sites of MHC molecules.
- *Immunotherapy* Monoclonal antibody (anti-CD4) treatment specific to CD4 is done to deplete all Th cells, thus preventing autoimmune diseases.

11.8 MODELS OF AUTOIMMUNE DISEASES

Inbred mouse strains are used to study immunological defects caused by autoimmune diseases. For example, F1 hybrids of New Zealand white (NZW) mice spontaneously develop an autoimmune disease similar to SLE and haemolytic anaemia. Various types of autoantibodies against DNA, nuclear proteins, erythrocytes, and T-lymphocytes were detected in these animals after two months to four months of age.

The mice model for SLE is the MRL/*lpr*/*lpr* strain. In such mice strain, *lpr* is detected as defective Fas gene. But this is not a true model for SLE study because there is no correlation between Fas protein and SLE of human. Non-obese diabetic (NOD) mice are very good models for the study of human IDDM.

AT A GLANCE

- There are two types of autoimmune diseases in humans: organ-specific and systemic diseases. In organ-specific autoimmune diseases, a single organ or gland is affected, while a variety of tissues or organs are affected in cell-mediated immune responses in which damage is caused by autoantibodies or their complexes.
- Autoimmune diseases may be due to spontaneous responses. For example, systemic lupus erythematosus is similar to the autoimmune disease that develops automatically in New Zealand white (NZW) mice. An experimental animal model can be generated by injecting mice with self-antigens along with adjuvants.
- $CD4^+$ Th cells play a pivotal role in the development of autoimmunity. Clones of T cells isolated from autoimmune-diseased mice cause disease in normal mice. The specific ratio of the number of Th-1 to Th-2 cells determines whether an autoimmune disease will occur or not. Th-1 cells promote, while Th-2 cells block the development of new autoimmune diseases and the progression of developed autoimmune diseases. Autoantigens presented to Th cells depend on the major histocompatibility complex (MHC) haplotypes of experimental mice.
- Autoimmune diseases have many pathways: release of segregated antigens, mimicry of molecules, inappropriate expression of class II MHC due to IFNγ stimulation, and activation by polyclonal B cells.
- Autoimmune therapies include (1) use of immunosuppressive drugs, (2) thymectomy and plasmapheresis in autoimmune diseases as a result of formation of immune complex, (3) specific T-cells vaccination, (4) use of autoantigen competing synthetic peptides that can bind to MHC molecules, (5) use of monoclonal antibody that binds with specific autoimmune disease-causing components and enhances tolerance to those components and autoantigens.

QUESTIONS

1. What is autoimmune disease? How does it occur? Discuss with suitable examples.
2. What are the main characteristics and causes of the occurrence of the following diseases?

a. Goodpasture's syndrome
b. Graves' disease
c. Systemic lupus erythematosus
d. Insulin-dependent diabetes mellitus
e. Rheumatoid arthritis
f. Hashimoto's thyroiditis
g. Myasthenia gravis
h. Pernicious anaemia
i. Multiple sclerosis
j. Autoimmune haemolytic anaemia

3. How is the animal model for EAE made and used to prove the role of T cells in the development of autoimmunity?
4. What is molecular mimicry? How does it cause autoimmune diseases? Discuss the role of myelin basic protein (MBP) in the development of autoimmune disease based on molecular mimicry.
5. How does localized viral infection cause organ-specific autoimmune diseases? Discuss.
6. What is the role of IFNγ in the development of autoimmune diseases? Discuss with suitable examples. Why is insulin promoter used to make a construct of IFNγ to develop transgenic IDDM mice model?
7. What kinds of cells express MHC? Discuss the role of unusual expression of MHC in the development of autoimmune diseases with suitable examples.
8. What are the different approaches used to cure or prevent autoimmune diseases? Discuss with reasons.

OBJECTIVE QUESTIONS

A. Fill in the blanks.

a. Two types of autoimmune diseases are _______ and _______.
b. Clone of _______ lymphocyte cells cause autoimmune disease in normal mice.
c. _______ cell promote while _______ cells block the development of new autoimmune diseases.
d. Experimental autoimmune disease mice model could be developed by injecting _______ antigens with _______ to normal mice.

e. Inappropriate expressions of class-II MHC in autoimmune disease occur due to _______ stimulation and activation by polyclonal _______ lymphocyte cells.
f. Autoantigens in IDDM is _______.
g. Linear deposition of IgG and C3b occur in _______ autoimmune disease.
h. Opsonisation and Phagocytosis of RBC occur in _______ autoimmune disease.
i. Treatment of autoimmune anaemia is done by vitamin _______.
j. Antibody _______ mimic thyroid stimulating hormone in .... disease.
k. A complement C5a mediated increase in the expression of type-3 complement receptor (CR3 on neutrophilsoccur in _______ autoimmune disease.

B. Match the following.

i. Autoantigen presentation to Th	a. Promote autoimmune disease
ii. Th-1 cells	b. Prevent autoimmune disease
iii. Th-2 cells	c. MHC haplotypes
iv. Horror autotoxins	d. New Zealand White
v. Systemic lupus erythematosus	e. Paul Ehrlich
vi. Self-tolerance	f. Goitre
vii. Hashimoto thyroiditis	g. Protection from self-reactive lymphocytes
viii. Autoimmune anaemia	h. Glomerular and alveolar cell damage
ix. Goodpasture syndrome	i. Increased urination
x. Keto-acidosis	j. Vit-B12

CHAPTER TWELVE 12

Immunity to Infections and Tumours

12.1 INTRODUCTION

Infection is circumvented initially by innate immunity and later by adaptive immunity. Innate immunity is mainly involved in preventive measures adopted by the cells or tissues or organs after entry of microbes. Adaptive immunity involves more aggressive measures to kill or remove the microbes, including detoxification of their by-products or degraded products. This is done either by complement activation, such as production of C3b, which helps in opsonization to enhance phagocytosis, or by the production of cytokines, such as IL-1, IL-6, and TNFα, by endothelial cells or macrophages.

Viral infection is controlled either by production of interferon, which inhibits viral replication, or by natural killer (NK) cells, which remove infected cells.

The infecting pathogens adopt various measures to counter adaptive immunity. Pathogens can reduce antigenicity after invading cells. They can remove their membrane antigens and can mimic the cell surfaces of host. They can cover themselves by molecules similar to the host membrane. They can also mutate their membrane antigens to ensure variation, thus escaping host immune attack.

12.2 CONTROLLING VIRAL INFECTION

Defence cells such as macrophages, monocytes, and fibroblasts produce IFNα/β, which activates the cells' JAK–STAT (Janus kinase–signal transducer and activator of transcription) pathway to induce 2′5′-oligoadenylate synthetase. This activates the ribonuclease RNaseL to split viral RNA into pieces. The JAK–STAT pathway can also activate dsRNA-dependent protein kinase to inhibit viral-specific protein synthesis (Figure 12.1).

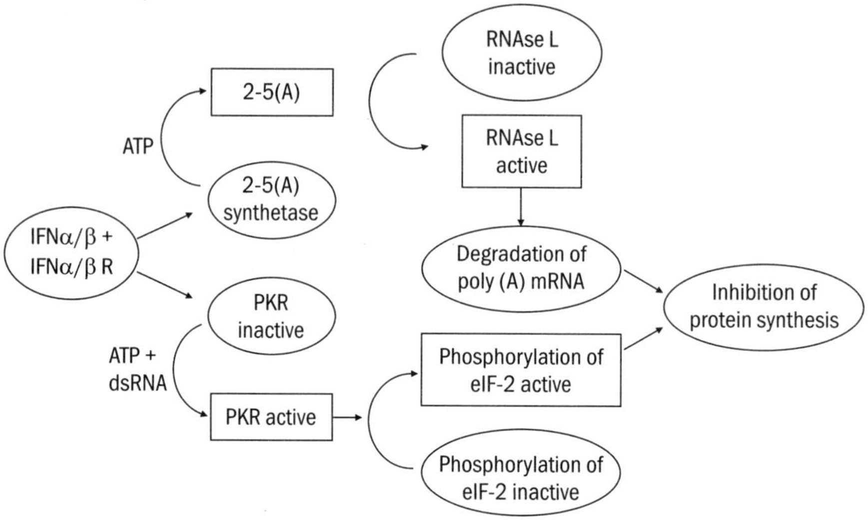

Figure 12.1 dsRNA-dependent activation pathway protein kinase to inhibit viral-specific protein synthesis

IFNα/β can bind to NK cells activating them to kill virus-infected target cells. The killing activity of NK cells is enhanced by the cytokine IL-12. Most viruses spread infection by producing receptors on their surface that bind to host cell membrane. Antibodies produced against these receptors will bind to them, thus preventing the viral particle from binding to the host cells. Soluble IgA (sIgA) produced in mucous prevents attachment of virus to the host mucosal cells. Polio vaccine works by stimulating the production of more sIgA, which prevents the attachment of virus to gastrointestinal cells (Lamm 1997). Vaccine against small pox was developed using these methodologies (Rosenthal, Merchlinsky, Kleppinger, *et al.* 2001). The viral antibody produced against the viral envelope proteins either binds to the viral envelope to prevent their attachment to the host cells or activates the complement to lyse the virion envelope. Either the Fc part of the antibody or the C3b complement fragment helps in the phagocytosis of virus by agglutination. The transfer of cytotoxic T-lymphocyte (CTL) clones, called adaptive transfer, is one of the mechanisms to control specific influenza virus but not the mutated virus (Alcami and Koszinowski 2000; Lachmann and Davies 1997).

Well-established viral infection is controlled by the cell-mediated lysis of viral cells. $CD8^+$ Tc cells and $CD4^+$ Th-1 cells are the cells mostly involved in the elimination of virus-infected cells. Activated Th-1 cells produce IL-2, IFNγ, and tumour necrosis factor (TNF). Initially, IL-2 and IFNγ control virus by activating NK cells. In the later stage, IL-2 acts indirectly by recruiting CTL precursor cells. CTLs eliminate virus-infected cells, thus eliminating the chances of further production of new viruses. These CTLs are specific to a particular type of viral strain. However, IFNγ are also directly involved in the elimination of viral particles. Some viruses such as hepatitis C virus interfere in the activation of RNA-dependent protein kinase (called PKR) to overcome the antiviral effect. Herpes simplex virus (HSV) strains block TAP molecules by producing an early protein called ICP47. Blocking leads to the culmination of antigen presentation with class I MHC on infected host cell membrane. This causes empty presentation of class I MHC. Empty class I MHC can participate in $CD8^+$ T-cells-mediated elimination of HSV-infected cells. Other viruses such as cytomegalovirus (CMV) inhibit the expression of class I MHC molecules, thus inhibiting antigen presentation. Measles virus and HIV both inhibit the expression of class II MHC molecules, thus preventing Th cells-mediated action (Biron 2001; Finkelman, Shea-Donohue, Goldhill, *et al.* 1997; Ramshaw, Ramsay, Karupiah, *et al.* 1997; Sher and Coffman 1992).

Some viruses inhibit complement-mediated lysis of cells by producing proteins that bind to either C4b or C3b. This blocks the classical or antibody-mediated alternative complement pathway (Knodler, Celli, and Finlay 2001; Lane, Montagne, and Fauci 2001; Lorenzo, Ploegh, and Tirabassi 2001).

12.3 VIRAL COMPONENT TAKING PART IN INFECTION

Two envelope glycoproteins haemagglutinin or HA (a trimer structure) and neuraminidase or NA present on the membrane surface of the influenza virus take part in the infection process. Haemagglutinin binds to the sialic acid groups on the host cells, and neuraminidase cleaves N-acetylneuraminic acid, which is used to produce nascent viral glycoproteins. These nascent glycoproteins help in the budding of viral particles on host cells. Influenza virus contains eight single-stranded RNAs (ssRNAs) coding different proteins, including attachment of proteins and RNA polymerase. The eight ssRNAs encode 10 proteins–PB1, PB2, protective antigen, haemagglutinin, nucleoprotein, neuraminidase, M1, M2, NS1, and NS2. Two strains of viruses differ from each other on the basis of nucleoproteins and matrix proteins. A/Sw/Iowa/15/30 (H1N1) designates strain-A isolate number 15 of swine in Iowa in 1930 with antigenic subtypes 1 of haemagglutinin and neuraminidase (Knodler, Celli, and Finlay 2001).

12.3.1 Antigenic Drift and Antigenic Shift

Antigenic drift is a series of spontaneous *point mutations* in the glycoproteins haemagglutinin and neuraminidase, while antigenic shift happens due to *genetic reassortment* between human virus and animal virus. Both enable the sudden origin of new subtype of viruses (such as influenza or other viruses) containing earlier epidemic genes of human virus and animal virus. Reassortment is possible due to the eight separate ssRNA coding of the aforementioned proteins followed by the translocation of some of the ssRNAs of animal and some of the ssRNAs of human into a single virus subtype, which gives rise to new viral subtype strains. Both antigenic drift and antigenic shift enable viruses to escape from immune elimination by the host. The originated strains become new variants of influenza or other viruses originating another localized epidemic. A new strain is selected by culturing a strain mixed with a specific monoclonal antibody to eliminate or neutralize old strains and allow the growth of new strains (Knodler, Celli, and Finlay 2001).

12.4 BACTERIAL INFECTION AND IMMUNE RESPONSE

Bacteria that infect in lower-amount inoculums with less virulence get eliminated immediately after infection by the innate immune mechanism. Bacteria that infect in higher-amount inoculums with more virulence may not be countered by innate immunity, but may trigger an adaptive specific immune response. Bacteria enter either through natural routes such as respiratory tract, gastrointestinal tract, and genito-urinary tract or through cuts or breaks in the skin or mucous membranes. The entered bacteria are countered by delayed-type hypersensitivity.

12.4.1 Protective Mechanism Against Bacterial Infection

Bacteria produce two types of toxins—*exotoxins* and *endotoxins*. Exotoxins are secondary metabolites secreted by bacteria, such as diphtheria toxin, and endotoxins are components of the bacterial cell walls, such as lipopolysaccharides. Both exotoxins and endotoxins are countered by antibodies and C3b complement split products produced by host cells. Both components help in opsonization-mediated phagocytosis. Gram-negative strains are eliminated by complement-mediated lysis mechanism. Other complement split products such as C3a, C4a, and C5a work as anaphylatoxins to induce degranulation of mast cells, either to attract more lymphocytes, macrophages, and neutrophils from the blood to the site of infection in tissues, which is called extravasation, or to cause vasodilation. Antibodies against toxins bind to the toxins to neutralize their toxic effects first and then to remove them by phagocytosis. Intracellular bacterial infections are cleared by cell-mediated immune responses, such as NK cells. In this mechanism, $CD4^+$ T cells secrete cytokines, including IFNγ, to remove the bacterium by macrophage-mediated phagocytosis. The following steps counter bacterial infection.

- Bacterial infection and production of toxin.
- Toxin neutralization by antibody or C3b complement-mediated lysis of bacteria.
- Opsonization and phagocytosis by macrophages.
- Mast cell degranulation by anaphylaxis, that is by C3a, C4a, and C5a.
- Extravasation, that is, the movement of lymphocyte, neutrophils, and macrophages.
- Chemotaxis.

12.4.2 Defence Mechanism of Bacteria Against Host Immune Response

Bacteria must first attach to host cells to proliferate. After proliferation, they invade the host tissue to cause toxin-mediated damage of the tissue. They develop hair-like projections called pili, which enable them to attach to intestinal or genito-urinary tract, causing infection. The bacterial cell wall surface molecules or adhesion molecules secreted may also help in the attachment to the host cell.

Many bacteria such as *Neisseria gonorrhoeae, Neisseria meningitides,* and *Haemophilus influenzae* secrete protease to cleave the hinge region of secretory IgA and produce Fab and Fc fragments, which prevent agglutination-mediated killing of microbes.

The pili are made up of protein pilin expressed by one/two genes and 10–20 silent genes. The arrangement of each gene into six regions is known as a mini-cassette. Due to gene conversion in each mini-cassette, silent genes are replaced by expression genes to generate antigenic variation, which enhances pathogenicity or the attachment potential. Change in the nature of pili prevents IgA-mediated neutralization, thus allowing the bacteria to evade immunization.

Some bacteria are able to escape the host immunity by producing multiple serotypes; for example, *Streptococcus pneumoniae* has 84 serotypes. The M-protein projection present on the surface of bacteria such as *Streptococcus pyogenes* prevents phagocytosis. Certain bacteria prevent phagocytosis by secreting coagulase enzymes that precipitate fibrin coat around the bacterial cell wall. Lipid-A moiety polysaccharide of the cell wall of certain Gram-negative bacteria prevent complement-mediated lysis. Elastase enzyme secreted by *Pseudomonas* inactivates the complement split products C3a and C5a, thus preventing localized inflammation. For some bacteria such as *Listeria monocytogenes,* phagolysosomes provide a suitable environment for their growth rather than killing them. *Mycobacterium avium* prevents fusion of phagolysosomes with lysosomes. These strains are also resistant to the oxidative attack of phagolysosome (Lamm 1997; Robertson and Meyer 1992).

12.4.3 Immunological Controlling Mechanism of Bacteria

Toxoid is an inactivated toxin obtained after treating with iodine trichloride. The fully non-toxic toxoid with intact antigenicity is developed by treating toxin with heat and formalin. It can also be developed by treating with formaldehyde. Toxoid was first developed by Ramon in 1923. It can be used for the development of antibody to protect against bacterial toxin, such as the diphtheria toxin produced by the bacterium *Corynebacterium diphtheriae.* Diphtheria is spread by airborne respiratory droplets. The exotoxin secreted by this bacterium destroys the underlying tissue of respiratory mucosa and also forms a tough fibrinous membrane composed of fibrin, white blood cells, and dead epithelial cells of the respiratory tract called *pseudo-membrane.* It can cause myocardial damage leading to heart attack, neurological damage leading to paralysis, and so on. Toxin is composed of two chains: (1) ganglioside receptors binding chain to help in internalization, and (2) toxin chain inhibiting protein synthesis. Both chains are linked to each other by disulphide bonds. Monoclonal antibody can be used to conjugate with toxin to prepare immunotoxin for targeting tumour cells.

Tuberculosis, a pulmonary infectious disease, is caused by bacterial strain *Mycobacterium tuberculosis*. Infection spreads through inhalation of small droplets containing a few bacilli. Bacilli multiply into the cells by preventing the merger of infected cells to lysosome. Immunity to *M. tuberculosis* is mediated by $CD4^+$ T cells and is the basis for the tuberculin skin test against purified protein derivative (PPD). Infection of bacterium leads to the formation of tubercles. Tubercles consist of activated macrophages and lymphocytes, which form multinucleated cells or epithelioid. The macrophages release lytic enzymes that destroy nearby healthy cells to cause tissue necrosis and prevent the proliferation of bacilli. These necrotic tissues form caseous lesions, called Ghon's complex, which become calcified and are visible by X-rays. These bacilli are killed by cytokines, primarily IFNγ produced by activated macrophages. Tuberculosis is treated by a combination of drugs such as isoniazid, rifampin, streptomycin, pyrazinamide, and ethambutol. The vaccine available to control this disease is BCG (Bacillus Calmette-Guérin), an attenuated strain of *Mycobacterium bovis* (Bloom and Murray 1992; Kaufmann 2001; Krause, Dimmock, and Morens 1997).

12.5 PROTOZOAN INFECTIONS

Protozoans are eukaryotic unicellular organisms. Infection by protozoans causes amoebiasis, Chagas' disease, sleeping sickness, malaria, leishmaniasis (Louis, Himmelrich, Parra-Lopez, *et al.* 1998; Scott 1998), and toxoplasmosis. Humoral immunity against these parasites works at the blood stream life cycle stages.

12.5.1 Malarial Disease by *Plasmodium* sp.

Among various strains of *Plasmodium* sp., *Plasmodium falciparum* is the most prevalent and virulent. *Anopheles* mosquitoes are the carriers or vectors of malaria. The situation becomes worse because many *Plasmodium* sp. are resistant to multiple drugs. Its vectors, *Anopheles* mosquitoes, are also becoming resistant to DDT.

12.5.2 Life Cycle of *Plasmodium* sp.

There are four stages in the life cycle of *Plasmodium* sp.—sporozoites, merozoites, gametocytes (male/female), and zygotes.

Sporozoites enter the blood stream after an infected mosquito bites the host. These sporozoites migrate from the blood stream to the liver to transform into merozoites in hepatocytes. Transformed hepatocytes are called multinucleate schizonts. Merozoites infect red blood cells to produce more merozoites. Merozoites further differentiate into male gametocytes and female gametocytes. These gametes reach the blood and pass into another mosquito when it bites the infected human. Two gametes fuse to form a zygote. Zygotes differentiate into sporozoites after reaching the mosquito's salivary gland. From the salivary glands, they get transferred to a human after a bite and the cycle continues.

Circumsporozoite antigen helps in the adhesion to hepatocytes. The adhesion site is a conserved region in the carboxy-terminal end called region-II due to its sequence homology with adhesion molecules.

After infection, these parasites trigger secretion of a large amount of cytokines from infected cells that might be the cause of malarial symptoms. The cytokine responsible for malarial symptom is TNF. Immune responses against these parasites are very low due to the frequent change in the surface antigens of *Plasmodium* sp.

The X-ray-irradiated sporozoite stage of *Plasmodium* sp. has been currently used to develop a vaccine that will prevent sporozoite multiplication and remove them after a mosquito bite. Encouraging results have been obtained after vaccination. But the main problem is to collect enough sporozoite to mass-produce the vaccine; it is difficult to culture sporozoites. Development of vaccines against synthetic subunit epitopes that are supposed to be recognized by T cells and B cells is in progress (Cox 1997; Good 2001; Hollingdale, McCormick, Heal, *et al.* 1998).

Trypanosoma **sp.**

Due to antigenic shifts in the parasites by the variant surface glycoprotein (VSG), it is very difficult for the host to immunologically control second infection with the antibody developed during the first infection. Antigenic shift occurs due to the duplication and translocation of gene segments encoding VSG molecules to the expression site (Borst, Bitter, Blundell, *et al.* 1998). Although most parasites escape the immune system, some of them are removed by two mechanisms—complement-mediated lysis and opsonization-mediated phagocytosis—after infection.

12.5.3 Parasitic Worms or Helminths

Helminths are large multicellular organisms residing outside the cells. The host develops very poor immunity against these worms, and hence they seem to escape the immune surveillance system. These organisms mainly infect the small intestine, such as *Schistosoma* sp., which causes the fatal disease schistosomiasis. The immune system adopts two mechanisms as a control measure against these organisms: (1) humoral immunity and (2) cell-mediated immunity. Humoral immunity is due to the production of a high level of anti-schistosome IgE antibodies, increase in the number of mast cells, which increases degranulation, and increase in the number of eosinophils. Th-2 cells secrete many cytokines after activation, such as IL-4, IL-5, and IL-3. IL-4 enables B cells to switch class and produce IgE. IL-5 enables bone marrow cells to differentiate into eosinophils. Eosinophils carry receptors for Fc with which the antibody attached to the worm surface causes antibody-dependent cell-mediated cytotoxicity (ADCC). IL-3 and IL-4 together help in the growth of mast cells. Mediators released by mast cells promote infiltration of macrophages and eosinophils to the worm-infected site. The mediators, mainly basic proteins released from eosinophils are toxic to the worm.

12.6 CANCER

Cancer cells are altered self-cells whose growth-regulating mechanisms have either collapsed or failed. Typically, normal cells are produced from earlier normal cells with a growth-control mechanism to keep the number of cells constant. Due to failure in growth control, a tumour is produced. This is also known as *neoplasm*. A tumour that does not invade healthy surrounding tissues is called *benign*. Tumours that keep on growing are called *malignant*. The term cancer is normally used for malignant tumours. A malignant primary tumour has the property of *metastasis* in which clusters of cancerous cells released from the tumour invade blood vessels to reach other tissues, where they proliferate. This gives rise to secondary tumours. Malignant tumours originating from hematopoietic cells of the bone marrow may be *leukaemia* or *lymphoma*. Leukaemia is a malignant cancer that proliferates as single cells, while lymphoma is a malignant cancer that forms tumour masses. The cancers that arise from mesodermal connective tissues such as bones, fat, and cartilage are known as *sarcomas*.

12.6.1 Cause of Malignancy

Malignancy in normal cells may be caused by chemical carcinogens (DNA-alkylating agent), physical agents (ionizing irradiation, ultraviolet rays), and viruses (SV40, polyoma). SV40 produces two early proteins called large T and little T, while polyoma produces three early proteins, namely large T, middle T, and little T, which play a crucial role in malignancy. Rous sarcoma virus is a retrovirus that transforms cells by oncogene or cancer gene called *v-src* (60 kDa protein kinase). *V-src* adds phosphate to tyrosine residues on proteins. In culture, transformation of plasmid-containing *v-src* induces malignancy in normal cells. Malignant cells/transformed cells can grow indefinitely with a decrease in the requirement of growth factors and serum. They are able to grow independent of anchorage and density.

12.6.2 Cellular Oncogenes and Viral Oncogenes

Cellular oncogenes (*c-onc*) were first reported by Howard Temin in 1971. He argued that upon infection, viruses may acquire oncogenes from the genome of normal cells.

Cellular oncogenes are also known as proto-oncogenes (Kufe 2000; Tindle 1996). Based on their activities, cancer genes are divided into three categories: (1) genes responsible for proliferation, (2) proliferation inhibitor or tumour suppressor genes, and (3) conversion of proto-oncogenes into oncogenes.

12.6.2.1 Genes responsible for proliferation

Genes responsible for proliferation include the following.

- Growth factor genes, such as *sis* gene (platelet-derived growth factor).
- Growth factor receptors genes, such as *tms* gene (colony-stimulating factor-1 receptor), *erb*-B gene (epidermal growth factor receptor), *neu* gene (epidermal growth factor related protein-HER2), and *erb*-A gene (receptor for thyroid hormone).
- Signal transducer genes, such as *src* gene (tyrosine kinase), *abl* gene (tyrosine kinase), Ha-*ras* gene (GTPase activity), N-*ras* gene (GTPase activity), and K-*ras* gene (GTPase activity).
- Transcription factor genes, such as *jun* gene (AP1 transcription factor component), *fos* gene (AP1 transcription factor component), and *myc* gene (DNA-binding protein).

12.6.2.2 Proliferation inhibitor or tumour suppressor genes

Proliferation inhibitor or tumour suppressor genes include *Rb* gene (suppresses retinoblastoma), *p53* gene (suppresses formation of lung and colon cancer), *DCC* gene (suppresses colon cancer), *APC* gene (suppresses adenomatous polyposis), *NF1* gene (suppresses neurofibromatosis), and *WT1* gene (suppresses Wilms' tumour). Mutation in these genes leads to their altered function and subsequently to cancer. Programmed cell death regulating genes *Bcl*-2 (suppresses apoptosis) also belong to this category.

12.6.2.3 Conversion of proto-oncogenes into oncogenes

How do proto-oncogenes form oncogenes? This question was first answered by R J Huebner and G J Todaro in 1972. They suggested that carcinogens or virus-mediated mutation or rearrangements of proto-oncogenes give rise to oncogenes. Mutation could be the basis of single or multiple copies of oncogenes leading to the production of many times higher amounts of hyperactive oncogenic products. Overall, it leads to the transformation of normal cells into cancerous cells. Even single-point mutation in responsible genes can lead to the conversion of normal cells into cancerous cells. The conversion of normal cells into cancerous cells is a multi-step process involving a series of somatic mutation.

As mentioned before, there are two types of tumours—lymphoma and leukaemia. Lymphomas develop and proliferate in the lymphoid tissue such as lymph nodes, thymus, and bone marrow, while leukaemias develop and proliferate as a single cell causing increase in their numbers either in the lymph or in the blood. Leukaemia may be acute or chronic. Examples of acute leukaemia include acute lymphocytic leukaemia (ALL) and acute myelogenous leukaemia (AML). Chronic leukaemias include chronic myelogenous leukaemia (CLL) and chronic myelogenous leukaemia (CML). Metastasis events occur in a sequence of genetic transformations, that is, starting from loss of antigen-presenting cells (APCs) in chromosome number 5q causing hyperproliferative epithelium tissue to finally DNA hypomethylation. This causes early adenoma to activate K-*ras* in 12p causing intermediate adenoma; loss of *DCC* in 18q causes late adenoma; loss of *p53* in chromosome 17p causes carcinoma, finally culminating in metastasis (Aisenberg 1993; Weinberg 1996).

12.6.3 Tumour Antigens

The study of the types, production, and properties of tumour antigens is called tumour immunology. Antigens are divided into two types—tumour-specific transplantation antigens (TSTAs) and tumour-associated transplantation antigens (TATAs). TSTAs are unique to tumour-associated cells resulting in mutations leading to the production of altered cellular proteins. These altered proteins are processed in the cytosol and presented in association with class I MHC molecules for recognition by CTLs. Tumour-associated proteins might be proteins expressed in immature cells but not in normal adult cells or proteins expressed in lower amount in normal cells but in higher amount in tumorous cells (Sahin, Türeci, and Pfreundschuh 1997). These proteins are recognized by T cells and categorized into the following five groups.

1. Expression of tumour-specific antigen-encoded genes.
2. Expression of various forms of mutated antigens of normal genes.
3. Expression of genes at specific differentiation stage or lineages.
4. Overexpression of tumour-specific antigen genes.
5. New mutated self-proteins or peptides.

Either TSTA or TATA is processed and expressed in association with class I MHC molecules for recognition by either CTLs or T-lymphocytes (Cournoyer and Caskey 1993; Srivastava 2002).

Tumour antigens and subsequently tumour cell lines are chemically induced by methylcholanthrene and ultraviolet light. Same carcinogens can trigger tumours at two different sites in the same host. The immune response developed against one tumour at one site does not protect the host against the other tumour at the different site. Immune responses develop when killed tumour cells are injected into the host to protect against the same live tumour cell lines.

Specific virus-induced tumour antigens are expressed and shared by other tumour antigens expressed by the same virus.

Tumour antigens are mostly not unique to tumour cells; they are also expressed by normal cells. The amount of tumour antigens on tumour cells will be manifold greater than normal cells. These antigens might be growth factors (such as epidermal growth factor) or growth factor receptors (such as p97, a transferrin growth factor). The transferrin growth factor transports iron into the cell and is expressed in normal cells up to 8000 molecules, while in tumour cells, the expression of the same protein is up to 50,000 to 500,000 molecules.

Tumour antigens also include oncofetal tumour antigens present on early embryonic developing cells as well as on cancerous cells such as alphafetoprotein (AFP) and carcinoembryonic antigen (CEA). An elevated level of AFP has been found in liver cancer in comparison to non-liver cancer. Since both marker proteins, AFP and CEA, are present on normal as well as cancerous cells, both cannot be used as tumour diagnostic markers; although they can be used as tumour growth monitor markers. Neu-proteins are used as a diagnostic marker for breast cancer cells (Baselga, Tripathy, Mendelsohn, *et al.* 1996; Boon, Coulie, and van den Eynde 1997; Coulie, Brichard, van Pel, *et al.* 1994; Sahin, Türeci, and Pfreundschuh 1997; van den Eynde and van der Bruggen 1996).

MAGE-1, MAGE-3, BAGE, GAGE-1, and *GAGE-2* are TATAs that are expressed in significant amounts on human melanoma tumours. Apart from the above five, other TATA proteins include tyrosinase, *gp100, Melan*-A or *MART-1,* and *gp75.* These proteins are also expressed on normal cells, but they are overexpressed on melanoma cells.

Tumour cells elimination mechanisms

Apart from CTLs that recognize other involved cells and eliminate tumour cells presenting antigens in association with class I MHC, NK cells and macrophages also help to get rid of tumours. NK cells can bind to antibody-associated tumour cells to eliminate them by ADCC mechanism. Similar to NK cells, macrophages are not MHC restricted. They recognize and bind the antibody of tumour cells by Fc receptors to follow an ADCC-mediated elimination of altered cells. Macrophages also eliminate tumour cells by secreting lytic enzymes, oxygen-free radicals, and nitrogen intermediates. The activated macrophages secrete cytokines, such as TNFα, which have anti-tumour activities (Houghton, Gold, and Blachere 2001; Rosenberg 2001; Rosenberg, Yannelli, Yang, *et al.* 1994).

12.6.4 Paul Ehrlich's Theory

Paul Ehrlich's theory, also known as the immune surveillance theory, states that frequently arising cancerous cells in the body are recognized as foreign cells and eliminated by the active immune system. Elimination is done by the cell-mediated immune surveillance mechanism as suggested by Lewis Thomas. The development of tumour can only be possible when cancerous cells escape or bypass the surveillance system. The escape mechanism involves decrease in the expression of tumour antigens and immune response compromise to these cancerous cells.

12.6.5 Cause of Failure of Immunization Against Tumour/Cancer

The serum used to immunize animals against tumour antigens has tumour growth promoting effect and also modulates tumour antigens. The actual mechanism is still not known, but K E Hellström and I Hellström suggested that serum or antibody acts as a blocking factor. It was postulated that the antibody binds to antigens to mask them from CTL-mediated and ADCC-mediated elimination (Hsu, Caspar, Czerwinski, *et al.* 1997; Kufe 2000; Pardoll 1996; Peterson and Ikonomidis 1996; Tindle 1996).

12.6.5.1 Poor expression of costimulatory signals

Costimulatory signals CD28 present on T cells and B7 present on APCs are required for IL-2 production and proliferation of T cells. Since both costimulatory signals are poorly expressed on T cells and APCs respectively, very less immunogenicity develops against tumour cells. This partial activating signal due to poor expression of costimulatory molecules causes clonal anergy (Allison, Hurwitz, and Leach 1995).

12.6.6 Immunotherapy to Cure Cancer

Chemotherapy and many anti-cancer drugs kill immune-governing cells. This weakens the immune system and exposes the host to various kinds of bacterial or viral infections. To overcome this problem, immunotherapy is the only method left to fight against cancer (Rosenberg 2001; Scott and Welt 1997). Research in this area is still going on for final application. Immunotherapy works in the following ways.

- The expression of costimulatory signals on APCs and T cells can be increased.
- Cytokines such as IFNα, IFNβ, IFNγ, TNFα, and TNFβ can be used to enhance immune responses against tumours.

12.7 DEVELOPMENT OF MODERN VACCINES: DIFFICULTIES AND FUTURE PROSPECT

Immunity is induced in animals by passive or active immunization. It is done by inducing serums or biomolecules called vaccines. The process of immunization is known as vaccination. Passive immunization is done by inducing a preformed antibody called serum in an animal. However, the serum can immunize an animal only for a short period. On the other hand, active immunization is done for a longer period of time by infecting or inoculating with the following types of biomolecules or vaccines: (1) attenuated microbes, which are avirulent microbial strains, (2) heat-killed or inactivated microbes, and (3) macromolecules (for example proteins) purified from microbes or pathogens.

Ideal vaccines are those that are heat stable, effective in single doses and against many diseases, can be administered within six years of age through mucosa, and are applicable worldwide.

There are two important factors to be taken care of when designing vaccines. Vaccines must activate immune system branches, that is, humoral and cell-mediated immunity, and they should develop immune memory B cells and T cells.

12.7.1 Passive Immunization

Passive immunization is effective against diseases such as botulism, diphtheria, hepatitis A and hepatitis B, and bite of black widow spider. It is achieved by either horse antivenin or human immune gamma globulin. Passive immunization becomes an essential requirement where proper antibody synthesis does not occur due to defective congenital or acquired B cells along with immunodeficient diseases. It is effective when a child is exposed to complicated diseases such as chickenpox (varicella) or measles where active immunization is ineffective. This is because active immunization takes some time to synthesize antibody, while in passive immunization it is already done by a preformed serum. It is also required to urgently ameliorate the effects of certain pathogens (such as *Clostridium tetani*) to protect from tetanus, diphtheria, botulism, measles, rabies, and hepatitis by injecting horse antibody.

However, passive immunization has some drawbacks. It can produce IgE antibody against injected serum to make a complex with vaccine antigen to cause systemic mast cell degranulation causing systemic anaphylaxis. It can also produce IgG or IgM antibodies that can form complement-activating immune complex to cause type III hypersensitive reaction on the site of tissues where it attaches.

12.7.2 Active Immunization

Active immunization is effective against measles, rabies, respiratory diseases, snake bites, tetanus, and respiratory syncytial virus. In this case, the immune system performs actively by the proliferation of antigen-reactive T cells and B cells leading to the formation of memory cells. This type of vaccination is mostly done in children starting from two months to six years of age to protect them from infectious diseases such as hepatitis B; diphtheria, whooping cough (pertussis), and tetanus (DPaT vaccine); measles, mumps, and rubella (MMR vaccine); polio (three-in-one inactivated [Salk or killed] polio vaccine [IPV] and oral [Sabin or live] polio vaccine [OPV]); *Haemophilus influenza* (Haemophilus influenzae type B [HiB] vaccine); chickenpox and shingles (varicella zoster vaccine [VZV]); pneumonia (pneumococcal conjugate vaccine [PCV]), and hepatitis A (hepatitis A vaccine at 18 months). In Table 12.1 the vaccination chart for children is given.

Table 12.1 Immunization chart recommended for the children (in USA)

Child vaccination type	1 month	2 months	4 months	6 months	12 months	15 months	4–6 years
Hepatitis B	+	+		+			
Diphtheria, tetanus, pertussis		+	+	+		+	+
Haemophilus influenzae type B		+	+	+	+		
Inactivated polio		+	+	+			+
Pneumococcal conjugate			+	+	+	+	
Measles, mumps, rubella					+		+
Varicella					+		

12.7.3 Modern Vaccine Types

12.7.3.1 Complete organism vaccine

Complete organism vaccines are of the following types: (1) inactivated (killed) bacterial cells or virus particles by using chemicals or γ-ray irradiation and (2) attenuated (avirulent) living bacterial cells or virus particles, for example, attenuated strains of *Mycobacterium bovis* (trade name BCG [Bacillus Calmette-Guérin]). The attenuated strain of *M. bovis* was developed by growing the bacterial strain on bile-gradient plate with increasing concentration of bile. Poliovirus, which is used as Sabin vaccine is attenuated by growing the virus on epithelial cells of monkey kidney, while rubella virus, which causes German measles, is attenuated by growing on duck embryo cells followed by human cell lines. The sabin vaccine consists of three attenuated viral strains of poliovirus.

The advantages of this type of vaccination include (1) single-dose usage against repeated booster dose; (2) production of sIgA (secreted-IgA), which provides defence against poliovirus; and (3) induction of IgM and IgG subtypes of antibodies. The Sabin poliovirus has a disadvantage, that is it requires booster dose.

Vaccines used for bacterial cells include that of anthrax, cholera, pertussis (whooping cough), plague, tuberculosis, typhoid, and against many virus particles such as hepatitis A (inactivated), influenza (inactivated), measles, mumps, poliovirus (Sabin-attenuated or Salk-inactivated virus), rabies, rotavirus, rubella (inactivated), chickenpox (varicella zoster), and yellow fever. All are live attenuated except a few, which are inactivated vaccine.

12.7.3.2 Macro-biomolecules

Macro-biomolecules include toxoids of diphtheria and tetanus, which are inactivated exotoxin; capsular polysaccharides of *Haemophilus influenzae* type B, *Neisseria meningitidis*, and *Streptococcus pneumoniae*; and hepatitis B surface antigen (HBsAg), which is a recombinant protein antigen.

12.7.3.3 DNA vaccine

DNA vaccine is an example of modern genetic vaccine that activates both innate and adaptive immune systems. It has many advantages over attenuated or inactivated vaccine. It is easily manufactured and purified. A single dose of the vaccine is sufficient, and hence multiple doses or booster dose is not required. It has the moderately highest level of stability. The stability level of various vaccine molecules in decreasing order is as follows: DNA vaccine > inactivated > attenuated. Attenuated vaccine can revert to virulent but DNA and chemical-treated vaccines cannot. Like attenuated vaccine, the DNA vaccine can trigger both humoral and cell-mediated immunity. Plasmid DNA vectors containing DNA sequence coding an antigen under promoter-based expression has the potential to develop potent vaccines used for therapeutic purpose to cure diseases occurring due to virus, bacteria, parasitic infections, and hereditary diseases. Naked plasmid DNA is safe and stable and can be advantageously used for the expression of antigen in cells for longer time periods than RNA or protein vaccines. The expression cassette comprises eukaryotic enhancer, promoter, and terminator/polyadenylation signal. An antigen expression in vector is improved by codon optimization, inclusion of an intron, a strong promoter, and exclusion of sequences causing DNA helical instability. These are the main factors that affect the immunogenicity of the DNA vaccine. Processing of antigens is improved for enhanced presentation of peptides in association with class I or class II major histocompatibility complex (MHC) to enhance adaptive immune responses. The potency of DNA vaccine could be increased by utilizing unmethylated CpG from bacterial DNA to enhance T-lymphocyte recruitment via TLR9. By engineering DNA vaccine to maximize epitope-specific immunity, it is possible to develop the so-called T-cell epitope-driven vaccine design. This vaccine is currently under development for its effectiveness against cancer/tumor.

12.7.3.4 Recombinant protein

Recombinant proteins are proteins (surface antigen of virus/bacteria/pathogen) of microbes and pathogens produced by using recombinant technology. The first kind of recombinant vaccine was developed against hepatitis B virus by cloning and expressing the gene of hepatitis B surface antigen (HBsAg).

12.7.3.5 Synthetic peptide

Synthetic peptide is not very successful because it is unable to develop immunity alone and requires adjuvants and conjugates, which are not safe. Theoretically, only those peptides are successful vaccines that are able to form immunodominant B-cell epitopes. It should also have the ability to generate memory T-helper cells (Th cells), that is, the ability to form immunodominant T-cell epitopes.

12.7.3.6 Vector vaccine

Vector vaccine is based on recombinant technology in which recombinant protein is produced inside the cell by inserting recombinant plasmid encoded with the desired antigen-encoding gene adjacent to the promoter. The foreign gene-encoded plasmid later gets incorporated into the vector genome. For example, the plasmid containing antigen gene is inserted between the thymidine kinase (TK) gene of vaccinia virus when transfected into the tissue culture cells along with vaccinia virus and is later incorporated into vaccinia virus genome by homologous recombination. Recombinant vaccinia vector is selected by bromodeoxyuridine selection, which kills TK^+ cells leaving TK cells containing foreign gene (Moss 1985).

Vector vaccine has some disadvantages. Most of the vaccines are poor immunogenic and are likely to trigger cell-mediated immune system rather than humoral immune system.

12.7.3.7 Multivalent vaccine

Multivalent vaccine is produced by mixing multiple copies of peptides. Multivalent antigens are prepared by two techniques. In the first method, T-cell and B-cell immunodominant epitopes are attached on the matrix opsonized with monoclonal antibodies specific to T-cell and B-cell epitopes. Thus, the solid matrix forms an antibody–antigen complex of B-cell and T-cell antigens. In the second method, the immunostimulating complex (ISCOM) is prepared, which is a mixture of liposomes (phospholipid bilayer), micelle (detergent), and antigens. ISCOM constitutes centrally located antigen liposome complex surrounded by micelles. In liposomes and micelles, the hydrophilic parts are oriented outward. Thus antigens are delivered into the cells by either liposomes or ISCOMs to be processed and presented on the cell's surface along with class I MHC molecules. In solid matrix antibody–antigen complexes, first the monoclonal antibody is attached to the matrices and then the desired immunodominant antigens of T cells and B cells are allowed to attach with the antibody. The complex formed could later be used as a vaccine.

12.7.3.8 HIV vaccine preparation and trial

There are a number of barriers in the testing of prepared HIV vaccine. Some of the barriers present are as follows.

- *Personal risks* which include side effects, lethality, injury, increased susceptibility to HIV, becoming antibody positive, side effects from placebo, distrust of institutions, problems to travel to foreign countries, temptation to have unsafe sex, less information about vaccine, unknown vaccine efficacy, effects on insurance and getting a job, problems in getting medical coverage, becoming infection prone.

- *Social risks* such as fear of discrimination in the society, avoiding sex and social contact, cultural and religious issues, effect on marriage, dominance of male over female, personal and transport costs, time and inconvenience, repeated vaccine injection, fear of needles and frequent blood draws, requirement to avoid/delay pregnancy, injecting a piece of virus.
- *Social costs* such as family commitments, misconceptions, vaccine-mediated HIV infection, ethics of study, fear of receiving untested vaccine, injection of HIV during trials, violation of confidentiality, worried about rectal administrations of the vaccine, and non-surety about the usefulness of the research.

AT A GLANCE

- Various infections are controlled initially by innate immune responses consisting of various kinds of barriers. If the infectious microbes cross these barriers, they are controlled by the production of non-specific components, including cytokines.
- Both humoral and cell-mediated immunities are involved in the control of viral infection. Viral receptors present on host cells can be blocked by specific antibodies, thus preventing infection. It is very difficult to control infection caused by viruses that can mutate their receptors frequently, such as influenza virus. Cell-mediated immunity acts to eliminate virus-infected host cells.
- Bacterial infections are controlled by antibodies produced against bacterial components. The produced antibody triggers the production of immune effector molecules. These molecules in turn take part in complement system activation to initiate local inflammation. Antibody-mediated mechanisms include complement activation for the lysis of bacterium, opsonization to enable better phagocytosis, and toxin neutralization by binding.
- Bacterial defence mechanisms include (1) production of protease to decrease IgA effectiveness by cleaving IgA dimmers; (2) secretion of capsules or proteins to prevent opsonization; (3) elimination of secreting toxins by phagocyte cells; and (4) formation of cysts to survive after phagocytosis.
- $CD4^+$ T cell-mediated immune response is the only defence mechanism adopted by the host cells for intracellular bacteria.
- For protozoan infection, initial control is mediated by humoral immune response in the blood stream. Cell-mediated immunity response occurs after infection of host cells.
- Escape mechanisms adopted by protozoan include (1) constantly changing nature of its glycoprotein coating by genetic switch procedure, such as *Trypanosoma brucei*; and (2) sloughing off antibody-attached glycoprotein by malarial parasites, such as Plasmodium^{spp}.
- Helminths are multicellular, but they do not multiply inside the cell. Helminth infections are controlled by an antibody-mediated defence mechanism promoted by cytokines secreted by Th-1 cells. *Schistosoma* infection is controlled by $CD4^+$ T cell-mediated immune response.
- Chemical carcinogens, physical carcinogens, or virus-mediated transformation of normal cells give rise to cancerous cells. Transformed cells have altered unlimited growth and can cause cancer when injected into normal animals.
- The proto-oncogenes present in normal cells get mutated to become oncogenes to cause cancer. Mutations might be of two types—translocation (such as B cell and T cell leukemia and lymphoma) or amplification.

- Tumour antigens or oncofetal antigens are either chemically induced or generated by oncovirus. They might be tumour specific, mutation specific, or differential lineages specific.
- Tumour antigens are identified by genetic, biochemical, and immunological techniques.
- Tumour-based immune responses are cytotoxic T-lymphocyte (CTL) mediated, natural killer (NK) cell mediated, macrophage mediated, and antibody-dependent cell-mediated cytotoxicity (ADCC) lysis or destruction of tumour cells.
- Tumour cells escape killing mediated by the immune system by modulation of their antigens, which reduces the expression of class I major histocompatibility complex (MHC) molecules, and by escaping CTL-mediated killing.
- Cancer immunotherapy is done by the following measures: (1) increasing costimulatory signals for the activation of T cells; (2) using genetic engineering techniques to secrete more cytokines, which enhance immune responses and antigen-presenting cell (APC) activity.
- Use of monoclonal antibody, immunotoxins, chemotherapy, and radiation are other methods currently in use to control cancer.
- Development of vaccines for cancer is still under progress.

QUESTIONS

1. What is antigenic drift and antigenic shift? Discuss the non-specific defence mechanisms adopted to control microbial infections.
2. What are the specific defence mechanisms to control various infections? Discuss.
3. How do plasmodium, bacteria, trypanosomes, and influenza virus survive the immune response of the host? Discuss.
4. What factors contribute to the emergence of new pathogens or re-emergence of earlier infected pathogens? Discuss.
5. Discuss the cytokine-based defence mechanism to control microbial and pathological infections.
6. What is the role of CTL in controlling infections? Discuss different factors that play a crucial role in CTL-mediated infection control.
7. What are the sources of tumour antigens? Discuss in brief.
8. What are the roles of various cytokines in tumour control? Discuss the mechanism involved in cytokine-mediated anti-tumour effects.

OBJECTIVE QUESTIONS

A. Fill in the blanks.

a. Virus infection is controlled by blocking viral receptor present on ________ using ________.

b. Host cells infected with virus are eliminated by ________ -mediated immunity.
c. Certain virus infections cannot be controlled easily because ________ of virus get mutated frequently.
d. Antibody produced after bacterial infection activates ________ to initiate local inflammation.
e. Antibody raised against infected bacteria helps phagocytosis by ________ and neutralizes ________ produced by bacteria.
f. Proteases produced by bacteria cleave dimmers of ________ antibody.
g. Opsonization of bacteria is prevented by the secretion of ________ or ________.
h. Some bacteria survive even after phagocytosis due to the formation of ________.
i. Protozoan infection is controlled initially by ________ immune response and later by ________ immune response.
j ________ are multicellular protozoan that do not multiply inside host cells.
k. Schistosoma infection is controlled by ________ T-cell-mediated immune response.
l. Tumour antigens are also known as ________ antigens.

B. Match the following.

i. Intracellular bacteria	a. Plasmodium
ii. Sloughing of antibody	b. $CD4^+$ T-cell immunity
iii. Genetic switch to change glycoprotein	c. Oncovirus
iv. Cancer immunotherapy	d. Trypanosoma
v. Tumor antigens	e. Immunotoxins
vi. Enhancement of APCs activity	f. Radiation
vii. Killing cancer cells	g. Cytokines
viii *Tms* gene	h. Receptor for thyroid hormone
ix *Neu* gene	i. Epidermal growth factor
x *Erb*-A gene	j. Protein-HER2

CHAPTER THIRTEEN 13

Immunology of Graft Rejection or Transplant Rejections

13.1 INTRODUCTION TO TRANSPLANTATION

Transfer of cells, tissues, and organs from one site to another in the same individual or different individuals of the same species or different species or genus is known as transplantation. The cells or tissues or organs transferred are known as graft. The graft is taken from the donor individual and transplanted to the recipient individual. The following types of transplants are possible.

- *Autograft* In autograft, one part of body tissues is transferred from one site to another site in the same individual. These grafts are usually accepted grafts.
- *Isograft* In isograft, tissues are transferred between two genetically identical twins, that is, monozygotic twins. These grafts are sometimes rejected because they are treated as foreign by the recipient individual.
- *Allograft* In allograft, tissues are transferred between two genetically different individuals but of the same species.
- *Xenograft* In xenograft, tissues are transferred between two individuals of different species.

Some commonly transplanted organs include kidney, bone marrow (to cure leukaemia, anaemia, and immunodeficiency), heart, pancreas (to cure diabetes mellitus), lung (to treat cystic fibrosis), liver (to treat congenital defects), and skin (to treat burn victims).

13.2 IMMUNOLOGY OF GRAFT REJECTION

Autografts and isografts are immunologically accepted transplanted grafts due to their identical genetic composition. Allografts and xenografts that are genetically dissimilar with the host tissues are treated as foreign and are thus rejected after transplantation.

Rejection of allografts and xenografts depends on the types of tissues and the presence of memory cells. Skin grafts get rejected faster than kidney and heart grafts (Kirk, Harlan, Armstrong, *et al.* 1997; Markees, Phillips, Noelle, *et al.* 1997). Allograft rejection could be primary or secondary (Adams 2000; Harlan and Kirk 1999; Hirose and Vincenti 1999; Lenschow, Zeng, Thistlethwaite, *et al.* 1992; Mollnes and Fiane 1999).

13.2.1 Primary Allograft Rejection

If the skin of a strain-B mouse is grafted into a strain-A inbred mouse, the first set rejection (Plate 1), called primary allograft rejection, takes place. It occurs in the following steps: revascularization occurs in 3–7 days, followed by infiltration of lymphocytes and inflammatory cells such as monocytes and neutrophils; loss of vascularization occurs in 7–10 days; necrosis occurs in 10 days; and complete rejection of the transplant occurs in 12–14 days.

13.2.2 Secondary Allograft Rejection

Secondary allograft rejection occurs due to the presence of memory cells when the skin graft of strain-B mouse is transplanted into a strain-A mouse bearing an earlier graft. Due to the presence of memory cells, the graft rejection is faster and completed within 5–6 days.

13.2.3 Role of T cells in Rejection

The role of T cells in rejection was proved in the 1950s by Avrion Mitchison. It was experimentally demonstrated that thymus knockout mice, devoid of functional T cells, were unable to reject allografts and might even be able to accept xenografts. Naïve strain-B mice injected with spleen T cells taken from earlier allograft-rejected strain-B mice showed faster second set rejection (Figure 13.1).

It was experimentally proved that both $CD4^+$ T cells and $CD8^+$ T cells have a crucial role in allograft rejection. Mice injected with monoclonal antibodies against $CD4^+$ T cells or $CD8^+$ T cells or both were able to knockout the function of corresponding cells. $CD8^+$ T cells alone had no effect on graft rejection, and the removal of the graft was similar to the control. The removal of $CD4^+$ T cells by treatment with monoclonal antibody prolonged allograft survival. The transplant survived from 15 days to 30 days. When both $CD4^+$ T cells and $CD8^+$ T cells were removed, the allograft survival increased up to 60 days proving the role of both $CD4^+$ T cells and $CD8^+$ T cells in graft rejection (Plate 2).

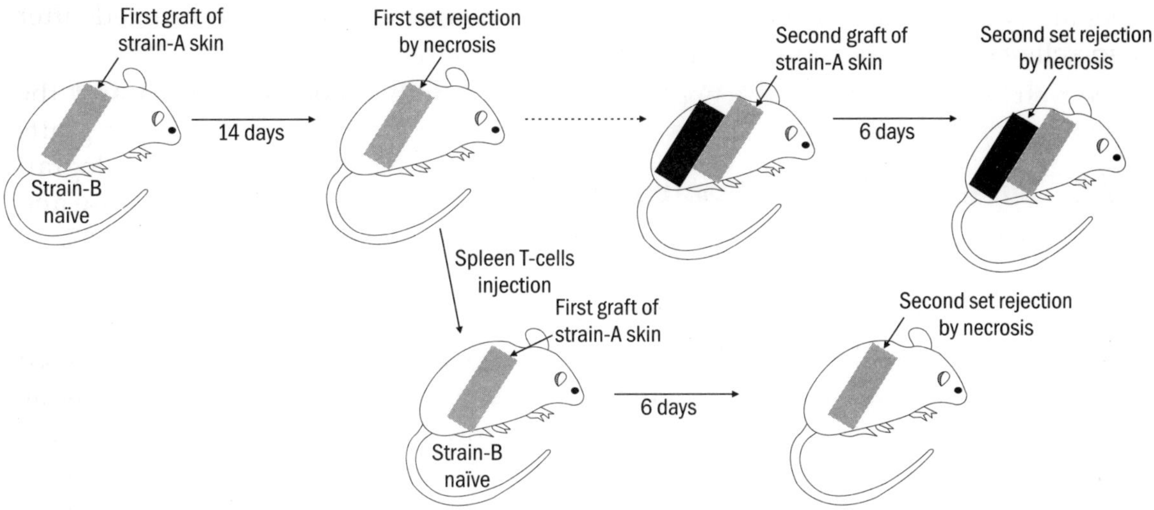

Figure 13.1 Experiment showing T cells-mediated graft rejection

Note T cells are derived from allograft mice and transferred to unprimed mice. Second set graft rejection occurs by original allogenic strain.

13.2.4 Role of Major Histocompatibility Complex in Graft Rejection

Antigenically similar cells or tissues are *histocompatible* and cannot trigger immune response to reject transplanted tissues. If cells and tissues are antigenically different, they are *histoincompatible* and trigger immune response to reject transplanted tissues. In major histocompatibility complex (MHC), more than 40 different loci encode histocompatible antigens. MHCs are known as H-2 complex in mice and human leukocyte antigen (HLA) complex in humans as described earlier in Chapter 6. MHC loci are inherited together from each parent as a complete set called *haplotype*. Inbred mice are homozygous in MHC loci. If two different haplotypes are crossed in the F1 generation, the progeny will be heterotype and will accept grafts from both haplotypes. None of the parents can accept grafts from F1 offspring due to the lack of one of the haplotypes.

13.2.5 RBC-dependent Graft Rejection

The ABO blood group is initially checked for recipient and donor compatibility. Antigens of blood groups are expressed on epithelial and endothelial cells. Antibody raised against any of these antigens will cause lysis of donor cells mediated by complement and antibody, which is incompatible for recipient cells.

In humans, HLA typing is done to check the compatibility of the donor and recipient cells and tissues. HLA typing is done by antibody-mediated microcytotoxicity to denote the presence and absence of MHC alleles. Antibodies specific to class I and class II MHC alleles are added to the well of the microtitre plate containing donor and recipient white blood cells (WBCs). The complement is then added after a few minutes of incubation to check the cytotoxicity of cells by trypan blue exclusion assay. Dead cells retain dye, while live cells do not.

13.2.6 Mixed Lymphocyte Reaction

The mixed lymphocyte reaction is used to quantify the class II MHC compatibility between donor cells and recipient cells. X-ray-irradiated or mitomycin-C-treated donor cells are used to determine T-cell activation by testing proliferation. Proliferation is measured by the uptake of $[^3H]$thymidine into cell DNA. The $[^3H]$thymidine uptake will depend on the class II MHC differences between the T-lymphocytes of donors and recipients, indicating very less or poor chances of graft survival (Plate 3).

The advantage of mixed lymphocyte reaction is that it gives a better degree of T-helper cell (Th cell) activation than microcytotoxicity typing. The disadvantage of the reaction is that it takes several days to run the assay.

13.2.7 Alloreactivity

The phenomenon of Th cells and T-cytotoxic cells (Tc cells) directly recognizing MHC antigens is known as alloreactivity. The MHC antigens are recognized after presentation with self-MHC molecules. Graft rejection occurs even if the MHC antigens of donors and recipients are identical due to differences in the antigens of minor histocompatibility. Graft rejection due to minor histocompatibility is less vigorous than MHC. Yet, because of this it is essential to use immune suppression drugs even in HLA-identical individuals.

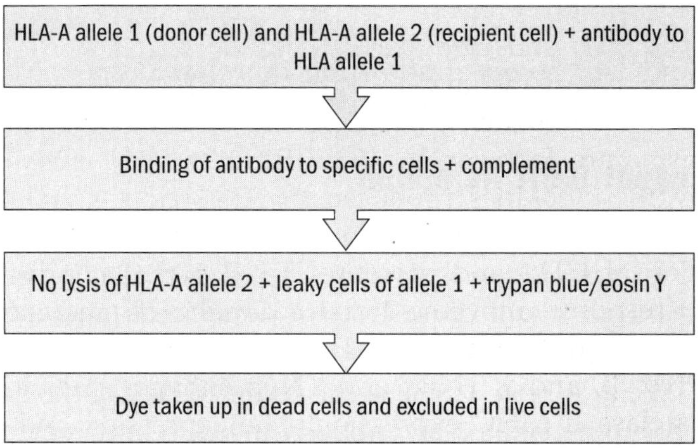

13.2.8 Cell-mediated Graft Rejection

Cell-mediated graft rejection occurs in two stages—sensitization stage and effector stage.

13.2.8.1 Sensitization stage

In the recipient, $CD4^+$ and $CD8^+$ lymphocytes that react with allograft antigens (alloantigens) proliferate and recognize MHC and minor histocompatibility complex (MiHC) alloantigens. The response to MHC is more pronounced than to MiHC, but sometimes the collective response to several minor differences is collectively more than the MHC response. Dendritic cells are present in most tissues of the recipient. They migrate to the grafting site to endocytose alloantigens, to process and present them with MHC molecules. In the grafts of kidney, thymus, and pancreatic islets, passenger leukocytes that are donor antigen-presenting cells (APCs) (basically dendritic cells) have been shown to migrate from the grafts to the nearby lymph nodes to induce immune responses mediated by T-lymphocytes. The exception is the brain. These passenger leukocytes express both class II and class I MHC molecules and are widely distributed in mammalian tissues. Sometimes these passenger cells induce tolerance to antigens in the absence of T-cell populations leading to the acceptance of graft. Apart from passenger cells, other cells that take part in the presentation of alloantigens are Langerhans cells and endothelial cells of blood vessel lining. These cells trigger alloantigen-induced T-cell proliferation in the host. The majority of the cells are $CD4^+$ T cells, which are known to recognize alloantigen peptides presented on class II MHC to induce allograft rejection.

13.2.8.2 Effector stage

The mechanism involved in graft rejection in the effector stage is cell-mediated delayed-type hypersensitivity (DTH) and cytotoxic T-lymphocyte (CTL)-mediated cytotoxicity. Other less common mechanisms taking part in graft rejection include antibody-dependent cell-mediated cytotoxicity (ADCC) and antibody-dependent complement lysis. These mechanisms are followed by the infiltration of T cells and macrophages, which are guided by the cytokines produced by T_{DTH} cells. Cytokines such as IL-2, IFNγ, and TNFβ secreted by Th cells play a central role in effector mechanism. IL-2 helps in the generation of CTLs and promotes T-cell proliferation. IFNγ helps in the development of DTH response, movement and activation of macrophages to the graft site. TNFβ is cytotoxic to graft cells. Apart from the aforementioned cytokines, other cytokines such as IFNα, β, and γ, TNFα and TNFβ enhance class I MHC expression, while IFNγ enhances class II MHC expression. The expression and level of all these cytokines increase the generation and infiltration of other cell types such as dendritic cells, endothelial cells, and myocytes. Dendritic cells are the first cells that express class II MHC, followed by endothelial cells and myocytes to induce CTL attack (Plate 4).

13.2.9 Role of Pre-existing Antibody in Graft Rejection

Pre-existing antibodies in the organs of the host cause hyperacute rejections before vascularization. Hyperacute rejection is carried out by the infiltration of neutrophils, which secrete lytic enzyme and antigen–antibody plus complement-mediated graft rejection. Vascularization is not initiated due to blood clotting in the capillaries.

13.3 IMMUNOTHERAPY AND DRUGS FOR GRAFT ACCEPTANCE

Most immunosuppressive drugs are non-specific but important for the transplant to survive. They are used to slow down the proliferation of T-lymphocytes, such as azathioprine (Imuran) used just after transplantation to inhibit mitotic division at the S phase of the cell cycle. It blocks inosinic acid synthesis. Inosinic acid is a precursor of purines, that is, adenylic (A) acid and guanylic (G) acid. This drug can block the proliferation of both B cells and T cells.

Other mitotic inhibitor drugs include cyclophosphamide, methotrexate, corticosteroids, cyclosporin-A (CsA), tacrolimus (FK506), and rapamycin (sirolimus). Cyclophosphamide is an alkylating agent that breaks the DNA chain by intercalating into the DNA helix. Methotrexate is used to block purine biosynthesis since it acts as a folic acid antagonist. Corticosteroids is anti-inflammatory and is given to graft recipients along with the aforementioned drugs.

Cyclosporin-A (CsA), tacrolimus (FK506), rapamycin (sirolimus) are fungal metabolites and used for immunosuppression during graft. CsA and FK506 block the transcription of IL-2 cytokine gene, thus blocking the activation of T cells. Both drugs also block the calcineurin phosphatase activity, thus preventing NFATc translocation to the nucleus and subsequently preventing the transcription of genes responsible for T-cell activation. FK506 forms a complex with immunophilin to inhibit calcineurin activity leading to the prevention of proliferation and differentiation of Th cells. The arrest occurs in G1 phase. More or less, all three drugs block Th cell proliferation and thus block the expression of cytokine, which prevents the proliferation of Th cells, Tc cells, natural killer (NK) cells, macrophages, and B cells involved in graft rejection. Moreover, CsA prolongs graft survival in kidney, liver, heart, and lung transplantation (Grover, Fullerton, Zamora, *et al.* 1997). The drawback of CsA is that it is toxic to kidneys causing kidney failure and death (Rayhill, Heimes, Kirk, *et al.* 1996).

Prolonged immunosuppressive therapy leads to the risk of cancer, hypertension, and metabolic bone disease (Hong and Kahan 2000; Woo, Lee, and Schubert 1997).

13.3.1 Use of X-ray Irradiation

The transplanted organ is X-ray irradiated just before transplantation to remove lymphocytes. The thymus, spleen, and lymph nodes of the recipient are also irradiated just before transplantation. Treatment doses are 200 rads per day, with a total of 3400 rads. Grafting is done during the immunosuppressed state. The only problem comes from the non-irradiated bone marrow stem cells, which renew recirculating lymphocytes.

13.3.2 Monoclonal Antibody Therapy

Monoclonal antibody therapy is used to block costimulatory signals and thus deplete immunity-promoting cells. It also causes anergy to T cells that react with allograft–antigens. Such monoclonal antibodies act against the CD3 molecule of the T-cell receptor (TCR) complex, IL-2 receptor (anti-TAC), CD4, ICAM-1, LFA-1, and cytokines (such as TNFα, IFNγ, and IL-2). The mode of action of CD3 monoclonal antibody involves the formation of a coat on T cells that cause binding to the Fc receptors on phagocytic cells. This removes T cells by phagocytosis. Antibody-tagged toxin (diphtheria toxin) is phagocytosed by cells causing the death of phagocytic cells. The monoclonal antibody against IL-2 prevents proliferation and activation of T cells. The anti-$CD4^+$ antibody promotes T cells to go into immunosuppressed state without its depletion.

Some monoclonal antibody shows redundancy; for example, LFA-1 monoclonal antibody binds to ICAM-2 and ICAM-1, while ICAM-1 monoclonal antibody binds to Mac-1, CD43, and LFA-1. The blocking of all these adhesion molecules and their pairs causes blocking of signal transduction through the ligand. The clinical problem in using monoclonal antibody is that all of them are of mice origin and are not effective because they are cleared by the raised human antibody. The solution to this problem is the development of mouse–human chimeric antibodies.

13.3.3 Blocking Costimulatory Signals

Blocking costimulatory signals can make the T cells anergic and inhibit the chances of graft rejection. Blocking the interaction between B7 molecules present on APCs to the CD28 or CTLA-4 molecule present on T cells prevents T-cells activation, thus graft rejection. CTLA-4 is expressed on activated T cells only, while CD28 is expressed on both activated and resting T cells. Other interacting costimulatory molecules are CD40 on APC and CD40 ligand or CD154 present on T cells (Harlan and Kirk 1999; Hong and Kahan 2000; Kirk, Harlan, Armstrong, *et al.* 1997). Lenschow, Zeng, Thistlethwaite, *et al.* (1992) demonstrated that blocking these costimulatory signals makes T-cells anergic, thus prolonging graft survival (Plate 5).

13.4 INDUCTION OF ALLOANTIGEN TOLERANCE

Induction of alloantigen tolerance was experimentally proved by Ray Owen in 1945. He found that non-identical twins in cattle can accept cells or tissues from genetically distinct cattle. This does not happen in non-identical twins of other mammals. It is possible in cattle because a shared placenta allows free circulation of cells from one twin to the other and thus does not allow the recognition of these placental partners as foreign. This allows the acceptance of non-identical twins.

It has been experimentally proved that neonates of mouse strain-A, when injected with cells from strain-C, are able to accept grafts from adult strain-C. This can be possible in human also, although so far it has not been proved experimentally (Fox and Harrison 2000; Woo, Lee, and Schubert 1997).

AT A GLANCE

- Immunological responses against transplanted organs/tissues/cells cause their rejection due to mismatch in specificity, memory, and self/non-self. The rejection of a transplant is called graft rejection.
- There are three types of graft rejection: (1) hyperacute rejection mediated by antigens of graft and antibodies of host; (2) acute rejection mediated by T-helper cells (Th cells) and cytotoxic T-lymphocytes (CTLs); and (3) chronic rejection mediated by humoral and cell-mediated immune components.
- Major histocompatibility complex (MHC) antigens present in the tissues are the major cause of graft rejection. For acceptance, the donor class I and class II MHC grafts should match with the acceptor class I and class II MHC grafts.
- Graft rejection is a process involving two steps: (1) sensitization stage (stimulation of T cells) and (2) effector stage (attack on the graft by activated T cells).
- Rejection of graft can be suppressed by immunosuppressive agents, monoclonal antibody, or irradiation by X-rays.
- Monoclonal antibodies act by removing reactive cell population and inhibiting costimulatory signals, which cause anergy in the reactive cells.
- Lymphocytes present in the bone marrow are the main cause of rejection of the bone marrow.
- Xenotransplantation, that is, transplantation of non-human organs could be the solution to the shortage of available organs.

QUESTIONS

1. Discuss the test(s) to be used to find a suitable organ for transplantation. What should be the possible MHC match needed between donors and recipients?
2. Discuss the mechanisms involved in the graft-versus-host disease (GVHD). What are the conditions that cause GVHD? Discuss.
3. How can CTL4A or anti-CD40L be used to block allograft rejection? How is this method better than treatment with CsA or FK506?

4. What is graft rejection? Discuss different types of graft rejection. How can graft rejection be prevented? Discuss in brief.
5. How can alloantigen tolerance be generated? Discuss.

OBJECTIVE QUESTIONS

A. Fill in the blanks.

a. Pre-existing host ________ is the main cause of hyperacute rejection of graft.
b. ________ and ________ cells are responsible for acute graft rejection.
c. The strongest force of graft rejection is encoded within ________ of host tissues.
d. Typing of ________ tissue antigens is done between a recipient and a donor. Graft rejection stages includes ________ and ________ stage.
e. Graft rejection is prevented by using ________ .
f. ________ antibodies could be used to suppress graft rejection by eliminating ________ cells.

B. Match the following.

i. Chronic rejection	a. Matching of tissues of donor with recipient
ii. Acute rejection	b. Th cells activation
iii. Tissue antigen class I and class II typing	c. Cellular and humoral immune components
iv. Sensitization stage of graft	d. Th cells and CTLs
v. Effector stage of graft	e. Host antibody
vi. Hyperacute rejection	f. Lymphoid X-ray irradiation
vii. Suppression of graft rejection	g. Attack on graft tissues
viii. Anergy	h. Exposure to utero or neonates injection
ix. Tolerance to alloantigens	i. Xenotransplants
x. Solution to shortage of organs	j. Inhibiting costimulatory signals

CHAPTER FOURTEEN 14

Cytokines

14.1 INTRODUCTION

Cytokines or chemokines are low-molecular weight proteins or glycoproteins normally present and produced by cells in soluble form. They are secreted by monocytes, macrophages, neutrophils, and dendritic cells in response to an antigen invasion. These molecules act as chemical messengers to connect and regulate both innate and adaptive immune systems. Cytokines are generally named on the basis of their sources and occasionally on the basis of their activity, such as lymphokines (produced by lymphocytes), monokines (produced by monocytes), chemokines (having chemotactic activities), and interleukins (produced by leukocytes and acting on other leukocytes). Functionally, based on the site where they act, they might be autocrine (acting on cells that secrete them), paracrine (acting on nearby cells), or endocrine (acting on distant cells) (Figure 14.1). Cytokines may be pleiotropic (acting on different cell types), redundant (two cytokines similar in function), synergistic (acting together), or antagonistic (opposing the activity of each other) based on their effects on cells (Plate 1).

Cytokines are used to establish and maintain immunological memory apart from attributing pleiotropy, redundancy, and antagonism to regulate interactive or coordinated cellular activity. Chemokines are cytokines that enable leukocytes, monocytes, and macrophages to migrate through the blood to the site of infection after they are produced by site-specific leukocytes, endothelial cells, epithelial cells, and fibroblasts cells. Chemokines include interleukin-8 (IL-8), MIP-1a, MIP-1b, MCP-1, MCP-2, MCP-3, GRO-a, GRO-b, GRO-g, RANTES, and eotaxin. Cytokine or chemokine molecules produced after infection at the site of entry drive inflammatory signals to regulate the residing and newly recruited phagocytic cells to destroy foreign molecules.

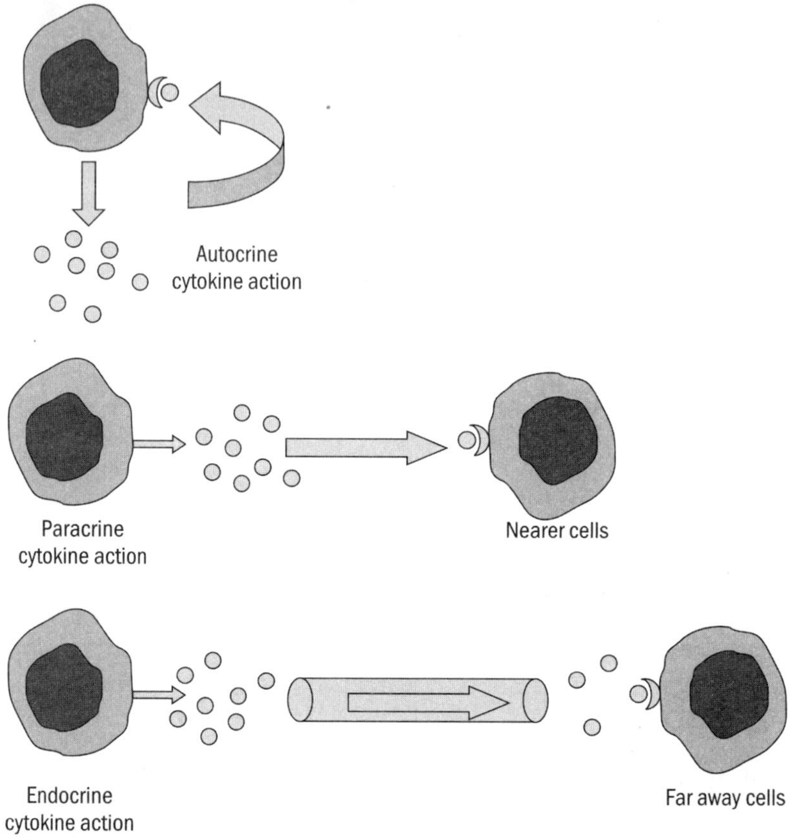

Figure 14.1 Functional differentiation of cytokines

Chemokines increase the affinity of integrins on leukocytes vascular wall ligands during diapedesis. They also regulate the polymerization and depolymerization of leukocytes actin, thus helping to move, migrate, and function as chemoattractants. Additionally, these molecules trigger white blood cells (WBCs) to release substances to kill and induce phagocytosis of invading pathogens and damaged tissues. They are also known to regulate B-lymphocytes, T-lymphocytes, and dendritic cells through lymph nodes and spleen. These cytokine-sensitive B-lymphocytes and T-lymphocytes are differentiated into memory cells by the cytokine milieu at the immunological synapse. Cytokines also play a crucial role in the effector functions and homing properties of T cells and B cells (Chabalgoity, Baz, Rial, *et al.* 2007).

14.2 CYTOKINE RECEPTORS

Cytokines bind to the specific receptor present on or expressed on the surface of target cells. Identification and characterization of these receptors are done by cloning cytokine receptor genes and using biochemical methods. Cytokine receptors are divided into five superfamilies based on their structures.

1. *Immunoglobulin receptor superfamily* Examples of members of this family include IL-1, M-CSF, and C-kit (Plate 2a).
2. *Haematopoietin receptor superfamily or class I cytokine receptor family* This superfamily's extracellular domain motifs have conserved sequences of cysteine amino-acid residues (CCCC) on the upper side and WSXWS motif on the lower side. Examples of these cytokines include IL-2, IL-3, IL-4, IL-5, IL-6, IL-7, IL-9, IL-11, IL-12, IL-13, IL-15, GM-CSF, G-CSF, OSM, LIF, CNTF, growth hormone, and prolactin (Plate 2b).
3. *Interferon receptor superfamily or class II receptor family* This receptor family has conserved CCCC motifs (C stands for cysteine) but does not have WSXWS motif. Apart from interferon-α (IFNα), IFNβ, and IFNγ, this group also contains IL-10 receptor (Plate 2c).
4. *TNF receptor family* This receptor family contains conserved cysteine in their motifs. Examples of this family include TNFα, TNFβ, CD40, nerve growth factor (NGF), and Fas (Plate 2d).
5. *Chemokine receptor superfamily* Chemokine receptors contain G-protein signalling. Examples of this family include IL-8, RANTES, MIP-1, PF4, MCAF, and NAP-2 (Plate 2e).

On the basis of affinity, these cytokine receptors can be divided into three types.

1. *Low-affinity receptor* Out of three subunits α, β, and γ, only one α-subunit is present (Plate 3). The affinity constant (K_a) of this receptor is 10^8 M, and the dissociation constant is 10^{-8} M. These receptors are expressed by $CD4^+$, $CD8^+$ T cells, and activated B cells.
2. *Intermediate-affinity receptor* Out of three subunits α, β, and γ, two subunits, α and β, are present (Plate 3). The affinity constant for intermediate affinity receptors is 10^7 M, while the dissociation constant is 10^{-9} M. These receptors are expressed by resting T cells and natural killer (NK) cells.
3. *High-affinity receptor* All three subunits α, β, and γ are present (Plate 3). The affinity constant for high-affinity receptors is 10^{11} M, while the dissociation constant is 10^{-11} M. These receptors are expressed by $CD4^+$, $CD8^+$ T cells, and activated B cells.

14.2.1 Antagonists of Cytokines

Antagonists are found in the soluble form in the bloodstream and extracellular fluid. These molecules arise from the extracellular domain of cytokine receptors after enzymatic cleavage. Cytokine receptors are not only membrane associated; rather, they are present in soluble form, such as sIL-2R, sIL-4R, sIL-6R, sIL-7R, sIFN-αR, sIFN-γR, sTNF-βR, and sLIFR. Among these soluble receptors, only sIL-2R released from chronic T-cell activation is well characterized. It is a proteolytically cleaved 45 kDa molecule of the amino terminal (192 amino acids) of the α-subunit. Soluble cytokine receptors are used as clinical markers for activated T cells in many diseases; for example, sIL-2R has been used to detect AIDS, autoimmune diseases, and rejection of transplants.

Cytokine antagonist proteins are secreted by viruses. They either bind to cytokines or mimic their activity to protect themselves from being killed by the host. For example, poxviruses and myxoma virus *(Leporipoxvirus)* produce soluble tumour necrosis factor (TNF)-binding protein, while small pox virus produces IL-1-binding protein to prevent the inflammatory response, thus giving viruses a selective advantage for survival. IL-10-like molecule produced by the Epstein–Barr virus, known as viral-IL-10 or vIL-10, binds with IL-10 receptors to suppress Th-1 type cell-mediated responses. These methods are adopted by viruses to overcome the formidable immunity barriers and escape immune-mediated suppression and removal.

14.2.2 Cytokine-based Diseases

Due to differential production of cytokines, certain diseases appear in the host, such as bacterial septic shock (BSS). BSS occurs due to the overstimulation of macrophages by the bacterial cell wall endotoxins to produce IL-1 and TNFα. High level of TNFα causes meningitis after bacterial infection. Such effects were also recorded after injection of recombinant TNFα, even when there is no bacterial infection. Immunotherapy by using antibodies against IL-1 and TNFα can counter BSS.

Superantigen is a bacterial toxin that can cause BSS. These superantigens can bind to class II major histocompatibility complex (MHC) molecules and the Vβ domain of T-cell receptor (TCR), thus activating T cells. A single molecule of a superantigen can interact with almost 20 T cells. Superantigens can cause food poisoning and bacterial toxic shock. Examples of superantigens include toxic shock syndrome toxin and exfoliating toxins produced by *Staphylococcus aureus,* pyrogenic toxins by *Streptococcus pyogenes* and *Mycoplasma arthritidis* supernatant (MAS).

14.3 CLASSIFICATION OF CYTOKINES

Cytokines act as messengers and have a crucial role in developing a micro-environment, which enables innate immunity to talk with adaptive immunity and boost the immune system. Such a micro-environment controls the growth and replication of pathogens by creating a proper balance between/among the cytokines, such as the effect of anti-IFNα on virus replication in leukocytes sensitive to viral infection. This may be attributed to changes in the cytokine profile balance, that is, a high production of TNF by leukocytes infected by vesicular stomatitis virus and a complete reduction in IL-6 production (Alcami and Koszinowski 2000; Bach, Aguet, and Schreiber 1998; Chabalgoity, Baz, Rial, *et al.* 2007; Orzechowska, Antoszków, Siemieniec, *et al.* 2007). On the basis of their functional ability, cytokines can be divided into three groups. The first group of cytokines is involved in the regulation of innate immune responses. The second group regulates adaptive immune responses. The third group involves in haematopoiesis. The cells of the innate immune system are haematopoietic cells, mast cells, basophils, monocytes, dendritic cells, and macrophages. The cells of the adaptive immune system include $CD4^+$ T cells, $CD8^+$ T cells, T-regulatory cells (Tr), and B cells. In the innate immune system, dendritic cells are the major antigen-presenting cells (APCs) to Th (o) $CD4^+$ T cells in lymph nodes that polarize into T-helper-1 (Th-1) and T-helper-2 (Th-2) cells. These cells subsequently produce various cytokines (Becker 2006).

14.3.1 Cytokines Regulating Innate Immune Responses

Cytokines regulating innate immunity are produced primarily by mononuclear phagocytes such as macrophages and dendritic cells. These cytokines can also be produced by T-lymphocytes, NK cells, endothelial cells, epithelial cells, and mucosal cells in response to pathogen-associated molecular patterns (PAMP), such as lipopolysaccharide (LPS), peptidoglycan monomers, teichoic acids, and double-stranded DNA. They act on leukocytes and endothelial cells to form blood vessels in order to promote and control early inflammatory responses at the site of production after infection. Some important cytokines in this category have been listed in the following sub-subsections.

14.3.1.1 Tumour necrosis factor

After being produced by pathogen-infected cells such as monocytes (Rydström and Wick 2007), macrophages, dendritic cells, and Th-1 cells, TNFα/β causes acute inflammation to take part in the first line of innate immune defence. Myeloid-associated immunoglobulin-like receptor IV (MAIR-IV), which expresses preferentially on neutrophils, physically associates with immunoreceptor tyrosine-based activating motif (ITAM)-associated adaptor FcRγ-chain to induce secretion of TNFα (Nakano, Tahara-Hanaoka, Nakahashi, *et al.* 2008). TNFα/β acts on endothelial cells to stimulate the inflammation and coagulation pathway. During diapedesis, stimulated cells produce selectins and ligands for leukocyte integrins. The chemokines produced after stimulation act as chemotaxis for the recruitment of leukocytes. Leukocytes activated by pathogen-associated molecular patterns (PAMPs) respond, in part, by secreting TNF, shedding the TNF receptor, and leading to chronic inflammatory diseases (Kepler and Chan 2007). TNFα/β in turn activates other neutrophils to promote extracellular killing of pathogens. At the same time, it stimulates liver cells to produce acute phase proteins and acts on muscle cells to stimulate catabolism of fat for energy. TNFα/β also has a cytotoxic effect on some tumour cells. It interacts with the hypothalamus to induce fever and sleep and stimulates the synthesis of collagen and collagenase for scar tissue formation. An intestinal bacterial pathogen, *Helicobacter hepaticus* significantly promotes mammary carcinoma in females and enhances intestinal adenoma multiplicity by a TNFα-dependent mechanism in Rag2-deficient mice. This shows the role of bacteria in causing cancer (Rao, Poutahidis, Ge, *et al.* 2006).

14.3.1.2 Interleukin-1

IL-1 is another important cytokine in the innate immune category having a function very similar to the function of TNFα/β. It mediates in acute inflammatory responses and works synergistically with TNFα/β to enhance inflammation. It activates the coagulation pathway, stimulates the liver to produce acute phase response proteins, initiates catabolism of fat for energy, and induces fever. It stimulates the synthesis of collagen and collagenase for scar tissue formation. It also stimulates the synthesis of adhesion factors on endothelial cells and leukocytes for diapedesis. IL-1 is largely produced by activated monocytes, macrophages, dendritic cells, and a variety of other cells in the body. LPS, which is a Gram-negative bacterial cell wall component, binds to Toll-like receptor 4 (TLR4) through LPS-binding protein (LBP) to activate monocytes and other cells to produce IL-1. Non-phosphorylated STAT1 and other transcription factors, that is, a special kind of STAT called LILRE, bind to the upstream induction sequence (UIS) of promoters and are an important factor in initiating and regulating the IL-1 cytokine gene (Unlu, Kumar, Waterman, *et al.* 2007). Initially, the IL-1 superfamily cytokines contained only IL-1α and IL-1β, but now there are six members, namely, IL-1F5, IL-1F6, IL-1F7, IL-1F8, IL-1F9, and IL-1F10, with structural homology. Accordingly, IL-1α, IL-1β, and IL-1RA have been named IL-1F1, IL-1F2, and IL-1F3, respectively (Sims, Nicklin, Bazan, *et al.* 2001).

14.3.1.3 Interleukin-12

IL-12 is a heterodimeric cytokine consisting of p35 and p40 chains encoded by two separate genes located on two different chromosomes in mammals. It is an inducer of cell-mediated immunity and regarded as a primary mediator of early innate immune responses against intracellular microbes (Trinchieri 1998). IL-12 is produced largely by macrophages and dendritic cells within a few hours of infection and acts as pro-inflammatory cytokine having the ability to enhance the phagocytic and bactericidal activity of phagocytic cells such as cytotoxic T-lymphocytes (CTLs) and NK cells by IFNγ production (Chan, Perussia, Gupta, *et al.* 1991). IL-12 constructs a plate to act as a stimulant in the differentiation of naive T4-lymphocytes into IFNγ-producing Th-1 T cells. At the same time, it is known for inhibiting the differentiation of Th-2 T cells. IL-12 functions by binding to specific cell surface heterodimeric receptors, namely, IL-12Rβ1 and IL-12Rβ2. Between the two receptors, IL-12Rβ2 is critical for the function of IL-12 because it is expressed only on activated T cells, unlike IL-12Rβ1, which is constitutively expressed on resting T cells having the nullifying effects of other cytokines. IL-12Rβ2 expression is also enhanced by the cytokines that promote Th-1 development and inhibited by the cytokines that promote Th-2 development. Since IL-12 cytokine function depends on STAT4 transcription factor in T cells and NK cells, IL-12 belongs to the STAT4 category cytokine.

14.3.1.4 Type I interferon

As a modulator, type I IFN is known to modulate the activity of virtually every component of the immune system. It is known to include more than 20 types of IFNs classified under IFNα, IFNβ, IFNω, and IFNτ, while type II IFN has only IFNγ.

Type I IFNs are produced by virus-infected cells to provide an early innate immune response. The IFN produced by the infected cells induces uninfected cells to produce enzymes capable of degrading both viral and cellular mRNA. These enzymes remain inactive in uninfected cells until it gets infected with the virus. Enzymes of infected cells then get activated and start degrading both viral and cellular mRNA inside the cells. This blocks viral protein synthesis, preventing viral replication and eventually eliminating virus-infected cells. At the same time, they promote body defences by enhancing the activities of macrophages, CTLs, dendritic cells, NK cells, and antibody-producing cells.

Type I IFNs also induce MHC-I antigen expression on the cells providing the specificity needed for recognition of antigens by CTLs. They also augment macrophages, NK cells, and B-lymphocytes to induce fever. IFNα is produced by T-lymphocytes, B-lymphocytes, NK cells, and monocytes/macrophages, while IFNβ is reported to be produced by virus-infected cells, fibroblasts, macrophages, epithelial cells, and endothelial cells.

14.3.1.5 Interleukin-6

IL-6 stimulates the liver to produce acute phase proteins. It also stimulates the proliferation of B-lymphocytes and increases neutrophils production. IL-6 is produced by MAIR-IV neutrophils after interacting with ITAM-bearing adaptor FcRγ-chain. The main source of IL-6 cytokine in the body is activated T-lymphocytes, macrophages, monocytes, endothelial cells, and fibroblasts. The level of IL-6 pro-inflammatory cytokine, which is involved in atherosclerosis and longevity after LPS activation, significantly lowers in carrier cells bearing TLR4 mutation or ASP299GLY-attenuated receptor TLR4 (Roelofs, Wenink, Toonen, *et al.* 2008). This suggests that the pathogen burden, by interacting with host genotype, determines the type and intensity of the immune-inflammatory responses accountable for pro-inflammatory status. Most importantly, IL-6 induces a neutrophil-dependent anti-tumour response, which results in cellular and anti-glioma immunity (Graf, Prins, and Merchant 2001).

14.3.1.6 Interleukin-10

IL-10 is an inhibitor of activated macrophages and dendritic cells to regulate innate immunity and cell-mediated immunity. It inhibits the production of IL-12 costimulator molecules and MHC-II molecules. These molecules are needed for cell-mediated immunity. IL-10 is produced mainly by macrophages and Th-2 cells. It plays a protective role as a downregulator of inflammation and also decreases the pathogen (such as, spirochetes of spirochetemia) load during infection with relapsing fever. Systemic injection of IL-10 reduces spirochetemial disease (Gelderblom, Schmidt, Londoño, *et al.* 2007) since it is responsible for the elimination of the responsible bacteria. Activated microglias are the main source of IL-10 in the brain. A lower level of IL-10 is a main cause of enhanced osteoclastogenesis. On the other hand, Bcl-3 gene inhibits the transcription of the IL-10 gene in macrophages, thus reducing the level of IL-10 (Riemann, Endres, Liptay, *et al.* 2005).

14.3.1.7 Interleukin-15

IL-15 is a pleiotropic cytokine having a crucial role in both innate and adaptive immunity (Diab, Cohen, Alpdogan, *et al.* 2005). It is produced by various leukocytes, including phagocytes such as macrophages. This cytokine is responsible for the proliferation and survival of NK cells, NKT-cells, and $CD8^+$ T-lymphocyte cells in Fus1 gene-dependent manner leading to the control of many autoimmune diseases, including inflammation and cancer. IL-15 is involved in lymphocyte homeostasis; it acts on memory $CD8^+$ T cells and also promotes the activation of neutrophils and macrophages. Neutrophils are known to express a functional IL-15 receptor (IL-15R) composed of α-, β-CD122 and γ-CD132 subunits. IL-15R is responsible for enhancement in neutrophils phagocytosis activity by Syk-dependent mechanism. IL-15 selectively inhibits tumour-promoting molecules. Thus it could be a new compound for the adjuvant in vaccines to treat solid tumours. Hence, IL-15 might be a potential therapeutic molecule to treat autoimmune inflammatory diseases and cancer (Diab, Cohen, Alpdogan, *et al.* 2005; Ivanova, Ivanov, Pascal, *et al.* 2007; Shanmugham, Petrarca, Frydas, *et al.* 2006; Ratthe and Girard 2004; Kim, Reznikov, Stuyt, *et al.* 2001).

14.3.1.8 Interleukin-18

IL-18 is a cytokine taking part in the pathogenesis of autoimmune diseases and is produced by macrophages, dendritic cells and Kupffer cells, including epithelial cells (skin, lung, renal tubules), intestinal epithelial, central nervous system (astrocytes, microglia) in response to stimuli of viral and bacterial pathogens, thus initiating the first line of defence, that is, innate immunity. It activates cells through IL-18R (IL-18 receptor). The receptor consists of IL-18Ra and IL-18Rb chains and comes under the group of IL-1R superfamily. First IL-18 binds to IL-18Ra and then recruits

IL-18Rb into the signalling complex for signalling. IL-18Ra is expressed on the surface of immune cells such as Th-1 lymphocytes and NK cells and on a variety of other cells, including macrophages and B cells, neutrophils and basophils, endothelial cells, smooth muscle cells, synovial fibroblasts, chondrocytes, and epithelial cells (Yoshimoto, Takeda, Tanaka, *et al.* 1998). This cytokine is known to stimulate IFNγ production by NK cells, T-lymphocytes, and other cells to induce cell-mediated immunity. After bacterial infection, IL-18 helps in maximal innate responsiveness of CD4 T cells to eliminate cells infected with bacteria (such as salmonella species) at an optimal level (Srinivasan, Salazar-Gonzalez, Jarcho, *et al.* 2007). Moreover, this cytokine along with IL-12 increases the survival and recruitment of monocytes at particular sites by activated T cells and chemokines CXCL8 from granulocytes producing chemokines CXCL9 and CXCL10 (Coma, Peña, Blanco, *et al.* 2006). The most important feature of IL-18 cytokine is its ability to bridge the gap between innate and adaptive immunity by IFNγ and CD134 pathway (Maxwell, Yadav, Rossi, *et al.* 2006). The anti-cancerous effects of this cytokine could be used for therapeutic purposes in the future (Acres, Gantzer, Remy, *et al.* 2005; Robertson, Young, Kitson, *et al.* 2006).

14.3.2 Cytokines Regulating Adaptive Immune Responses

Cytokines regulating adaptive immunity are produced primarily by T-lymphocytes. These cytokines function in the proliferation and differentiation of B-lymphocytes and T-lymphocytes. The function is dependent on antigen recognition to activate effector cells. Some of the important cytokines taking part in adaptive immunity have been discussed below.

14.3.2.1 Interleukin-2

IL-2 is secreted by $CD4^+$ T-lymphocytes and is used as a growth factor for NK cells and antigen-stimulated T-lymphocytes and B-lymphocytes. Initially, it was known as T-cell growth factor (TCGF) due to its role in maturation and proliferation of T-lymphocytes. IL-2 also plays a critical role in maintaining the immune system by enhancing the killing ability of NK cells and Fas-mediated apoptosis. While the exact mechanisms of how IL-2-based immunotherapy controls metastatic diseases and kills cancerous cells is not clear, it is obvious that the IL-2-activated lymphocyte, along with its secretory products such as IFN and/or TNF, plays an important role in the lysis of tumour cells. Based on recent findings, it has been suggested that IL-2 administration in multiple tumours of lungs, colon, stomach, pancreas, ovary, breast, liver, mesothelium, brain, urinary tracts, head, and neck is useful to control their growth (Grande, Firvida, Navas, *et al.* 2006). IL-2 also has pro-inflammatory effect and has been reported to induce vascular leakage syndrome by the release of TNFα, IL-1β, and nitric oxide (NO) from activated leukocytes. It inhibits neutrophils migrating to specific inflammatory sites by a mechanism dependent on nitric oxide (Moreno, Alves-Filho, Bertozi, *et al.* 2006) in an organ-specific manner but does not block LPS-induced leukocyte infiltration in the lung. This characteristic feature of IL-2 is the main cause of increased risk of infection.

14.3.2.2 Interleukin-4

IL-4 cytokine comes under the category of anti-inflammatory cytokines. It is produced mainly by mature Th-2 cells, dendritic cells, and mast cells to help naive Th-2 cells in differentiation and maturation. The initial cell that produces this cytokine in the beginning has not been identified yet. It is a regulator of humoral and adaptive immunity and stimulates proliferation of B cells and T cells as well as differentiation of $CD4^+$ T cells to Th-2 cells leading to the production of IgE as a defence against helminthes. IL-4 protects the eye from *Staphylococcus aureus* corneal infection (Cole, Hume, Khan, *et al.* 2007). It is known to antagonize the effects of IFNγ and thus inhibits cell-mediated immunity. It is also used to target hypoxic cells in tumours in hypoxia oncolytic adenovirus (HYPR-Ad) encoded gene therapies; that is, it has anti-tumour and anti-angiogenic activity (Hong, Cho, Min, *et al.* 2007; Post, Sandberg, Kyle, *et al.* 2007). Thus IL-4 is therapeutically used for all solid tumours that have hypoxia or active hypoxia-inducible factor-1 (HIF-1). It also protects against insulitis and diabetes (Wilson, Kent, Patton, *et al.* 1998) by reducing the cell death of pancreatic β cells in type I diabetes from Th-1 pro-inflammatory cytokines in the islet milieu (Kaminski, Kaminski, and Morgan 2007). IL-4 receptor (IL-4R) is of two types—type I (CD132/IL-4Rα heterodimer) and type II (IL-4Rα/IL-13Rα1 heterodimer). Type I (CD132/IL-4Rα heterodimer) expresses in neutrophils. IL-4R components express on pancreatic β cells and reveal the direct protective effect of IL-4 on viability in response to pro-inflammatory cytokines.

14.3.2.3 Interleukin-5

IL-5 is produced by mast cells upon bacterial infection. Its expression is correlated with the maturation, activation, survival, tissue recruitment, and differentiation of eosinophils, which cause allergic inflammation (Stein, Collins, Villanueva, *et al.* 2006). Thus the increased level of IL-5 is related to the increase in the amount of eosinophils, that is, hypereosinophilia and hyper-IgE. Hence, increase in the concentration of IgE during allergy is related to IL-5 concentration. Diseases related to the hyper concentration or imbalance of IL-5 cytokines include nasal polyps, eosinophilic esophagitis, Gleich's syndrome, Graves' disease, and carcinogenesis.

14.3.2.4 Interferon-γ

IFNγ is produced primarily by Th-1 cells and $CD8^+$ cells, that is, by activated T-lymphocytes and NK cells upon pathogen invasion. It promotes the activity of CTLs, macrophages, and NK cells. IFNγ is involved in cell-mediated immunity to control the pathogen immediately after infection. It is known to produce MHC-I molecules,

MHC-II molecules, and costimulatory molecules by APCs to activate and increase the antimicrobial and tumoricidal activity of monocytes, macrophages, neutrophils, and NK cells. IFNγ stimulates the differentiation of T4-lymphocytes into Th-1 cells. At the same time, it inhibits the proliferation of Th-2 cells. It also stimulates the production of IgG subclasses to promote opsonization, activate the complement pathway of the immune system, and augment or inhibit other cytokine activities. It also exerts antiviral activity.

14.3.2.5 Transforming growth factor-β

Transforming growth factor (TGF-β) is mainly produced by T-lymphocytes, macrophages, and other cells. It inhibits the proliferation and effector function of T-lymphocytes and the proliferation of B-lymphocytes and macrophages. Most importantly, it takes part in the repair of damaged tissues.

14.3.2.6 Lymphotoxin

Lymphotoxin is produced by activated T-lymphocytes and is known to play a role in the recruitment and activation of neutrophils and in lymphoid organogenesis. It has been reported that lymphotoxin is chemically very similar to TNF and acts as a mediator of acute inflammatory responses.

14.3.2.7 Interleukin-13

A cytokine, IL-13 has anti-inflammatory and pleiotropic immune regulatory nature and is secreted by Th-2 cells in response to infections. It regulates the cellular functions of innate immunity against infection by microbial pathogens and that of acquired immunity against helminthes and protozoa infection. This cytokine activates B-lymphocytes by enhancing the synthesis of B-cell DNA to augment the production of IgE and mucus. It plays an important role in Th-2 immunity, asthma, pathogenesis, and the maturity of dendritic cells. It also enhances the capacity of dendritic cells to suppress T-cell secretion of IFNγ (Webb, Cai, Matthaei, *et al.* 2007). The IL-13 receptor is of two different types—IL-13α1 and IL-13α2. The second type, IL-13α2, is over-expressed in different types of cancer cells, including liver cancer (Hou, Du, Wang, *et al.* 2010).

14.3.3 Haematopoiesis Stimulating Cytokines and Growth Factors

Many cytokines and growth factors are produced by the secondary lymphoid organs such as spleen, thymus, and lymph nodes to stimulate the growth and differentiation of haematoblasts (haematopoietic stem cell) into common lymphoid progenitor cells and common myeloid progenitor cells. In the presence of cytokines such as IL-2, IL-7, IL-12, SDF, FLT3, TNFα, and TGFβ, common lymphoid progenitor cells differentiate into small lymphocytes that differentiate further into T-lymphocytes and B-lymphocytes by specific cytokines. On the other hand, common myeloid progenitor cells differentiate into megakaryocytes (thrombocytes), erythrocytes, and myeloblasts in the presence and proper balance of specific cytokines. These groups of cytokines in a particular ratio or balance lead to the differentiation of myeloblast cells into monocytes, basophils, neutrophils, and eosinophils.

Colony-stimulating factors (CSF) are of various types and are classified on the basis of the cells they are promoting to differentiate, such as GM-CSF for granulocyte macrophage, G-CSF for granulocyte, and M-CSF for macrophage. These factors take part by binding to the receptor appearing on the particular cells, that is, they can stimulate only those cells having a specific type of receptor. GM-CSF binds and activates neutrophils, eosinophils, and monocytes. During diapedesis, activated cells have increased adhesive powers to adhere to the capillary walls. GM-CSF is known to enhance phagocytosis and extracellular killing by increasing superoxide anion generation and antibody-dependent cytotoxicity.

14.4 BALANCE OF IMMUNE REGULATOR CYTOKINES

Two types of cytokines regulate adaptive immunity—Th-1 cytokines and Th-2 cytokines. The micro-environments of cytokines are responsible for the activation of macrophages, which leads to their subsequent specific functions, thus boosting the immune power.

The activated (polarized) macrophages are normally categorized into two groups, namely, M-1 and M-2, in analogy with the Th-1/Th-2 dichotomy in T-cell responses. M-1 macrophage and M-2 macrophage differ from each other on the basis of receptor expression, cytokine/chemokines production, and effectors molecules function. M-1 macrophages produce IL-12 and TNFα, while M-2 macrophages produce IL-10, IL-1 receptor antagonist (IL-1Ra), and type II IL-1 receptor (IL-1RII). The differential production of chemokines creates genetic and environmental conditions that are thought to promote M-1/Th-1 macrophages polarization and inhibit M-2/Th-2 regulatory activity, thus establishing chronic inflammatory conditions. Such conditions might be responsible for developing autoimmune adaptive responses, such as an infection or trauma via mechanisms of molecular mimicry. Subsequently, this results in two types of T-cell subsets—Th-1 and Th-2. Th-1 and Th-2 also give rise to a third and fourth type of T-cell subsets, namely, regulatory T-cells (Tregs) and Th-17, with different membrane-bound markers (Table 14.1).

14.4.1 Th-1 Subsets

Th-1 subset cytokines are involved in many cell-mediated immunological functions (Table 14.2). They are also involved in the production of opsonization-promoting IgG antibodies. IgG antibodies bind to the Fc receptors of phagocytes with high affinity and

Table 14.1 Cytokines regulating T-cell subsets

T-cell subtype being targeted	*Cytokine taking part*
Differentiation of naive T cells into Th-1	IL-2 + IFNα and IFNβ
Differentiation of naive T cells into Th-2	IL-4
Differentiation of naive T cells into Th-7	TGFβ + IL-6
Differentiation of naive T cells into Treg	TGFβ + IL-10
Activation of Th cells	IL-27
Negative regulation of Th-1 to Th-2 and Th-17 cells	IFNγ
Negative regulation of Th-2 to Th-1 cells, and Th-2 to Th-17 cells	IL-4

Table 14.2 Th-1, Th-2, Tregs, and Th-17 cytokine subsets and their functions

Cytokine	*Function*
Th-1 cytokines:	
IFNγ, TNFα, IL-2, IL-12, IL-15, IL-18	Activates and enhances antimicrobial and tumoricidal activity of monocytes, macrophages, neutrophils, and NK cells
Th-2 cytokines:	
IL-4, IL-5, IL-6, IL-10	Stimulates the proliferation and differentiation of B-lymphocyte cells and production of neutrophils
Tregs cytokines:	
TGFβ, IL-10	Maintains peripheral tolerance and homeostasis
Th-17 cytokines:	
IL-17	Facilitates pro-inflammatory responses

interact with the complement system. This subset is also associated with the promotion of excessive inflammation and tissue injury.

A defining cytokine of the Th-1 subset is IFNγ, which activates macrophages. Stimulation of macrophage cells increases its microbicidal activity, upregulates the level of class II MHC, and secretes cytokines such as IL-12. This secretion induces Th cells to differentiate into Th-1 subset (Table 14.1). IFNγ secretion by Th-1 cells also induces antibody class switching to IgG classes (such as IgG2a in mouse) to support phagocytosis and fixation of complement. TNFβ and IFNγ are cytokines that mediate inflammation, and their secretion accounts for Th-1 cells with inflammatory activity such as delayed hypersensitivity. Th-1 cells produce IL-2 and IFNγ cytokines that promote the differentiation of fully cytotoxic Tc cells from $CD8^+$ precursors (Table 14.2). This pattern of cytokine production makes the Th-1 subset suitable for responding to viral infections and intracellular pathogens. Finally, IFNγ inhibits the expansion of Th-2 population.

14.4.2 Th-2 Subsets

Th-2 subset cytokines (Table 14.2) are known to stimulate eosinophils, differentiate B cells, and promote the production of relatively more amounts of antibodies, such as, IgM, IgE, and non-complement-activating IgG isotypes. The subset supports allergic reactions due to environmental factors. Secreted Th-2 cytokines support eosinophil-mediated attack on helminth (roundworm) infection, removing the helminths from the host body. They also control the extent of production of IgM antibody and pass signals to B cells to switch from IgM production to IgE production. Eosinophils are endowed with IgE-specific Fc receptors. Upon roundworm infection, Th-2 cells evoke production of IgE antibody to bind to the worm-specific antigen and Fc receptors of eosinophils, forming a bridge between the worm and eosinophils. This interaction leads to eosinophil-mediated killing and removal of infected worms. Although IgE activity is valuable to the host, it also has some non-beneficial actions that cause allergy to the host. Th-2 class of cytokines such as IL-4 and IL-10 are known to suppress the expansion of Th-1 cell populations.

14.4.3 Regulatory T-cell Subsets

Regulatory T-cell (Treg) cytokine (Table14.2) profiles are distinct from either Th-1 or Th-2 cells and are reported to have immunosuppressive function. Tregs can be differentiated from $CD4^+$ T cells on the basis of different cell surface markers and expression of cytokines (Wan and Flavell 2006). Tregs secrete high levels of IL-10 and/or TGFβ but do not secrete immune-protective cytokines such as IFNγ or IL-4. They also play a main role in the maintenance of peripheral tolerance and homeostasis of mucosal surfaces (Groux, O'Garra, Bigler, *et al.* 1997; Levings, Sangregorio, Galbiati, *et al.* 2001). Although Tregs get activated in vivo against bacterial, viral, and parasitic antigens, they may act to prevent infection-induced immunopathology or suppress Th-1 responses to sustain pathogen persistence in the host (McGuirk and Mills 2002; Doetze, Satoguina, Burchard, *et al.* 2000; McGuirk, McCann, and Mills 2002). Tregs suppress the immune system in cancer patients by blocking the conversion of dendritic cells to tumour antigen-specific T cells or CTLs, which are responsible for killing tumour cells. Normally, dendritic cells activated by the tumour antigen are continuously synthesized by the tumour cells. The activated dendritic cells get transformed into the tumour antigen-specific T cells or CTLs to remove the tumour cells (Antony, Piccirillo, Akpinarli, *et al.* 2005). At the same time, they suppress the differentiation of naive Th cells into Th-1 and Th-2 cells (Plate 4).

14.4.4 Th-17 Lineage Cells

Th-17 lineage cells are effector molecules (Table 14.2) that can be considered an expanded arm of $CD4^+$ T cells. These lineage cells are developed by distinct cytokine signals and antagonized by the products of the Th-1 and Th-2 lineages (Park, Li, Yang, *et al.* 2005; Harrington, Hatton, Mangan, *et al.* 2005; Harrington, Mangan, and Weaver 2006). TGFβ and IL-6, both cytokines, help in the differentiation of naive T cells into Th-17 lineage cells (Table 14.1), but at the same time, IFNγ and IL-4 potentially inhibit their development. The key immunomodulatory cytokine TGFβ is responsible for differentiation and development of immunosuppressive molecules such as Th-17 to facilitate pro-inflammatory responses. This can also upregulate the expression of IL-23 receptor. IL-23 signalling might not be required for IL-17 production and Th-17 commitment, but it is indispensable for stabilizing the Th-17 phenotype, which allows the long-term survival of Th-17 cells. Th-17 lineages have a role in the control of infections. The association of IL-23 and IL-17 in the host is required for protection against some extracellular bacteria (Happel, Dubin, Zheng, *et al.* 2005; Malley, Srivastava, Lipsitch, *et al.* 2006). Murine intranasal immunization with the C polysaccharide of *Streptococcus pneumoniae* in association with cholera toxin induces protection against a challenge with live capsulated *Pneumococci* (Malley, Srivastava, Lipsitch, *et al.* 2006).

14.5 ROLE OF CYTOKINES IN THE REGULATION OF B CELLS

B cells play a fundamental role in host defences since they work as sentinels, survey the body for the presence of non-self-antigens, act as inducers of the immune response by presenting antigens to antigen-specific T cells, and perform as effectors by differentiating into cells producing antigen-specific antibodies. They have two types of immune responses—B-1/MZB (marginal zone B-cells)-guided innate immune responses and B-2-guided conventional adaptive immune responses.

The whole life of a B cell, beginning from stem cells to mature plasma cells, is governed by cytokines and other factors present in the micro-environment. The most important stage in the life cycle is during the development of B-cell progenitors, which express receptors for different cytokines, known as B-cell receptors (BCRs).

In addition to cell-intrinsic signals and those provided by pre-BCRs, maturation of B-cell progenitors in the bone marrow is primarily regulated by the interaction of growth factors produced by the surrounding cells, that is, stromal cells. The feeder stromal cell layers normally required by early progenitor B cell for development are later replaced by a cytokines cocktail (Marshall, Fleming, Wu, *et al.* 1998). Among these cytokines, IL-7 (25 kDa protein) has an important role as a growth factor that exerts a major influence on the development of murine B-lineage cells (Acosta-Rodríguez, Merino, Montes, *et al.* 2007).

The survival and maturation of B cells in the periphery are tightly regulated by different mediators such as TNF family members, B-cell activating factor (BAFF) and a proliferation-inducing ligand (APRIL) and their receptors such as TACI, BCMA, and BAFF-R. BAFF, a B-cell survival/maturation factor, is secreted by IFNγ or IL-10-activated macrophages, monocytes, dendritic cells, and G-CSF-stimulated neutrophils (Mackay and Ambrose 2003). Generally, BAFF neither plays a critical role in the development of B-1 cells nor in the survival of T-1 B cells, but its over-expression causes B-cells neoplasia and autoimmune diseases. T-2 cells as well as the subsequent stages of B-cell maturation such as conventional B-2 cells and MZ B cells are BAFF dependent (Batten, Groom, Cachero, *et al.* 2000). The survival effect of BAFF is mediated, at least partially, by the activation of the NFkB2 pathway. The over-expression of cytokine IL-9 causes an increase in the proportion of mature B-1 cells.

14.5.1 Cytokine-mediated Modulation, Shaping, and Homing of B Cells in Secondary Lymphoid Organs

In contrast to primary lymphoid tissues where naïve T cells and B cells develop, the secondary lymphoid organs provide a micro-environment for the generation of immune responses mediated by T cells and B cells. Immune responses are developed by achieving a spatial structure in the secondary lymphoid organs with cellular interactions. The functional spatial structure in the secondary lymphoid organs is maintained by distinct cytokines in the proper ratio. For example, TNF family ligands and TNF receptors play a critical role in determining an optimal splenic architecture and consequently B-cell development. The presence of TNFα decides the formation of marginal sinuses and germinal centres, whereas lymphotoxin-α (LT-α) has a critical role in the development of lymph nodes and Peyer's patches. In addition, proper development of lymph nodes requires TNF-related activation-induced cytokine (TRANCE) (Kim, Mebius, MacMicking, *et al.* 2000). Apparently, TRANCE and LT-α β both do not control the same pathway; rather, they control separate critical pathways for lymph node development. Once the organ is developed, TNF/lymphotoxin cytokines get involved in the maintenance of lymph node architecture (Tumanov, Kuprash, and Nedospasov 2003).

14.5.2 Role of Cytokines in T-cell-dependent B-cell Responses

Priming of naïve B cells is a critical initiating event in the development of humoral effector response (antibody secretion) and immune cell memory by cognate Th cells. Thus circulating mature follicular B cells that encounter antigens and receive the appropriate T-cell help will subsequently mount either an extra follicular response or a germinal centre reaction by a process of class switch recombination (CSR) imprinted by cognate Th cells. Cytokines derived from Th cells are considered a major factor in regulating CSR in vivo. Cytokines IL-4 and IFNγ have been shown to reciprocally regulate immunoglobulin isotype production, that is, IL-4 drives the switch to IgG1 and IgE while IFNγ promotes IgG2a. IL-21 is known to influence class switching in mouse- and human-inducing IgG1 and IgG3 and inhibiting IgE (Pène, Gauchat, Lécart, *et al.* 2004). On the other hand, IL-10 is a switch factor for IgG1 and IgG2. TGFβs are particularly abundant in mucosa and have been implicated in the induction of IgA responses in vivo (Cazac and Roes 2000).

Cytokines secreted by Th cells and follicular dendritic cells are crucial to decide the final fate of germinal centre B cells either by favouring cell survival or by forcing centroblasts to a plasma cell or memory B-cell pathway. Here, IL-4 directs B cells to a memory phenotype via CD40-mediated signals, while IL-2 and IL-10 induce terminal plasma cell differentiation in the absence of CD40. IL-10 can also act synergistically with CD27/CD70 signals to induce plasma cell differentiation (Agematsu, Nagumo, Oguchi, *et al.* 1998). IL-6 enhances proliferation of myeloma cells by interacting with signals mediated by fibroblast growth factor receptor 3 (Ishikawa, Tsuyama, Liu, *et al.* 2005). Furthermore, virus-infected plasmacytoid dendritic cells secret IL-6, which in combination with IFNγ helps in B-cell differentiation into immunoglobulin-secreting plasma cells.

14.5.3 Role of Cytokines in T-cell-independent Polyclonal B-cell Responses

B-cell polyclonal activation can be modulated by cytokines produced by macrophages or dendritic cells infected with virus, bacteria, and parasite. Dendritic cells stimulated by intact bacteria and bacterial cell wall components via TLR secrete B-cell stimulatory cytokines such as IL-1 and IL-6. This is possible by TLR-4 and TLR-9 ligands that bind LPS and CpG, respectively, to signal B-cells. They secrete IL-6 and IL-10 in an autocrine fashion synergizing with TLR signals to promote B-cell mitogenesis. Another cytokine IL-15, produced by monocytes/macrophages, is able to stimulate human B-cell proliferation and immunoglobulin secretion. IL-15, in concert with other monocyte-derived cytokines, is thought to promote the polyclonal B-cell activation frequently associated with HIV-1 infection.

14.6 CYTOKINES IN THE DEVELOPMENT AND MAINTENANCE OF IMMUNOLOGICAL MEMORY

To achieve protection from infection, it is important that memory $CD8^+$ T cells be generated in enough numbers and persist as a functional long-lived population. They should circulate or migrate rapidly to sites where re-exposure to infection or tumour reoccurrence may occur. They should either constitutively express the necessary array of effector molecules or rapidly synthesize them on re-stimulation. In such a simplified scenario, cytokines might be used to facilitate differentiation, persistence, and functions of memory T cells. Three cytokines IL-12, IFNα, and IFNγ are important for optimal $CD8^+$ T-cells expansion and differentiation. Short-term exposure to antigens can drive naive $CD8^+$ T cells to differentiate into effectors, but longer stimulation is needed for differentiation of memory T cells. However, excess or chronic antigenic stimulation reduces the differentiation of memory T cells.

IL-7 plays a pivotal role in memory $CD8^+$ T-cell development, although the relationship between signalling through IL-7R and memory cell differentiation is not well known. There are substantial evidences that IL-15 is a homeostatic cytokine for the maintenance of memory $CD8^+$ T cells (Ma, Koka, and Burkett 2006). Other cytokines may also be involved in the maintenance of memory cells, such as IL-21, which acts synergistically with IL-15 to promote survival and proliferation of memory $CD8^+$ T cells.

14.7 USE OF CYTOKINES AS VACCINES

Cytokines have been currently tested as an immunomodulatory molecule to develop and formulate vaccine-induced immune responses, with the aim of eliciting protective immune-effector mechanisms to avoid the pathological effects of immune responses (Villinger 2003). Cytokines are added either as recombinant proteins or as cytokine-encoding plasmids to either skew the immune response phenotype, thus influencing the quality of memory cells or simply as a classical adjuvant to strengthen the immune responses elicited upon vaccination. This is done by the incorporation or codelivery of typical Th-1 or Th-2 cytokines with DNA vaccines. The magnitude and nature of the immune response to the DNA vaccine can be differentially regulated. For example, the coadministration of IFNγ or IL-12 with a DNA vaccine encoding hepatitis B virus (HBV) envelop protein favours the development of a Th-1 phenotype, whereas coinjection of IL-4 enhances the development of Th-2 cells.

The IL-2 encoding plasmid in herpes simplex virus (HSV) is used as a vaccine to effectively drive Th-1 responses and increase vaccine efficacy. The use of cytokine IL-7 as a molecular adjuvant for DNA vaccination drives immune responses to Th-2 phenotype but also augments antigen-specific CTLs. This suggests an alternative pathway for CTL development, where CTLs could be induced in a Th-2 environment in vivo.

Cytokines/chemokines such as GM-CSF, IL-2, IL-12, IL-18, IFNγ, TNFα, and others present in DNA-vaccine formulations increase the recruitment of APCs at the site of vaccine administration and also increases their potency. Although cytokines enhance the potency, there are some limitations in their use in immunotherapies. It has a very short half-life in vivo. To circumvent this problem, cytokines are covalently attached to monomethoxy polyethylene glycol (PEG) called *pegylated cytokines*. Pegylated cytokines have also shown extended in vivo biological activity. Thus lower doses could be used to reduce the risk of adverse reactions. Pegylated IFNα augments the cellular response to a DNA vaccine encoding hepatitis C virus (HCV) core protein and has been successfully tested for the treatment of HCV and HBV (Gehring, Gregory, Kuzushita, *et al.* 2005). Fused cytokine–immunoglobulin complex molecules can be used to increase the half-life and efficacy of exogenous cytokines and may be administered with DNA vaccine for better response.

14.8 CYTOKINES IN CANCER THERAPY

Systemic administration of cytokines, either alone or in combination with other drugs, has been extensively evaluated as adjuvant therapy. Tumour cells transfected with different cytokine-encoded plasmids (such as IL-1, IL-2, IL-4, IL-6, IL-7, IL-12, IL-18, TNFα, G-CSF, GM-CSF, and IFNγ) can induce effective anti-tumour immunity (Grande, Firvida, Navas, *et al.* 2006; Post, Sandberg, Kyle, *et al.* 2007; Hou, Du, Wang, *et al.* 2010 ; Atkins 2006). However, vaccines based on genetically modified autologous tumour cells have certain limitations as it requires the availability of tumour tissue from the concerned patient. Out of the aforementioned cytokines, IL-2 has been extensively tested and approved for the treatment of metastatic melanoma and renal cell carcinoma patients. However, in contrast to the above report, some clinical trials have shown no clinical impact on patients, which makes further research necessary. IL-2 has an ability to expand regulatory T-cell populations leading to immune tolerance rather than anti-tumour immunity (Atkins 2006). The cytokines IL-15 and IL-2 have common receptor subunits as that of IL-2 and share the same signalling pathways, but they do have different and even competing roles. IL-15 has the ability to activate T cells and NK cells. It also has a role in the persistence of $CD8^+$ memory T cells. This suggests that IL-15 might be a better choice than IL-2 for the treatment of cancer. IL-15 can be a better component of vaccines against infectious diseases than IL-2. IL-7, on the other hand, induces an increase in $CD4^+$ and $CD8^+$ lymphocytes with a decrease in the percentage of $CD4^+$ T-regulatory cells. Due to this property, it controls the renal cell carcinoma model and causes a 75% reduction in pulmonary metastases. It also increases the total body number of T cells, NK cells, B cells, and macrophages (Rosenberg, Sportès, Ahmadzadeh, *et al.* 2006). Over-expression of IL-13 and its receptor IL-13Rα2 has been found in many cancer cells. The IL-13Rα2 receptor has a specific chain that can be used for targeted killing of cancerous cells.

Since these key regulatory small protein molecules can guide and interlink innate and adaptive immune systems, their protective role for immunity, including avoidance of immunopathological effects, is important. By identifying the right balance of various key cytokines, immunity can be boosted to cure even chronic diseases. Better immune systems can be developed by keeping the immunological memory for longer periods for future protection. Immunological memory largely depends on the interaction between various Th-1 and Th-2 cytokines. Their proper ratio can control infectious diseases, even deadly cancer. The ability of a specific ratio of cytokines to control diseases has prompted the use of these protein molecules as adjuvants, specifically as immune modulators in vaccines to strengthen vaccine-induced immune responses against specific infection and cancer-like diseases. Cytokine receptors have proved a potential target for therapy and killing of cancerous cells.

AT A GLANCE

- Cytokines are defined as low molecular weight proteins produced by monocytes, macrophages, neutrophils, and dendritic cells.
- They can be classified as lymphokine (produced by lymphocytes), monokine (produced by monocytes), chemokine (based on chemotactic activities), and interleukin (produced by leukocytes).
- Based on their functionalities, cytokines can be pleiotropic, redundant, synergistic, or antagonistic in nature. A proper ratio of each type of cytokine is required to maintain the immunological potential of any organism to fight against diseases.
- Cytokines can also be classified based on their involvement in generating both innate and adaptive immunity and their interactive roles.
- The immunomodulatory role of cytokines in various diseases such as hepatitis B and hepatitis C, including cancer, has been well researched to find methods of prevention and cure.
- Their immunomodulatory roles at the basic level can be determined by understanding the problematic disorders that occur due to overall change in their ratio in diseased organs.
- Immunotherapy through cytokines and cytokine receptors can cure or prevent many diseases. Such therapies are currently in clinical trial.

QUESTIONS

1. Discuss in brief the similarities and differences between cytokines, chemokines, growth factors, and hormones.
2. Discuss the salient features of cytokines. What are Th-1 and Th-2 cytokines? Discuss.
3. What types of immune responses are induced and mediated by the Th-1 and Th-2 cytokine subset?

OBJECTIVE QUESTIONS

A. Fill in the blanks.

a. Cytokines are _______ having low molecular weight secreted by _______.
b. The total number of cytokines is approximately _______.
c. Cytokines bind to _______ present on the cell surface.
d. Cytokine receptors are classified in the _______ superfamily receptors.
e. Cytokine-induced class I receptor activates _______ signal transduction pathway.
f. Th cells get differentiated into Th-1 and Th-2 subsets by _______ environment.
g. Cytokines secreted from Th-1 cells act as _______ and _______ cells activation to eliminate pathogens.
h. Th-2 cytokines help in the production of _______ isotypes and _______ antibody-mediated immune response.
i. Interferon-α is used in the treatment of _______.
j. Antigen reacting simultaneously with class II MHC molecule and Vβ domain of TCR is called _______.

B. Match the following.

i. Pleiotropy	a. Roferon and intron-A
ii. Redundancy	b. Hodgkin lymphoma treatment
iii. Synergy	c. Multiple sclerosis treatment
iv. Antagonism	d. Haematopoietin receptor family
v. Class I cytokine receptor	e. One cytokine acting on many cells
vi. Class II cytokine receptor	f. Many cytokines acting on one cell
vii. TNF receptor family	g. Two cytokines having accumulative effect
viii. Interferon-β	h. One cytokine having opposite activity to the other
ix. Interferon-α	i. Nerve growth factor
x. Interferon-γ	j. Interferon receptor family

Bibliography

Acosta-Rodríguez, E. V., M. C., Merino, C. L., Montes, C. C., Motrán, and A., Gruppi. 2007. Cytokines and chemokines shaping the B-cell compartment. *Cytokine and Growth Factor Reviews* 18: 73

Acres, B., M., Gantzer, C., Remy, N., Futin, N., Accart, O., Chaloin, J., Hoebeke, J. M., Balloul, and S., Paul. 2005. Fusokine interleukin-2/interleukin-18, a novel potent innate and adaptive immune stimulator with decreased toxicity. *Cancer Research* 65: 9536

Adams, D. H. 2000. Cardiac xenotransplantation: clinical experience and future direction. *The Annals of Thoracic Surgery* 70: 320

Agematsu, K., H., Nagumo, Y., Oguchi, T., Nakazawa, K., Fukushima, K., Yasui, S., Ito, T., Kobata, C., Morimoto, and A., Komiyama. 1998. Generation of plasma cells from peripheral blood memory B cells: synergistic effect of interleukin-10 and CD27/CD70 interaction. *Blood* 91: 173

Ahearn, J. M., and D. T., Fearon. 1989. Structure and function of the complement receptors CR1 (CD35) and cR2 (CD21). *Advances in Immunology* 46: 183

Aisenberg, A. C. 1993. Utility of gene rearrangements in lymphoid malignancies. *Annual Review of Medicine* 44: 75

Akdis, C. A., and K., Blaser. 1999. IL-10-induced anergy in peripheral T-cell and reactivation by micro-environmental cytokines: two key steps in specific immunotherapy. *The FASEB Journal* 13: 603

Alcami, A., and U. H., Koszinowski. 2000. Viral mechanisms of immune evasion. *Immunology Today* 9: 447

Alcami, A., and U. S., Koszinowski. 2000. Viral mechanisms of immune evasion. *Trends in Microbiology* 8: 410

Allison, J. P., A. A., Hurwitz, and D. R., Leach. 1995. Manipulation of costimulatory signals to enhance antitumor T-cell responses. *Current Opinion in Immunology* 7: 682

Antony, P. A., C. A., Piccirillo, A., Akpinarli, S. E., Finkelstein, P. J., Speiss, D. R., Surman, D. C., Palmer, C. C., Chan, C. A., Klebanoff, W. W., Overwijk, S. A., Rosenberg, and N. P., Restifo. 2005. CD8+ T cell immunity against a tumor/self-antigen is augmented by CD4+ T helper cells and hindered by naturally occurring T regulatory cells. *Journal of Immunology* 174: 2591

Atkins, M. B. 2006. Cytokine-based therapy and biochemotherapy for advanced melanoma. *Clinical Cancer Research* 12: 2353

Aubry, J. P., S., Pochon, P., Graber, K. U., Jansen, and J. Y., Bonnefoy. 1992. Cd21 is a ligand for CD23 and regulates IgE production. *Nature* 358: 505

Bach, E. A., M., Aguet, and R. D., Schreiber. 1998. The IFNγ receptor: a paradigm for cytokine receptor signaling. *Annual Review of Immunology* 15: 563

Banchereau, J., F., Briere, C., Caux, J., Davoust, S., Lebecque, Y. J., Liu, B., Pulendran, and K., Palucka. 2000. Immunology of dendritic cells. *Annual Review of Immunology* 18: 767

Barnes, K. C., and D. G., Marsh. 1998. The genetics and complexity of allergy and asthma. *Immunology Today* 19: 325

Baselga, J., D., Tripathy, J., Mendelsohn, S., Baughman, C. C., Benz, L., Dantis, N. T., Sklarin, A. D., Seidman, C. A., Hudis, J., Moore, P. P., Rosen, T., Twaddell, I. C., Henderson, and L., Norton. 1996. Phase II study of weekly intravenous recombinant humanized anti-p185HER2 monoclonal antibody in patients with HER2/neu-overexpressing metastatic breast cancer. *Journal of Clinical Oncology* 14: 737

Batten, M., J., Groom, T. G., Cachero, F., Qian, P., Schneider, J., Tschopp, J. L., Browning, and F., Mackay. 2000. BAFF mediates survival of peripheral immature B lymphocytes. *Journal of Experimental Medicine* 192: 1453

Beaven, M. A., and H., Metzger. 1993. Signal transduction by Fc receptors: the FcεRI case. *Immunology Today* 14: 222

Becker, Y. 2006. Molecular immunological approaches to biotherapy of human cancers: a review, hypothesis, and implications. *Anticancer Research* 26: 1113

Benoist, C., and D., Matis. 2001. Autoimmunity proved by infection how good the case for T-cell epitope mimicry is? *Nature Immunology* 2: 797

Benschop, R. J., and J. C., Cambier. 1999. B-cell development: signal transduction by antigen receptors and their surrogates. *Current Opinion in Immunology* 11: 43

Bentley, G. A., and R. A., Mariuzza. 1996. The structure of the T-cell antigen receptor. Annual *Review of Immunology* 14:563

Berke, G. 1994. The binding and lysis of target cells by cytotoxic lymphocytes: molecular and cellular aspects. *Annual Review of Immunology* 12: 735

Berland, R., and H. H., Wortis. 2002. Origins and functions of B-1 cells with notes on the role of Cd5. *Annual Review of Immunology* 20: 253

Biron, C. A. 2001. Interferons alpha and beta as immune regulators: a new look. *Immunity* 14: 661

Bjorkman, P. J., and P., Parham. 1990. Structure and function and diversity of Class I major histocompatibility complex molecules. *Annual Review of Biochemistry* 59:253

Bjorkman, P. J., M. A., Saper, B., Samraoui, W. S., Bennett, J. L., Strominger, and D. C., Wiley. 1987a. Structure of the human class I histocompatibility antigen HLA-A2. *Nature* 329: 506

Bjorkman, P. J., M. A., Saper, B., Samraoui, W. S., Bennett, J. L., Strominger, and D. C., Wiley. 1987b. The foreign antigen binding site and T-cell recognition regions of Class I histocompatibility antigens. *Nature* 329: 512

Blackwell, T. K., and F. W., Alt. 1989 Mechanism and development program of immunoglobulin gene rearrangement in mammals. *Annual Review of Genetics* 23: 605

Bleackley, R. C., C. G., Lobe, B., Duggan, N., Ehrman, C., Fregeau, M., Meier, M., Letellier, C., Havele, J., Shaw, and V., Paetkau. 1988. The isolation and characterization of a family of serine protease genes expressed in activated cytotoxic T lymphocytes. *Immunological Reviews* 103: 5

Bloom, B. R., and C. J. L., Murray. 1992. Tuberculosis commentary on a re-emergent killer. *Science* 257: 1055

Bodmer, J. G., S. G. E., Marsh, E. D., Albert, W. F., Bodmer, B. Dupont, H. A., Erlich, B. Mach, W. R., Mayr, P. Parham, T. Sasazuki, G. M. Th. Schreuder, J. L. Strominger, A. Svejgaard, and P. I. Terasaki. 1994. Nomenclature for factors of the HLA system. *Tissue Antigens* 322: 145

Bonnefoy, J. Y., S., Pochon, J. P., Aubry, P., Graber, J. F., Gauchat, K., Jansen, and L., Flores-Romo. 1993. A new pair of surface molecules involved in human IgE regulation. *Immunology Today* 14: 1

Boon, T., P. G., Coulie, and B., van den Eynde. 1997. Tumor antigens recognized by T-cells. *Immunology Today* 18: 267

Borris, L., and L., Rosenwasser. 1997. TH1/ TH2 lymphocytes: doubt some more (editorial). *Journal of Allergy and Clinical Immunology* 99: 161

Borst, P., W., Bitter, P. A., Blundell, I., Chaves, M., Cross, H., Gerrits, F., van Leeuwen, R., McCulloch, M., Taylor, and G., Rudenko. 1998. Control of VSG gene expression sites in Trypanosoma brucei. *Molecular and Biochemical Parasitology* 91: 67

Brenner, M. B., J., McLean, D. P., Dialynas, J. L., Strominger, J. A., Smith, F. L., Owen, J. G., Seidman, S., Ip, F., Rosen, and M. S., Krangel. 1986. Identification of a putative second T-cell receptor. *Nature* 322: 145

Brown, J. H., T. S., Jardetzky, J. C., Gorga, L. J., Stern, R. G., Urban, J. L., Strominger, and D. C., Wiley. 1993. Three-dimensional structure of the human class II histocompatibility antigen HLA-DR1. *Nature* 364: 33

Burdin, N., and M., Kronenberg. 1999. CD1-mediated immune responses to glycolipids. *Current Opinion in Immunology* 111:326

Burton, D. R. 1990. Antibody: the flexible adaptor molecule. *Trends in Biochemical Sciences* 15: 64

Busse, W., and W., Neaville. 2001. Anti-immunoglobulin E for the treatment of allergic disease. *Current Opinion in Allergy and Clinical Immunology* 1: 105

Carosella, E. D., J., Dausset, and M., Kirszenbaum. 1996. HLA-G revisited. *Immunology Today* 17: 407

Carroll, M. C. 2000. The role of complement in B-cell activation and tolerance. *Advances in Immunology* 74: 61

Cazac, B., and J., Roes. 2000. TGF-beta receptor controls B-cell responsiveness and induction of IgA in vivo. *Immunity* 13: 443

Chabalgoity, J. A., A., Baz, A., Rial, and S., Grille. 2007. The relevance of cytokines for development of protective immunity and rational design of vaccines. *Cytokine and Growth Factor Reviews* 18: 195

Chan, S. H., B., Perussia, J. W., Gupta, M., Kobayashi, M., Pospísil, H. A., Young, S. F., Wolf, D., Young, S. C., Clark, and G., Trinchieri. 1991. Induction of interferon-γ production by natural killer cell stimulatory factor: characterization of responder cells and synergy with other inducers. *Journal of Experimental Medicine* 173: 869

Chang, T. W. 2000. The pharmacological basis of anti-IgE therapy. *Nature Biotechnology* 18: 157

Charlton, B., and K. J., Lafferty. 1995. The Th-1/Th-2 balance in autoimmunity. *Current Opinion in Immunology* 7: 793

Chien, Y. -H., R., Jores, and M. P., Crowley. 1996. Recognition by γ/δ T cells. *Annual Review of Immunology* 14: 511

Cole, N., E. B., Hume, S., Khan, L., Garthwaite, T., Schubert, V., Reeve, and M. D., Willcox. 2007. The corneal response to infection with Staphylococcus aureus in the absence of interleukin-4. *Immunology and Cell Biology* 85: 333

Coma, G., R., Peña, J., Blanco, A., Rosell, F. E., Borras, J. A., Esté, B., Clotet, L., Ruiz, R. M., Parkhouse, and M., Bofill. 2006. Treatment of monocytes with interleukin IL-12 plus IL-18 stimulates survival, differentiation and the production of CXC chemokine ligands (CXCL)8, CXCL9, and CXCL10. *Clinical and Experimental Immunology* 145: 535

Conrad, D. H. 1990. The low affinity receptor for IgE. *Annual Review of Immun*ology 8: 623

Coulie, P. G., V., Brichard, A., van Pel, T., Wölfel, J., Schneider, C., Traversari, S., Mattei, E., de Plaen, C., Lurquin, J. P., Szikora, J. C., Renauld, and T., Boon. 1994. A new gene coding for a differentiation antigen recognized by autologous cytolytic T lymphocytes on HLA-A2 melanomas. *Journal of Experimental Medicine* 180: 35

Cournoyer, D., and C. T., Caskey. 1993. Gene therapy of the immune system. *Annual Review of Immunology* 11: 297

Cox, F. E. 1997. Designer vaccines for parasitic diseases. *International Journal of Parasitology* 27: 1147

CSGA (Collaborative Study on the Genetics of Asthma). 1997. A genome-wide search for asthma susceptibility loci in ethnically diverse populations. *Nature Genetics* 15: 389

Daser, A., N., Meissner, U., Herz, and H., Renz. 1995. Role and modulation of T-cell cytokines in allergy. *Current Opinion in Immunology* 7: 762

Davies, A. C., and M. J., Schulman. 1989. IgM-molecular requirements for its assembly and function. *Immunology Today* 10: 118

Davies, D. R., and H., Metzger. 1984. Structural basis of antibody function. *Annual Review of Immunology* 1: 87

Davis, M. M., and P. J., Bjorkman. 1988. T-cell antigen receptor genes and T-cell recognition. *Nature* 334: 395

Davis, M. M., J. J., Boniface, Z., Reich, D., Lyons, J., Hampl, B., Arden, and Y. -H., Chien. 1998. Ligand recognition by αβ T-cell receptors. *Annual Review of Immunology* 16: 523

Demotz, S., H. M., Grey, E., Appella, and A., Sette. 1989. Characterization of a naturally processed MHC class II restricted T-cell determinant of hen egg lysozyme. *Nature* 342: 682

Diab, A., A. D., Cohen, O., Alpdogan, and M. A., Perales. 2005. IL-15: targeting CD8+ T-cells for immunotherapy. *Cytotherapy* 7: 23

Doetze, A., J., Satoguina, G., Burchard, T., Rau, C., Löliger, B., Fleischer, and A., Hoerauf. 2000. Antigen-specific cellular hyporesponsiveness in a chronic human helminth infection is mediated by Th3/Tr1-type cytokines IL-10 and transforming growth factor-beta but not by a T (h) 1 to T (h) 2 shift. *International Immunology* 12: 623

Erikson, J., L., Mandik, A., Bui, A., Eaton, H., Noorchashm, K. A., Nguyen, and J. H., Roark. 1998. Self-reactive B-cells in nonautoimmune and autoimmune mice. *Immunologic Research* 17: 49

Feinstein, A., N., Richardson, and M. I., Taussig. 1986. Immunoglobulin flexibility in complement activation. *Immunology Today* 7: 169

Finkelman, F. D., I. M., Katona, J. F., Urban Jr, J., Holmes, J., Ohara, A. S., Tung, J. V., Sample, and W. E., Paul. 1988. IL-4 is required to generate and sustain in vivo IgE response. *Journal of Immunology* 141: 2335

Finkelman, F. D., T., Shea-Donohue, J., Goldhill, C. A., Sullivan, S. C., Morris, K. B., Madden, W. C., Gause, and J. F., Urban Jr. 1997. Cytokine regulation of host defense against parasitic gastrointestinal nematodes: lessons from studies with rodent models. *Annals of Tropical Medicine and Parasitology* 15: 505

Fox, A., and I. C., Harrison. 2000. Innate immunity and graft rejection. *Immunological Reviews* 173: 141

Francis, T. Jr. 1960. On the doctrine of original antigenic sin. *Proceedings of the American Philosophical Society* 104: 572

Fugmann, S. D., I. I., Lee, P. E., Shockett, I. J. Villey, and D. G., Schatz. 2000. The RaF protein and V(D)J recombination: complexes, ends and transposition. *Annual Review of Immunology* 18: 495

Garcia, K. C., L., Teyton, and I. A., Wilson. 1999. Structural basis of T-cell recognition. *Annual Review of Immunology* 17: 369

Garcia, K. C., M., Degano, R. L., Stanfield, A., Brunmark, M. R., Jackson, P. A., Peterson, L., Teyton, and I. A., Wilson. 1996. An αβ T-cell receptor structure at 2.5 Å and its orientation in the TCR-MHC complex. *Science* 274: 209

Garrett, T. P., M. A., Saper, P. J., Bjorkman, J. L., Strominger, and D. C., Wiley. 1989. Specificity pockets for the side chains of peptide antigens in HLA-w68. *Nature* 342: 692

Gehring, S., S. H., Gregory, N., Kuzushita, and J. R., Wands. 2005. Type 1 interferon augments DNA-based vaccination against hepatitis C virus core protein. *Journal of Medical Virology* 75: 249

Gelderblom, H., J., Schmidt, D., Londoño, Y., Bai, J., Quandt, R., Hornung, A., Marques, R., Martin, and D., Cadavid. 2007. Role of interleukin 10 during persistent infection with the relapsing fever Spirochete Borrelia turicatae. *American Journal of Pathology* 170: 251

Good, M. F. 2001. Towards a blood-stage vaccine for malaria: are we following all the leads? *Nature Reviews Immunology* 1: 117

Graf, M. R., R. M., Prins, and R. E., Merchant. 2001. IL-6 secretion by a rat T9 glioma clone induces a neutrophil-dependent antitumor response with resultant cellular, antiglioma immunity. *Journal of Immunology* 166: 121

Grande, C., J. L., Firvida, V., Navas, and J., Casal. 2006. Interleukin-2 for the treatment of solid tumors other than melanoma and renal cell carcinoma. *Anticancer Drugs* 17: 1–12

Grey, H. M., A., Sette, and S., Buus. 1989. How T-cells see antigen. *Scientific American* 261(5): 56

Groux, H., A., O'Garra, M., Bigler, M., Rouleau, S., Antonenko, J. E., de Vries, and M. G., Roncarolo. 1997. CD4+T-cell subset inhibits antigen-specific T-cell and responses and prevents colitis. *Nature* 389: 737

Grover, F. L., D. A., Fullerton, M. R., Zamora, C., Mills, B., Ackerman, D., Badesch, J. M., Brown, D. N., Campbell, P., Chetham, A., Dhaliwal, M., Diercks, T., Kinnard, K., Niejadlik, and M., Ochs. 1997. The past, present, and future of lung transplantation. *The American Journal of Surgery* 173: 523

Haas, W., P., Pereira, and S., Tonegawa. 1993. Gamma/delta cells. *Annual Review of Immunology* 11: 637

Haddad, E., F., Le Deist, S., Blanche, M., Benkerrou, P., Rohrlich, E., Vilmer, C., Griscelli, and A., Fischer. 1995. Treatment of Chediak–Higashi syndrome by allogenic bone marrow transplantation: report of 10 cases. *Blood* 11: 3328

Hanson, L. A., and M., Korotkova. 2002. The role of breast feeding in prevention of neonatal infection. *Seminars in Neonatology* 7: 275

Happel, K. I., P. J., Dubin, M., Zheng, N., Ghilardi, C., Lockhart, L. J., Quinton, A. R., Odden, J. E., Shellito, G. J., Bagby, S., Nelson, and J. K., Kolls. 2005. Divergent roles of IL-23 and IL-12 in host defense against Klebsiella pneumoniae. *Journal of Experimental Medicine* 202: 761

Hardy, R. R., and K., Kayakawa. 2001. B-cell development pathways. *Annual Review of Immunology* 19: 595

Harlan, D. M., and A. D., Kirk. 1999. The future of organ and tissue transplantation: can T-cell costimulatory pathway modifiers revolutionize the prevention of graft rejection? *Journal of the American Medical Association* 282: 1076

Harriman, W., H., Volk, N., Defranoux, and M., Wabl. 1993. Immunoglobulin class switch recombinations. *Annual Review of Immunology* 11: 361

Harrington, L. E., P. R., Mangan, and C. T., Weaver. 2006. Expanding the effector CD4T-cell repertoire: the Th17 lineage. *Current Opinion in Immunology* 18: 349

Harrington, L. E., R. D., Hatton, P. R., Mangan, H., Turner, T. L., Murphy, K. M., Murphy, and C. T., Weaver. 2005. IL-17- producing CD4+ effector T cells develop via a lineage distinct from the T helper type 1 and 2 lineages. *Nature Immunology* 6: 1123

Hausmann, S., and K. W., Wucherpfennig. 1997. Activation of autoreactive T-cells by peptides from human pathogens. *Current Opinion in Immunology* 9: 831

Henderson, A., and K., Calame. 1998. Transcriptional regulation of B-cell development. *Annual Review of Immunology* 16: 163

Hesslein, D. G., and D. G., Schatz. 2001. Factors and forces controlling V(D)J recombination. *Advances in Immunology* 78: 169

Hirose, R., and F., Vincenti. 1999. Review of transplantation-1999. *Clinical Transplantation* 1999: 295

Hollingdale, M. R., C. J., McCormick, K. G., Heal, A. W., Taylor-Robinson, P., Reeve, R., Boykins, and J. W., Kazura. 1998. Biology of malarial liver stages: implications for vaccine design. *Annals of Tropical Medicine and Parasi*tology 92: 411

Holt, P. G. 1994. Immunoprophylaxis of atopy: light at the end of the tunnel? *Immunology Today* 15: 484

Hong, J. C., and B. D., Kahan. 2000. Immunosuppressive agents in transplantation: past, present, and future. *Seminars in Nephrology* 20: 108

Hong, K. H., M. L., Cho, S. Y., Min, Y. J., Shin, S. A., Yoo, J. J., Choi, W. U., Kim, S. W., Song, and C. S., Cho. 2007. Effect of interleukin-4 on vascular endothelial growth factor production in rheumatoid synovial fibroblasts. *Clinical and Experimental Immunology* 147: 573

Horwitz, M. S., L. M., Bradley, J., Harbertson, T., Krahl, J., Lee, and N., Sarvetnick. 1998. Diabetes induced by Coxsackie virus: initiation by bystander damage and not molecular mimicry. *Nature Medicine* 4: 781

Hou, L., J., Du, J., Wang, Y., Liu, W., Sun, Y., Zheng, L., Zhang, H., Hu, X., Dai, W., Guan, Y., Ma, and T., Hong. 2010. Expression of IL-12R alpha in liver cancer cells and its effects on targeted therapy of liver cancer. *Journal of Cancer Research and Clinical Oncology* 136: 839

Houghton, A. N., J. S., Gold, and N. E., Blachere. 2001. Immunity against cancer: lessons learned from melanoma. *Current Opinion in Immunology* 13: 134

Hoyne, G. E., N. M., Kristensen, H., Yssel, and J. R., Lamb. 1995. Peptide modulation of allergen-specific immune responses. *Current Opinion in Immunology* 7: 757

Hsu, F. J., C. B., Caspar, D., Czerwinski, L. W., Kwak, T. M., Liles, A., Syrengelas, B., Taidi-Laskowski, and R., Levy. 1997. Tumor-specific idiotype vaccines in the treatment of patients with B-cell lymphoma. *Blood* 89: 3129

Hunkapillar, T., and L., Hood. 1989. Diversity of the immunoglobulin gene superfamily. *Advances in Immunology* 44: 1–63

Ishikawa, H., N., Tsuyama, S., Liu, S., Abroun, F. J., Li, K., Otsuyama, X., Zheng, Z., Ma, Y., Maki, M. S., Iqbal, M., Obata, and M. M., Kawano. 2005. Accelerated proliferation of myeloma cells by interleukin-6 cooperating with fibroblast growth factor receptor 3 mediated signals. *Oncogene* 24: 6328

Ivanova, A. V., S. V., Ivanov, V., Pascal, J. M., Lumsden, J. M., Ward, N., Morris, L., Tessarolo, S. K., Anderson, and M. I., Lerman. 2007. Autoimmunity, spontaneous tumourigenesis, and IL-15 insufficiency in mice with a targeted disruption of the tumour suppressor gene Fus1. *Journal of Pathology* 211: 591

Jacob, J., G., Kelsoe, K., Rajewsky, and U., Weiss. 1991. Intraclonal generation of antibody mutants in germinal centers. *Nature* 354: 389

Kägi, D., B., Ledermann, K., Bürki, P., Seiler, B., Odermatt, K. J., Olsen, E. R., Podack, R. M., Zinkernagel, and H., Hengartner. 1994. Cytotoxicity mediated by T-cells and natural killer cells is greatly impaired in perforin-deficient mice. *Nature* 369: 31

Kägi, D., F., Vignaux, B., Ledermann, K., Bürki, V., Depraetere, S., Nagata, H., Hengartner, and P., Golstein. 1994. Fas and perforin pathways as major mechanisms of T-cell-mediated cytotoxicity. *Science* 265: 528

Kaminski, A., E. R., Kaminski, and N. G., Morgan. 2007. Pre-incubation with interleukin-4 mediates a direct protective effect against the loss of pancreatic beta-cell viability induced by proinflammatory cytokines. *Clinical and Experimental Immunology* 148: 583

Kaufmann, S. H. 2001. How can immunology contribute to the control of tuberculosis? *Nature Reviews of Immunology* 1: 20

Keegan, A. D., and W. E., Paul. 1992. Multichain immune recognition receptors: similarities in structure and signaling pathways. *Immunology Today* 13: 63

Kepler, T. B., and C., Chan. 2007. Spatiotemporal programming of a simple inflammatory process. *Immunological Reviews* 216: 153

Kerr, M. A. 1990. The structure and function of human IgA. *Biochemical Journal* 271: 285

Kim, D., R. E., Mebius, J. D., MacMicking, S., Jung, T., Cupedo, Y., Castellanos, J., Rho, B. R., Wong, R., Josien, N., Kim, P. D., Rennert, and Y., Choi. 2000. Regulation of peripheral lymph node genesis by the tumor necrosis factor family member TRANCE. *Journal of Experimental Medicine* 192: 1467

Kim, S. H., L. L., Reznikov, R. J., Stuyt, C. H., Selzman, G., Fantuzzi, T., Hoshino, H. A., Young, and C. A., Dinarello. 2001. Functional reconstitution and regulation of IL-18 activity by the IL-18R beta chain. *Journal of Immunology* 166: 148

King, C., and N., Sarvetnick. 1997. Organ-specific autoimmunity. *Current Opinion in Immunology* 9: 863

Kirk, A. D., D. M., Harlan, N. N., Armstrong, T. A., Davis, Y., Dong, G. S., Gray, X., Hong, D., Thomas, J. H., Fechner, Jr., and S. J., Knechtle. 1997. CTLA4-Ig and anti-CD40 ligand prevent renal allograft rejection in primates. *Proceedings of the National Academy of Sciences USA* 94: 8789

Knodler, L. A., J., Celli, and B. B., Finlay. 2001. Pathogenic trickery: deception of host cell processes. *Nature Reviews Molecular Cell Biology* 2: 578

Kohler, G., and C., Milstein. 2005. Continuous cultures of fused cells secreting antibody of predefined specificity. 1975. *Journal of Immunology* 174: 2453

Kraehenbuhl, J. P., S., Kerneis, J., Berdoz, B., Corthésy, L., Rindisbacher, I., Corthésy-Theulaz, A., Phalipon, M. R., Neutra, and E., Pringault. 1995. Molecular events in production and transport of secretory IgA: a basis for design of reagents for passive immune protection. *Veterinary Research* 26: 231

Krause, R. M., N. J., Dimmock, and D. M., Morens. 1997. Summary of antibody workshop: the role of humoral immunity in the treatment and prevention of emerging and extant infectious diseases. *Journal of Infectious Diseases* 176: 549

Kufe, D. W. 2000. Smallpox, polio and now a cancer vaccine? *Nature Medicine* 6: 252

Kühn, R., K., Rajewsky, and W., Müller. 1991. Generation and analysis of interleukin-4 deficient mice. *Science* 254: 707

Lachmann, P. J., and A., Davies. 1997. Complement and immunity to viruses. *Immunological Reviews* 159: 69

Lamm, M. E. 1997. Interaction of antigens and antibodies at mucosal surfaces. *Annual Review of Microbiology* 51: 311

Lane, H. C., J. L, Montagne, and A. S., Fauci. 2001. Bioterrorism: a clear and present danger. *Nature Medicine* 7: 1271

Lane, P. 1996. Development of B-cell memory and effector function. *Current Opinion in Immunology* 8: 331

Lanier, L. L., and J. H., Phillips. 1996. Inhibitory MHC class I receptors on NK cells and T cells. *Immunology Today* 17: 86

Laurent, J., and M. T., Guinnepain. 1999. Angioedema associated with C1 inhibitor deficiency. *Clinical Reviews in Allergy and Immunology* 17: 513

Laver, W. G., G. M., Air, R. G., Webster, and S. J., Smith-Gill. 1990. Epitopes on protein antigens: misconceptions and realities. *Cell* 61: 553

Lefranc, M. -P., and T. H., Rabbitts. 1989. The human T-cell receptor γ-gene (TRG) genes. *Trends in Biochemical Sciences* 14: 214

Leiden, J. M. 1993. Transcriptional regulation of T-cell receptor genes. *Annual Review of Immunology* 11: 539

Lekstorm-Himes, J. A., and J. I., Gallin. 2000. Advances in immunology: immunodeficiency diseases caused by defects in phagocytes. *The New England Journal of Medicine* 343: 1703

Lenschow, D. J., Y., Zeng, J. R., Thistlethwaite, A., Montag, W., Brady, M. G., Gibson, P. S., Linsley, and J. A., Bluestone. 1992. Long-term survival of xenogeneic pancreatic islets induced by CTLA4-Ig. *Science* 257: 789

Levin, M. C., S. M., Lee, F., Kalume, Y., Morcos, F. C., Dohan Jr, K. A., Hasty, J. C., Callaway, J., Zunt, D., Desiderio, and J. M., Stuart. 2002. Autoimmunity due to molecular mimicry as a cause of neurological disease. *Nature Medicine* 8: 509

Levings, M. K., R., Sangregorio, F., Galbiati, S., Squadrone, R., de Waal Malefyt, and M. G., Roncarolo. 2001. IFN-α and IL-10 induce the differentiation of human type 1 T regulatory cells. *Journal of Immunology* 166: 5530

Lewis, S. M. 1994. The mechanism of V(D)J joining: lessons from molecular, immunological, and comparative analyses. *Advances in Immunology* 56: 27

Liblau, R. S., S. M., Singer, and H. O., McDevitt. 1995. TH-1 and TH-2 CD4+ T-cells in the pathogenesis of organ-specific autoimmune diseases. *Immunology Today* 16: 34

Lindahl, G., U., Sjöbring, and E., Johnsson. 2000. Human complement regulators: a major target for pathogenic microorganisms. *Current Opinion in Immunology* 12: 44

Liu, Y. J. 2001. Dendritic cell subsets and lineages and their functions in innate and adaptive immunity. *Cell* 106: 259

Liu, Y. J., O., de Bouteiller, and I., Fugier-Vivier. 1997. Mechanisms of selection and differentiation in germinal centers. *Current Opinion in Immunology* 9: 256

Lockshin, M. D. 1998. Why women? *Journal of the American Medical Women's Association* 53: 4

Lodish, H., A., Berk, S. L., Zipursky, P., Matsudaira, D., Baltimore, and J., Darnell. 2000. *Molecular Cell Biology*. New York: Freeman

Lokki, M. L., and H. R., Colten. 1995. Genetic deficiencies of complement. *Annals of Medicine* 27: 451

Long, E. O. 1999. Regulation of immune responses through inhibitory receptors. *Annual Review of Immunology* 17: 875

Lopez-Botet, M., and T., Bellon. 1999. Natural killer cell activation and inhibition by receptors for MHC class I. *Current Opinion in Immunology* 11: 301

Lorenzo, M. E., H. L., Ploegh, and R. S., Tirabassi. 2001. Viral immune evasion strategies and the underlying cell biology. *Seminars in Immunology* 13: 1–9

Louis, J., H., Himmelrich, C., Parra-Lopez, F., Tacchini-Cottier, and P., Launois. 1998. Regulation of protective immunity against Leishmania major in mice. *Current Opinion in Immunology* 10: 459

Lympany, P., K. I., Welsh, G. M., Cochrane, D. M., Kemeny, and T. H., Lee. 1992. Genetic analysis of the linkage between chromosome 11q and atopy. *Clinical and Experimental Allergy* 22: 1085

Ma, A., R., Koka, and P., Burkett. 2006. Diverse functions of IL-2, IL-7 and IL-15 in lymphoid homeostasis. *Annual Review of Immunology* 24: 657

Mackay, F., and C., Ambrose. 2003. The TNF family members BAFF and APRIL: the growing complexity. *Cytokine and Growth Factor Reviews* 14: 311

Madden, D. R., J. C., Gorga, J. L., Strominger, and D. C., Wiley. 1991. The structure of HLA-B27 reveals nonamer self-peptides bound in an extended conformation. *Nature* 353: 321

Malley, R., A., Srivastava, M., Lipsitch, C. M., Thompson, C., Watkins, A., Tzianabos, and P. W., Anderson. 2006. Antibody-independent, IL-17A–mediated crossserotype immunity to pneumococci in mice immunized intranasally with the cell wall polysaccharide. *Infection and Immunity* 74: 2187

Manis, J. P., M., Tian, and F. W., Alt. 2002. Mechanism and control of class-switch recombination. *Trends in Immunology* 23: 31

Manning, T. C., and D. M., Kranz. 1999. Binding energetics of T-cell receptors: correlation with immunological consequences. *Immunology Today* 20: 417

Markees, T. G., N. E., Phillips, R. J., Noelle, L. D., Shultz, J. P., Mordes, D. L., Greiner, and A. A., Rossini. 1997. Prolonged survival of mouse skin allografts in recipients treated with donor splenocytes and antibody to CD40 ligand. *Transplantation* 64: 329

Marshall, A. J., H. E., Fleming, G. E., Wu, and C. J., Paige. 1998. Modulation of the IL-7 dose-response threshold during pro-B cell differentiation is dependent on pre-B cell receptor expression. *Journal of Immunology* 161: 6038

Matsuuchi, L., and M. R., Gold. 2002. New views of BCR structure and organization. *Current Opinion in Immunology* 13: 270

Maxwell, J. R., R., Yadav, R. J., Rossi, C. E., Ruby, A. D., Weinberg, H. L., Aguila, and A. T., Vella. 2006. IL-18 bridges innate and adaptive immunity through IFN-gamma and the CD134 pathway. *Journal of Immunology* 177: 234

McGuirk, P., and K. H., Mills. 2002. Pathogen-specific regulatory T cells provoke a shift in the Th1/Th2 paradigm in immunity to infectious diseases. *Trends in Immunology* 23: 450

McGuirk, P., C., McCann, and K. H., Mills. 2002. Pathogen-specific T regulatory 1 cell induced in the respiratory tract by a bacterial molecule that stimulates interleukin 10 productions by dendritic cells: a novel strategy for evasion of protective T helper type 1 responses by Bordetella pertussis. *Journal of Experimental Medicine* 195: 221

McHeyzer-Williams, M. G., and R. B., Ahmed. 1999. B-cell memory and the long lived plasma cell. *Current Opinion in Immunology* 11: 172

Meffre, E., R., Casellas, and M. C., Nussenzweig. 2000. Antibody regulation of B-cell development. *Nature Immunology* 1: 379

Moiller, G (ed.). 1977. Immunoglobulin D: structure, synthesis, membrane representation and function. *Immunological Reviews* 37: 287

Moiller, G (ed.). 1978. Immunoglobulin E. *Immunological Reviews* 41

Moiller, G (ed.). 1992. Fc receptors. *Immunological Reviews* 125

Moiller, G (ed.). 1993. B-cell antigen receptor complex. *Immunological Reviews* 132

Molina, H., and V. M., Holers. 1996. Markedly impaired humoral immune response in mice deficient in complement receptors 1 and 2. *Proceedings of the National Academy of Sciences USA* 93: 8720

Mollnes, T. E., and A. E., Fiane. 1999. Xenotransplantation: how to overcome the complement obstacle? *Molecular Immunology* 36: 269

Moreno, S. E., J. C., Alves-Filho, G., Bertozi, T. M, Alfaya, J., Thèze, S. H., Ferreira, and B. B., Vargaftig. 2006. Systemic administration of interleukin-2 inhibits inflammatory neutrophil migration: role of nitric oxide. *British Journal of Pharmacology* 148: 1060

Moss, B. 1985. Vaccinia virus expression vectors: a new tool for immunologists. *Immunology Today* 6: 243

Muller-Eberhard, H. J. 1988. Molecular organization and function of the complement system. *Annual Review of Biochemistry* 57: 321

Nakano, T., S., Tahara-Hanaoka, C., Nakahashi, I., Can, N., Totsuka, S., Honda, K., Shibuya, and A., Shibuya. 2008. Activation of neutrophils by a novel triggering immunoglobulin-like receptor MAIR-IV. *Molecular Immunology* 45: 289

Natarajan, K., N., Dimasi, J., Wang, R. A., Mariuzza, and D. H., Margulies. 2002. Structure and function of natural killer cell receptors: multiple molecular solutions to self, non-self discrimination. *Annual Review of Immunology* 20: 853

Nathan, C., and M. U., Shiloh. 2000. Reactive oxygen and nitrogen intermediates in the relationship between mammalian hosts and microbial pathogens. *Proceedings of the National Academy of Sciences USA* 97: 8841

Nielsen, C. H., E. M., Fischer, and R. G. Q., Leslie. 2000. The role of complement in the acquired immune response. *Immunology* 100: 4

Nonaka, M. 2000. Origin and evolution of the complement system. *Current Topics in Microbiology and Immunology* 248: 37

Novak, N., S., Kraft, and T., Bieber. 2001. IgE receptors. *Current Opinion in Immunology* 13: 721

O'Garra, A., L., Steinman, and K., Gijbels. 1997. Cd4+ T-cell subsets in autoimmunity. *Current Opinion in Immunology* 9: 872

Orzechowska, B., Z., Antoszków, I., Siemieniec, M., Lorenc, B., Jatczak, Z., and Błach-Olszewska. 2007. Cytokine production by human leukocytes with different expressions of natural antiviral immunity and the effect of antibodies against interferons and TNF-alpha. *Archivum Immunologiae et Therapiae Experimentalis (Warsz)* 55: 111

Papavasiliou, F. N., and D. G., Schatz. 2002. Somatic hypermutation of immunoglobulin genes. Merging mechanisms for genetic diversity. *Cell* 109: S35

Pardoll, D. M. 1996. Cancer vaccines: a road map for the next decade. *Current Opinion in Immunology* 8: 619

Park, H., Z., Li, X. O., Yang, S. H., Chang, R., Nurieva, Y. H., Wang, Y., Wang, L., Hood, Z., Zhu, Q., Tian, and C., Dong. 2005. A distinct lineage of CD4 T cells regulates tissue inflammation by producing IL-17. *Nature Immunology* 6: 1133

Pascual, V., and J. D., Capra. 1991. Human immunoglobulin heavy chain variable region genes: organization, polymorphism, and expression. *Advances in Immunology* 49: 1–74

Paul-Eugene, N., J. P., Kolbe, A., Calenda, J., Gordon, K., Hitoshi, K., Tadamidsu, J. M., Mencia-Huerta, P., Braquet, and B., Dugas. 1993. Functional interaction between β2-adrenoceptor agonists and interleukin-4 in the regulation of CD23 expression and release and IgE production in humans. *Molecular Immunology* 30: 157

Peiser, L., S., Mukhopadhyay, and S., Gordon. 2002. Scavenger receptors in innate immunity. *Current Opinion in Immunology* 14: 123

Pène, J., J. F., Gauchat, S., Lécart, E., Drouet, P., Guglielmi, V., Boulay, A., Delwail, D., Foster, J. C., Lecron, and H., Yssel. 2004. Cutting edge: IL-21 is a switch factor for the production of IgG1 and IgG3 by human B cells. *Journal of Immunology* 172: 5154

Peng, Chun-Wei and Yan, Li. 2010. Application of quantum dots-based biotechnology in cancer diagnosis: Current status and future perspectives. *Journal of Nanomaterials* 2010 <Details available at www. hindawi.com/journals/jinm/2010/676839

Peterson, Y., and G., Ikonomidis. 1996. Recombinant Listeria monocytogenes cancer vaccines. *Current Opinion Immunology* 8: 651

Pickering, M. C., and M. J., Walport. 2000. Links between complement abnormalities and systemic lupus erythematosus. *Rheumatology* 39: 133

Porcelli, S. A., B. W., Segelke, M., Sugita, I. A., Wilson, and M. B., Brenner. 1998. The CD1 family of lipid antigen presenting molecules. *Immunology Today* 19: 362

Post, D. E., E. M., Sandberg, M. M., Kyle, N. S., Devi, D. J., Brat, Z., Xu, M., Tighiouart, and E. G., van Meir. 2007. Targeted cancer gene therapy using a hypoxia inducible factor dependent oncolytic adenovirus armed with interleukin-4. *Cancer Research* 67: 6872

Powis, S. H., and J., Trowsdale. 1994. Human major histocompatibility complex genes. *Behring Institute Mitteilungen* 94: 17

Przylepa, J., C., Himes, and G., Kelsoe. 1998. Lymphocyte development and selection in germinal centers. *Current Topics in Microbiology and Immunology* 229: 85

Ramshaw, I. A., A. J., Ramsay, G., Karupiah, M. S., Rolph, S., Mahalingam, and J. C., Ruby. 1997. Cytokines and immunity to viral infections. *Immunological Reviews* 159: 119

Rao, V. P., T., Poutahidis, Z., Ge, P. R., Nambiar, C., Boussahmain, Y. Y., Wang, B. H., Horwitz, J. G., Fox, and S. E., Erdman. 2006. Innate immune inflammatory response against enteric bacteria Helicobacter hepaticus induces mammary adenocarcinoma in mice. *Cancer Research* 66: 7395

Rathmell, J. C., and C. B., Thompson. 1999. The central effectors of cell death in the immune system. *Annual Review of Immunology* 17: 781

Ratthe, C., and D., Girard. 2004. Interleukin-15 enhances human neutrophil phagocytosis by a Syk-dependent mechanism: importance of the IL-15R alpha chain. *Journal of Leukocyte Biology* 76: 162

Raulet, D. H. 1994. How γδ T-cells make a living? *Current Biology* 4: 246

Rautemaa, R., and S., Meri. 1999. Complement resistance mechanisms of bacteria. *Microbes and Infection/Institute Pasteur* 1: 785

Ravetch, J. V., and D. H., Margulies. 1994. New tricks for old molecules. *Nature* 372: 323

Ravetch, J. V., and J. P., Kinet. 1991. Fc receptors. *Annual Review of Immunology* 9: 457

Rayhill, S. C., M., Heimes, A. D., Kirk, and H. W., Sollinger. 1996. Simultaneous pancreas–kidney transplantation: recent experience at the University of Wisconsin. *Experimental and Clinical Endocrinology and Diabetes* 104: 353

Razin, E., I., Pecht, and J., Rivera. 1995. Signal transduction in the activation of mast cells and basophils. *Immunology Today* 16: 370

Reth, M. 1995. The B-cell antigen receptor complex and co-receptor. *Immunology Today* 16: 310

Riemann, M., R., Endres, S., Liptay, K., Pfeffer, and R. M., Schmid. 2005. The I-kappaB protein Bcl-3 negatively regulates transcription of the IL-10 gene in macrophages. *Journal of Immunology* 175: 3560

Robertson, B. D., and T. E., Meyer. 1992. Genetic variation in pathogenic bacteria. *Trends in Genetics* 8: 422

Robertson, S. E., J. D., Young, S., Kitson, A., Pitt, J., Evans, J., Roes, D., Karaoglu, L., Santora, T., Ghayur, F. Y., Liew, J. A., Gracie, and I. B., McInnes. 2006. Expression and alternative processing of IL-18 in human neutrophils. *European Journal of Immunology* 36: 722

Roelofs, M. F., M. H., Wenink, E. J., Toonen, M. J., Coenen, L. A., Joosten, W. B., van den Berg, P. L., van Riel, and T. R., Radstake. 2008. The functional variant (Asp299gly) of Toll-like receptor 4 (TLR4) influences TLR4-mediated cytokine production in rheumatoid arthritis. *Journal of Rheumatology* 35: 558

Romagnani, S. 2001. T-cell responses in allergy and asthma. *Current Opinion in Allergy and Clinical Immunology* 1: 73

Rose, N. R. 1998. The role of infection in the pathogenesis of autoimmune disease. *Seminars in Immunology* 10: 5

Rosenberg, S. A. 2001. Progress in human tumour immunology and immunotherapy. *Nature* 411: 380

Rosenberg, S. A., C., Sportès, M., Ahmadzadeh, T. J., Fry, L. T., Ngo, S. L., Schwarz, M., Stetler-Stevenson, K. E., Morton, S. A., Mavroukakis, M., Morre, R., Buffet, C. L., Mackall, and R. E., Gress. 2006. IL-7 administration to humans leads to expansion of CD8+ and CD4+ cells but a relative decrease of CD4+ T-regulatory cells. *Journal of Immunotherapy* 29: 313

Rosenberg, S. A., J. R., Yannelli, J. C., Yang, S. L., Topalian, D. J., Schwartzentruber, J. S., Weber, D. R., Parkinson, C. A., Seipp, J. H., Einhorn, and D. E., White. 1994. Treatment of patients with metastatic melanoma with autologous tumor-infiltrating lymphocytes and interleukin 2. *Journal of the National Cancer Institute* 86: 1159

Rosenstreich, D. L., P., Eggleston, M., Kattan, D., Baker, R. G., Slavin, P., Gergen, H., Mitchell, K., McNiff-Mortimer, H., Lynn, D., Ownby, and F., Malveaux. 1997. The role of cockroach allergy and exposure to cockroach allergen in causing morbidity among inner-city children with asthma. *The New England Journal of Medicine* 336: 1356

Rosenthal, S. R., M., Merchlinsky, C., Kleppinger, and K. L., Goldenthal. 2001. Developing new smallpox vaccines. *Emerging Infectious Diseases* 7: 422

Rothenberg, E. V. 2000. Step wise specification of lymphocyte developmental lineages. *Current Opinion in Genetics and Development* 10: 370

Russell, J. H., and T. J., Ley. 2002. Lymphocyte mediated cytotoxicity. *Annual Review of Immunology* 20: 370

Rydström, A., and M. J., Wick. 2007. Monocyte recruitment, activation, and function in the gut-associated lymphoid tissue during oral Salmonella infection. *Journal of Immunology* 178: 5789

Sahin, U., O., Türeci, and M., Pfreundschuh. 1997. Serological identification of human tumor antigens. *Current Opinion in Immunology* 9: 709

Salter, R. D., R. J., Benjamin, P. K., Wesley, S. E., Buxton, T. P. J., Garrett, C., Clayberger, A. M., Krensky, A. M., Norment, D. R., Littman, and P., Parham. 1990. A binding site for the T-cell co-receptor CD8 on the α3 domain of HLA-A2. *Nature* 345: 41

San José, E., A. G., Sahuquillo, R., Bragado, and B., Alarcón. 1998. Assembly of the TCR/CD3 complex: CD3 epsilon/delta and CD3 epsilon/gamma dimmers associate indistinctly with both TCR alpha and TCR beta chains. Evidence for a double TCR heterodimer model. *European Journal of Immunology* 28: 12

Schiffer, M., R. L., Girling, K. R., Ely, and A. B., Edmundson. 1973. Structure of a lambda-type Bence-Jones protein at 3.5 angstrom resolution. *Biochemistry* 12: 4620–4631

Schroeder, H. W. Jr., and L., Cavacini. 2010. Structure and function of immunoglobulins. *Journal of Allergy and Clinical Immunology* 125: S41

Scott, A. M., and S., Welt. 1997. Antibody-based immunological therapies. *Current Opinion in Immunology* 9: 717

Scott, P. 1998. Differentiation, regulation, and death of T helper cell subsets during infection with Leishmania major. *Immunologic Research* 17: 229

Shanmugham, L. N., C., Petrarca, S., Frydas, J., Donelan, M. L., Castellani, W., Boucher, B., Madhappan, S., Tete', K., Falasca, P., Conti, and J., Vecchiet. 2006. IL-15 an immunoregulatory and anti-cancer cytokine. Recent advances. *Journal of Experimental and Clinical Cancer Research* 25: 529

Sher, A., and R. L., Coffman. 1992. Regulation of immunity to parasites by T-cells and T-cell-derived cytokines. *Annual Review of Immunology* 10: 385

Silverstein, A. M., and N. R., Rose. 1997. On the mystique of the immunological self. *Immunological Reviews* 159: 197

Sims, J. E., M. J., Nicklin, J. F., Bazan, J. L., Barton, S. J., Busfield, J. E., Ford, R. A., Kastelein, S., Kumar, H., Lin, J. J., Mulero, J., Pan, Y., Pan, D. E., Smith, and P. R., Young. 2001. A new nomenclature for IL-1 family genes. *Trends in Immunology* 22: 536

Sloan-Lancaster, J., and P. M., Allen. 1996. Altered peptide ligand induced partial T-cell activation: molecular mechanisms and role in T-cell biology. *Annual Review of Immunology* 14: 1–27

Smith, K. G., and D. T., Fearon. 2000. Receptor modulators of B-cell receptor signalling—CD19/CD22. *Current Topics in Microbiology and Immunology* 245: 195

Snapper, C. M., and J. J., Mond. 1996. A model for induction of T-cell independent humoral immunity in response to polysaccharide antigens. *Journal of Immunology* 157: 2229

Srinivasan, A., R. M., Salazar-Gonzalez, M., Jarcho, M. M., Sandau, L., Lefrancois, and S. J., McSorley. 2007. Innate immune activation of CD4 T-cells in salmonella-infected mice is dependent on IL-18. *Journal of Immunology* 178: 6342

Srivastava, S. 2002. Roles of heat-shock proteins in innate and adaptive immunity. *Nature Reviews Immunology* 2: 185

Stanfield, R. L., and I. A., Wilson. 1995. Protein peptide interactions. *Current Opinion in Chemical Biology* 8: 138

Stanfield, R. L., and I. A., Wilson. 1995. Protein-peptide interactions. *Current Opinion in Structural Biology* 5: 103

Stavnezer, J. 1996. Antibody class switching. *Advances in Immunology* 61: 79

Stein, M. L., M. H., Collins, J. M., Villanueva, J. P., Kushner, P. E., Putnam, B. K., Buckmeier, A. H., Filipovich, A. H., Assa'ad, and M. E., Rothenberg. 2006. Anti-IL-5 (mepolizumab) therapy for eosinophilic esophagitis. *Journal of Allergy and Clinical Immunology* 118: 1312

Steinman, L., J. R., Oksenberg, and C. C., Bernard. 1992. Association of susceptibility to multiple sclerosis with TCR genes. *Immunology Today* 13: 49

Stern, L. J., and D. C., Wiley. 1994. Antigenic peptide binding by class I and class II histocompatibility proteins. *Structure* 2: 245

Stern, L. J., J. H., Brown, T. S., Jardetzky, J. C., Gorga, R. G., Urban, J. L., Strominger, D. C., Wiley. 1994. Crystal structure of the human class II MHC protein HLA-DR1 complexed with an influenza virus peptide. *Nature* 368: 215

Streilein, J. W., M. R., Dana, and B. R., Ksander. 1997. Immunity causing blindness: five different paths to herpes stromal keratitis. *Immunology Today* 18: 443

Tainer, J. A., E. D., Getzoff, Y., Paterson, A. J., Olson, and R. A., Lerner. 1985. The atomic mobility component of protein antigenicity. *Annual Review of Immunology* 3: 501

Teixeira, M. M., T. J., Williams, and P. G., Hellewell. 1995. Mechanisms and pharmacological manipulation of eosinophil accumulation in vivo. *Trends in Pharmacological Sciences* 16: 418

Theofilopoulos, A. N. 1995a. The basis of autoimmunity: Part I. mechanisms of aberrant self-recognition. *Immunology Today* 16: 90

Theofilopoulos, A. N. 1995b. The basis of autoimmunity: Part II. genetic predisposition. *Immunology Today* 16: 150

Tindle, R. W. 1996. Human papillomavirus vaccines for cervical cancer. *Current Opinion in Immunology* 8: 643

Trinchieri, G. 1998. Immunobiology of interleukin-12. *Immunologic Research* 17: 269

Tumanov, A. V., D. V., Kuprash, and S. A., Nedospasov. 2003. The role of lymphotoxin in development and maintenance of secondary lymphoid tissues. *Cytokine and Growth Factor Reviews* 14: 275

Turner, M. W. 1998. Mannose binding lectin (MBL) in health and disease. *Immunobiology* 199: 327

Underhill, D. M., and A., Ozinsky. 2002. Toll-like receptors: key mediators of microbe detection. *Current Opinion in Immunology* 14: 103

Unlu, S., A., Kumar, W. R., Waterman, J., Tsukada, K. Z., Wang, D. L., Galson, and P. E., Auron. 2007. Phosphorylation of IRF8 in a pre-associated complex with Spi-1/PU.1 and non-phosphorylated Stat-1 is critical for LPS induction of the IL1B gene. *Molecular Immunology* 44: 3364

Van den Eynde, B., and P., van der Bruggen. 1996. T cell defined tumor antigens. *Current Opinion in Immunology* 9: 684

Villinger, F. 2003. Cytokines as clinical adjuvants: how far are we? *Expert Review of Vaccines* 2: 317

Vyse, T. J., and B. L., Kotzin. 1998. Genetic susceptibility to systemic lupus erythromatosus. *Annual Review of Immunology* 16: 261

Wallach, D., E. E., Varfolomeev, N. L., Malinin, Y. V., Goltsev, A. V., Kovalenko, and M. P., Boldin. 1999. Tumor necrosis factor receptor and Fas signaling mechanisms. *Annual Review of Immunology* 17: 331

Wan, Y. Y., and R. A., Flavell. 2006. The roles for cytokines in the generation and maintenance of regulatory T-cells. *Immunological Reviews* 212: 114

Ward, A. C., D. M., Loeb, A. A., Soede-Bobok, I. P., Touw, and A. D., Friedman. 2000. Regulation of granulopoiesis by transcription factors and cytokine signals. *Leukemia* 14: 973

Webb, D. C., Y., Cai, K. I., Matthaei, and P. S., Foster. 2007. Comparative roles of IL-4, IL-13, and IL-4Ralpha in dendritic cell maturation and CD4+ Th2 cell function. *Journal of Immunology* 178: 219

Wedemayer, G. J., P. A., Patten, L. H., Wang, P. G., Schultz, and R. C., Stevens. 1997. Structural insights into the evolution of an antibody combining site. *Science* 276: 1665

Weinberg, R. A. 1996. How cancer arises. *Scientific American* 275: 62

Weiner, H. L., G. A., Mackin, M., Matsui, E. J., Orav, S. J., Khoury, D. M., Dawson, and D. A., Hafler. 1993. Double-blind pilot trial of oral tolerization with myelin antigens in multiple sclerosis. *Science* 259: 1321

Weiss, A., and D. R., Littman. 1994. Signal transduction by lymphocyte antigen receptor. *Cell* 76: 263

Weissman, I. L. 2000. Translating stem and progenitor cell biology to the clinic: barriers and opportunities. *Science* 287: 1442

Welsh, R. M., M. Y., Lin, B. L., Lohman, S. M., Varga, C. C., Zarozinski, and L. K., Selin. 1997. Alpha beta and gamma delta T-cell networks and their roles in natural resistance to viral infections. *Immunological Reviews* 159: 79

Wentworth, P., and K., Janda. 1998. Catalytic antibodies. Current Opinion in Chemical Biology 8: 138

Wilder, R. L. 1995. Neuroendocrine-immune system interactions and autoimmunity. *Annual Review of Immunology* 13: 307

Wilson, I. A., and R. I., Stanfield. 1994. Antibody-antibody interactions: new structures and new conformational changes. *Current Opinion in Structural Biology* 4: 857

Wilson, S. B., S. C., Kent, K. T., Patton, T., Orban, R. A., Jackson, M., Exley, S., Porcelli, D. A., Schatz, M. A., Atkinson, S. P., Balk, J. L., Strominger, and D. A., Hafler. 1998. Extreme Th1 bias of invariant Valpha24JalphaQ T cells in type 1 diabetes. *Nature* 391: 177

Woo, S. B., S. J., Lee, and M. M., Schubert. 1997. Graft-vs-host disease. *Critical Reviews in Oral Biology and Medicine* 8: 201

Wucherpfennig, K. W. 2001. Mechanisms for the induction of autoimmunity by infectious agents. *Journal of Clinical Investigation* 108: 1097

Xing, Y., and J., Rao. 2008. Quantum dot bioconjugates for in vitro diagnostics and in vivo imaging. *Cancer Biomarkers* 4: 307

Yokoyama, W. M. 1998. Natural killer cell receptors. *Current Opinion in Immunology* 10: 298

Yoshimoto, T., K., Takeda, T., Tanaka, K., Ohkusu, S., Kashiwamura, H., Okamura, S., Akira, and K., Nakanishi. 1998. IL-12 up-regulates IL-18 receptor expression on T cells, Th1 cells, and B cells: synergism with IL-18 for IFN-gamma production. *Journal of Immunology* 161: 3400

Zeng, Z., A. R., Castaño, B. W., Segelke, E. A., Stura, P. A., Peterson, and I. A., Wilson. 1997. Crystal structure of mouse CD1: an MHC-like fold with a large hydrophobic binding groove. *Science* 277: 339

Zhao, Z. S., F., Granucci, L., Yeh, P. A., Schaffer, and H., Cantor. 1998. Molecular mimicry by herpes simplex virus type I: autoimmune disease after viral infection. *Science* 279: 1344

Glossary

AIDS: It is known as Acquired Immune Deficiency Syndrome. The deficiency occurs in the human immunity due to infection of human specific virus HIV-1 to CD4-T cells of human.

Acute phase proteins: Proteins that are secreted after infection and inflammation.

ADCC: It is known as antibody-dependent cell mediated cytotoxicity. In this kind of cytotoxic reaction, Fc containing killer cells identify target cells mediated by antibody.

Adhesion molecules: Molecules present on the cell surface to help them to bind to matrix or nearby cells without activating them, e.g. integrin and selectins.

Adjuvant: An adjuvant is a non-specific immune response enhancing substance used with antigens having low immunogenic capability.

Affinity: It is a measurement of binding strength between antigenic epitope to an antibody.

Affinity maturation: At the time of secondary immune response, the average antibody affinity increases to maximum which is known as affinity maturation.

Allelic exclusion: The genes from paternal and maternal side that do not allow the use of other genes is known as allelic exclusion, e.g. antibody and TCR genes.

Allergen: External non-parasitic or microbial agents that trigger the formation of IgE antibody to cause allergic or hypersensitivity reaction are known as allergen, e.g. pollen, dust, dandruffs, animal hairs, egg.

Allergy: An altered immune reaction that occurs after second contact of allergens, now known as type-I hypersensitivity reaction.

Allotype antigens: Antigens which are expressed as allelic protein detected by another member of same species.

Alternative pathway: Complement activation pathway to form a positive feedback reaction loop due to split of C3 component in the presence of factors B, D, P, H to generate C3 convertase alternate pathway.

Amplification loop: This is a positive feedback alternative pathway due to C3 convertase which is a split product due to factors B, D, P, and H.

Anaphylatoxins: Sub-units of complement, e.g. C3a and C5a, that cause degranulation of mast cells and contraction of smooth muscle.

Anaphylaxis: It is vasodilation and constriction of smooth muscles, e.g. bronchus, due to production of IgE against allergens, which leads to death.

Anchor residues: It is amino acid residues of antigenic peptides essential for interaction of MHC molecules pockets sites.

Anergy: This is the failure to make an immune response even stimulation with highly immunogenic antigens.

Antagonist peptides: Antigenic peptides analogous that binds to MHC molecules to prevent T-cells activation.

Antibody: Protein molecules produced by animals and binded with the same antigens in specific manner against which they are produced.

Antigen: Generally protein molecules that activate the immune cells B or T to synthesize antibody against which they can bind in specific manner. Antigen can also bind with B-cell receptor and T-cell receptor.

Antigen receptors: It is B-cell receptor (BCR) which is a B-cell surface immuno-globuline and T-cell receptor (TCR) having an ability to bind with antigens.

Antigenic determinants: It is also known as epitopes.

Antigenic peptides: Peptides that in association with MHC activate T cells.

Antigen presentation: Presentation of antigens after processing them on the cell surface in association with MHC for recognition by lymphocytes by certain specific cells of the body called APCs is known as antigen presentation.

Antigen processing: Making small pieces of antigen for presentation at the cell surface so it could be recognized by the lymphocyte cells.

Antiviral proteins: These are interferon proteins used to limit viral replication after infection into the body cells.

APCs: These are antigen presenting cells that process antigens in small fragments and present them on cell surface with MHC for recognition by the lymphocyte cells.

Apoptosis: It is also known as programmed cell death identified by DNA fragmentation, cell cytoplasm condensation, plasma membrane shrinkage, degeneration of cell organelles.

ARAMs: It is known as antigen receptor activation motifs of CD79 and CD3. It is also known as amino acid sequences of CD79 and CD3 extra cellular domain to be get phosphorylated by activated lymphocytes.

Arthus reaction: It is manifestation reaction of type-III hypersensitivity which appeared as inflammation on skin after a few hours of administrating antigens.

Atopy: It is manifestation reaction of type-I hypersensitivity reaction which appears as eczema, asthma, rhinitis, and food allergy. It is a hereditary genetic predisposition to the development of immediate hypersensitivity reaction.

Autocrine: A process in which cytokine act on same cells from which they are produced.

Autoimmunity: The immune reaction against the self cells or tissues in an individual.

Avidity: It is a functional combining strength and ability between antigen epitope and antibody paratope determined by the valencies between antibody and antigen.

β2-microglobulin: Protein or polypeptide molecule that constitute part of Class-I MHC molecules and membrane protein is known as β2-microglobulin.

B7-1 (CD80) and B7-2 (CD86): Co-stimulatory molecules present on APCs and responsible for ligation of CD28 on T cells.

B cells: Lymphocytes which are derived from the stem cells in the bone marrow and responsible for the production of antibody against antigen infection. The cells developed with minimum mutation in the receptor region bringing them close to germline sequence are known as B1 cells while mature cells with more number of mutation, taking part in immune system against antigen specific immunity by synthesizing antibodies are called B2 cells.

B-cell co-receptor complex: Group of surface molecules of mature B cells consisting of CD19, CD21 (complement receptor type 2) and CD81 to help in co-stimulation.

BCR: It is a B-cell receptor complex present on the B-cell surface along with CD79a and CD79b.

Basophil: A polymorphonuclear leucocyte with anti-inflammatory properties and stained with basic dyes is known as basophil.

BCG: It is Bacillus Calmette Guerin adjuvant, an attenuated strain of *Mycobacterium tuberculosis* used to vaccinate against *Mycobacterium* sp. infection.

Bcl-2: An anti-apoptotic gene gets activated and expressed transiently on B cells to rescue them from apoptosis.

Biozzi mice: Type of selectively bred mice that produces two types of antibody strains, low or high, against variety of antigens.

Blood groups: A set of genes, A and B, that gets expressed and their proteins appear on RBC membrane and gives them a group's assignment. Group-A has antigen protein A, group-B has antigen protein B, group-AB has antigen protein of both A and B, and group-O has none of the antigens.

Bradykinin: It is a vasoactive mediator nonapeptide generated by kinin system to stimulate pain receptors present on the skin.

Bursa of Fabricius: An organ present in birds where B cells mature. This lymphoepithelial organ is present at the junction of hind gut and cloaca.

Bystander lysis: It is a C5b67 complement complex mediated lysis of cells also known as innocent-bystander lysis. Here C5b67 complex is inserted into the phospholipid bilayer membrane.

C-domains: A little variable constant domains of antibody takes part in the binding of antigens.

C-genes: Genes that encode the constant region of the heavy and the light chain of an antibody.

c-Kit (CD117): Stem cell receptor taking part in the early development of leucocytes.

C1-C9: Complement system's classical or lytic pathways functionally involved in inflammation, opsonization, and cell lysis.

C3 convertases: Complement sub-unit complex of and which has an ability to cleave C3 complement component.

Carrier: Part of immunogenic molecule recognized by T cells is known as carrier.

Caspases: A set of proteases (enzymes) that helps to generate signal transduction for apoptosis.

CD markers: Cluster of differentiation, cell differentiation markers present on leucocytes and platelets that distinguish cells based on specific monoclonal antibody.

CDR: It is a complementarity determining regions present in the light and heavy chains within V-regions of antibody that constitute the antigen binding site.

Cell adhesion molecules (CAMs): A immunoglobulin super-family proteins taking part in cell adhesion includes ICAM-3, VCAM-1, MAdCAM-1, and PECAM.

Chemokines: A low molecular weight cytokines acting on 7-transmembrane pass receptors involved in chemotactic (chemotaxis) and other behaviour of leukocytes to mediate inflammatory response. The main families are: CC group, CXC group and includes IL-8, MCP-1, RANTES, MIP-1α, MIP-1β, MIP-II IP-10, etc.

Chemotaxis: Migration of cells in the direction of chemotactic factors along the concentration gradient.

Class I/II/III MHC molecules: Class-I molecules are one MHC-encoded peptide associated with β2-microglobulin; class-II molecules are two MHC-encoded peptide covalently attached to each other, and class-III molecules are complement components molecules.

Classical pathway: Soluble antigen-antibody complex activates complement system components C1, C2, C3, and C4 (named in the order to their discovery) in plasma to generate C3 convertase of classical pathway.

Class switching: It is a mechanism by which activated B cells link immunoglobulin heavy chain (C_H) genes to flanking recombined V-gene (V_H gene) producing different classes of specific antibodies with the help of *switch factors* (cytokines). It is also called isotype switching because it is basically a $V_HD_HJ_H$ rearrangement.

Clonal selection: It is an antigen-based selective activation, division, and differentiation of lymphocytes, that is B cells or T cells into a group of cells expressing specific antibodies.

Cell-mediated immunity: Immunity generated by cells or cell-based immunity rather than humoral factors.

Collectins: A surfactant protein including conglutinin and mannan binding lectin (MBL) able to kill bacteria directly by disrupting lipid membranes or by opsonization.

Complements: Serum glycoproteins that form pore in the cell membrane upon activation either by classical pathway or by lectin pathway.

Complement receptors (CR1-CR4 and C1qR): CR1, CR2, CR3, and CR4 expressed on various cells including phagocytic cells for the binding of sub-units of C3, that is C3b, C4b, or iC3b. C1qR binds to C1q. CR2 receptor binds to C3d. Binding helps in lysis, phagocytosis, etc.

Congenic: Two strains that genetically differ only at one genetic locus.

Constant regions: Invariant parts of heavy and light chain of immunoglobulin (Ig), and α, β, γ, and δ chains of T-cell receptor (TCR).

Co-stimulation: Signal that is required to stimulate B cells and T cells via co-stimulatory molecules CD40 and CD28, respectively. B-cell co-receptor is a complex comprising CD19, CD21 (CR2), and CD81 (TAPA-1). Signaling through CTLA-4 inhibits T-cell stimulation and CD22 acts as inhibitor of B-cell activation.

Cross reaction: Two antigens having similar types of antigenic determinants.

CSFs (colony stimulating factors): Cytokines that control stem cell differentiation.

CTLA-4 (CD-152): Compete with CD28 to bind B7-ligand of APCs to inhibit T-cell stimulation.

Cyclo-phosphamide: Immunosuppressive drug used to slow down lymphocyte proliferation.

Cyclosporin-A: It is also known as FK506 drug used to suppress T cells in autoimmunity and suppress graft rejection.

Cytokines: Soluble molecules which act as immunological mediator to mediate interaction between cells (lymphocytes).

CTLs: These are Cytotoxic T cells that kill virally infected target cells by binding to the presented antigens with Class-I MHC molecules.

D-gene: It is a gene encoding heavy chain present between V and J genes in immunoglobulin and β and δ chain genes of TCR.

Decay accelerating factor (DAF) or *CD55*: Mammalian cell surface associated protein prevents activation and deposition of C3b and also dissociates C3 convertase. Other similar nature proteins are C4bBP, CR1, and Factor-H.

Defensins: These are small antibacterial proteins produced by neutrophils.

Degranulation: The release of granules from mast cells and basophils is known as degranulation.

Dendritic cells: Cells derived either mononuclear phagocytic cells or lymphoid cells present in tissues. These cells phagocytose antigens and migrate into lymph nodes, e.g. spleen to present for T cells.

Domains: Overall final shape of the molecules or its part formed due to tertiary structure for rendering specific functional properties is known as domains.

DTH: It is delayed-type hypersensitivity. It is delayed skin reactions also known as type-IV hypersensitivity.

Effector cells: Short-lived lymphocytes and phagocytes derived from both naïve and memory cells to carry out specialized functions.

Endocytosis: Internalization of material present in extracellular environment by specific cells either by the process of phagocytosis (solid) or by pinocytosis (liquid).

Endothelium: It contains cells lining blood vessels and lymphatic vessels.

Endotoxin: Gram negative bacterial lipopolysaccharides (LPS) that can activate B cells and macrophages are known as endotoxin.

Eosinophils: Polynuclear granulocytes that get stained with acidic dyes and control parasitic worms infection in hypersensitivity.

Epithelioid cells: These are activated mononuclear phagocytes present in granuloma.

Epitopes: The part of the site of recognition of antigens, antigenic determinant that comes in contact with an antibody and TCR.

Epstein-Barr virus (EBV): Virus that causes Burkitt's lymphoma and infectious mononucleosis. It can transform B cells into stable-cell lines.

Exon: These are functional part of genes that encode proteins.

Fab: Antigen binding region of antibody consisting of light chain and part of heavy chain, identified after enzymatic digestion.

Factors B, P, D, H and I: These are sub-units or components present in the alternative pathway of complement system.

Fas (CD95): Molecule present on various cell types is a target for FasL binding on CTLs.

Fc: An antibody part able to bind antibody receptors on cells and C1q component of complement system.

Fc receptors: Cell surface molecules able to bind Fc part of the antibody in class specific manner.

Flow cytometry: Analysis or differentiation of cells based on their surface marker molecules.

Fluoresence activated cell sorter (FACS): A machine used to analyse or differentiate cells by flow cytometry.

Follicular dendritic cells (FDCs): APCs present in the area of B cells of lymphoid tissues.

Formyl-methionyl peptides: These are also known as fMet used by prokaryotes to initiate protein synthesis. fMet-Leu-Phe peptides are chemotactic for mononuclear neutrophils and phagocytes.

Framework segments (FR): V-part of antibody present in hyper-variable region.

Freund's adjuvant: Aqueous antigen emulsified in oil is called incomplete Freund's antigen. Incomplete Freund's antigen combined with killed *Mycobacterium tuberculosis* is called complete Freund's adjuvant.

γδ TCR: It is γδ T-cell receptor of T cells that have limited diversity for antigens.

GALT (gut-associated lymphoid tissue): Gut-associated lymphoid tissues are peripheral or secondary lymphoid organs which trap antigens to provide them an opportunity to interact with mature lymphocytes.

Genome: The total genetic material present within a cell.

Genotype: Genetic constitution of any animal inherited from their parents.

Germ line: The genetic material of initial cell line, that is eggs and sperm passed down to gametes before any modification due to recombination.

Germinal centres: Part of secondary lymphoid tissues, where B cell differentiation, that is, affinity maturation, class switching, and formation of plasma cell and memory cells takes place.

Giant cells: Multinucleated cells formed by the fusion of activated macrophages and supported by activated lymphocytes in the centre of the granuloma.

Granuloma: A tumour-like mass of cells, centre of which consists of activated macrophages surrounded by activated lymphocytes.

Granulocytes: Neutrophils, eosinophils, and basophils cells are called granulocytes.

Granulomatous reactions: Visible intense inflammatory response (destructive to host) due to non-clearance of antigen in delayed-type hypersensitivity.

Granzyms: Serine proteases also called fragmentins secreted during CTLs.

Grave's disease: It is a disease found in patients who produce auto-antibodies that bind TSH receptor.

H-2: Major histocompatibility complex found in mice.

Haemagglutination: It is a specific type of agglutination in which antibodies mediate the clumping of erythrocytes.

Haplotype: A complete set of genetic loci which is closely linked and is inherited from a single parent.

Hapten: These are small molecules that are antigenic but none-immunogenic in themselves.

Helper T cells (Th cells): Functional T-cells sub-class that helps the activity of cytotoxic T cells (CTLs) and B cells in antibody production and recognizes class-II MHC associated antigens.

Heterologous: Two different forms of one molecule, each derived from both paternal side and maternal side.

HEV (high endothelial venule): An area in blood vessel used by lymphocytes for extravasation.

Hinge: It is a part of an antibody which is present between Fc and Fab region that gives flexibility to the antibody during interaction with antigen. It is encoded by a separate exon.

Histamine: It is a vasoactive amine compound released during degranulation from mast cells and basophils.

Histocompatibility: It is the ability to accept graft.

HIV (human immunodeficiency virus): A virus which is a causative agent of AIDS and targets host T cells.

HLA: It is MHC found in human beings.

HLA typing: It is an antibody mediated microcytotoxicity assay that detects the presence and absence of various MHC alleles.

Homing receptors: These are receptors that give direction to lymphocytes to move to particular lymphoid and inflammatory tissues.

hnRNA: It is heteronuclear RNA from which mRNA originate after the splicing of intron.

Homologous restriction factor (HRF): HRF protect cells from non-specific lysis with the help of complement complex by binding to C8 and stopping the formation of poly-C9 complex and its insertion into the cell membrane.

Humoral: It is extracellular fluids that also include serum and lymph.

Humoral immune response: Immunity generated against bacteria and macromolecules due to the presence of antibody in the extracellular fluid including serum and lymph.

Hybridoma: Clone of hybrid cells in which each cell nucleus contains chromosome from each of the fused cells, that is, from myeloma (cancerous) cells and normal B cells to retain the properties of both immortal growth and monoclonal antibody production.

Hypersensitivity: An inappropriate strong increased immune response deleterious to an individual is termed as allergy. It is either antibody mediated or Th cells mediated.

Hypervariable region: The most variable three regions within the V-domain (variable domain) of each chain of immunoglobulin which show high variability in the sequence and provide most to the antigen-antibody binding site.

ICAM-1 (CD54), ICAM-2 (CD102), ICAM-3 (CD50): These are intracellular adhesion molecules present on leukocytes, target cells, and APCs which bind to CD2 and integrin LFA-1 present on T cells.

Iccosomes: These are immune complex coated bodies released from the follicular dendritic cells.

Idiotope: A single antigenic determinant present on V-region of antibody or TCR.

IDDM: Insulin-dependent diabetes mellitus, a kind of autoimmune disease caused due to the development of antibodies against cells of islets of Langerhans present in the pancreas.

Idiotype: A set of idiotopes or antigenic determinant of immunoglobulin to characterize unique antibody and TCR.

IELs (intra-epithelial lymphocytes): These are lymphocytes which are found in the epithelial layer of mammalian mucosal lining and characterized by the high numbers of γδT cells.

Immune-complex: These are antigen-antibody-complement complex in type-III hypersensitivity.

Immunity: The state of resistance against infectious pathogens and macromolecules.

Immunization: It is a state of immunity generated in organisms.

Immunoblotting or Westernblotting: A technique used to identify protein antigen with the help of specific antibody.

Immunodominant: Subunits of epitopes which produce more pronounced immune response than others.

Immunofluorescence: The technique of staining cells with fluorescent tagged antibody to visualize under fluorescent microscope.

Immunogen: The molecules that elicit immune response is an immunogen. All immunogens are antigens but all antigens are not immunogens.

Immunogenicity: The capability of immunogen to induce immunity under certain condition.

Immuno tolerance: It is immunologic unresponsiveness to self, low doses antigens induced in B cells and T cells during development in bone marrow and thymus.

Immunoglobulins: Serum antibodies are called immunoglobulilns.

Immunoglobulin super gene family (IgSF): Group of proteins, the domains of which are similar or homologous to that of immunoglobulins. IgSFs include Class-I and Class-II MHC, TCR, CD2, CD3, CD4, CD8, ICAMs, VCAM, Fc receptors and other molecules.

Immunotoxin: Antibody conjugated toxic agent usually protein, e.g. ricin, to produce agents highly cytotoxic for specific cells.

Incomplete antibody: An antibody that binds antigen but is unable to induce agglutination.

Induced fit: These are adjustments of antigen with respect to antibodies by displacing amino acids at the binding sites.

Inflammation: A group of many interdependent reactions to bring immunological cells to the site of infection or damage. The symptoms include increase in blood supply, vascular permeability, and transendothelial migration of leukocytes.

Inflamation: It is a condition which is determined by symptoms such as redness, pain, heat, and swelling.

Integrins: A cell surface heterodimeric adhesion molecule (LFA-1, VLA-4 and Mac-1) that separately interacts with Ig superfamily, e.g. CAMs, complement fragments, and extracellular matrix.

Intrferons (IFNs): Glycoproteins secreted by cells against viral infections to generate immunity.

Interleukins (IL-1 – IL-22): Cytokines secreted by leukocytes to generate immune signals that help in the differentiation of hematopoietic stem cells and the cells of immune system.

Intron: It is the non-coding segment of genes situated between two exons.

In vitro: Experiments with living cells or cellular components that are done outside the living organism.

In vivo: The experiments which are carried out in or on the animals.

Isotype: These are antibody types determined by the constant region sequence of heavy chain to generate five types of antibodies, e.g., IgA, IgD, IgE, IgG, and IgM.

Isotype switching: It is the conversion of one isotype antibody into another isotype due to genetic rearrangements of the constant region sequence of heavy chain in B cells. The process is also known as class switching.

ITAMs: Immunoreceptor tyrosine-based activation motifs that are phosphorylated by kinases for cell activation.

ITIMs: Immunoreceptor tyrosine-based inhibition motifs that are phosphorylated by kinases for cell inhibition.

JAKs: These are enzymes with two catalytic domains activated by cross phosphorylation.

J-chain: It is a monomeric polypeptide chain called joining-chain present in polymeric IgA and IgM between two heavy chains of antibody to form polymeric structure.

J-gene segment: It is a segment of genes that joins variable region to the constant region in rearranged immunoglobulin and TCR and also encodes hypervariable region. Recombination takes place during lymphocyte ontogeny to contribute genes for variable domains.

κ (kappa) chains: One of the two types of light chains. The other type of light chain is λ (lambda).

Karyotype: It is the chromosomal constitution of a cell.

Kinins: A group of vasoactive mediators, peptide in nature, produced during inflammatory response after injury.

Knockout: A transgenic animal in which normal gene for a particular protein has either been deleted or mutated so that the animal either makes no such protein or make a non-functional protein.

Kuffer cells: Macrophages or phagocytic cells present in liver sinusoids are called kuffer cells.

λ (lambda) chains: These are one of the light chains of immunoglobulin.

Langerhan's cells: APCs found in the skin, migrate to lymph nodes to become dendritic cells for presenting antigens to T cells.

Leader (L) peptide: It is a peptide segment present at the N-terminus of newly synthesized immunoglobulins to help its transport to the cell surface by inserting it into the lipid bilayer. It is removed from mature Ig by proteolysis.

Lectin pathway: A complement pathway initiated by mannan-binding-lectin (MBL) to intersect the classical complement pathway.

Leukotrienes: These are archidonic acid derivatives produced in leukocytes that mediate inflammation and type-I hypersensitivity. It is also called as *slow reactive substance of anaphylaxis (SRS-A).*

Leukocyte: Also called white blood cells, it includes lymphocytes, granulocytes, platelets, monocytes, and macrophages.

LCAs (leukocyte common antigens): It is CD45 having phosphatase activity present on all leukocytes. It is present in different forms.

LFAs (leukocyte functional antigens): It mediates adhesion between leukocytes and antigen of other cells in a non-specific way. It is of three types: LFA-1 is CD11a/CD18, LFA-2 is CD2, and LFA-3 is CD58.

Ligand: Any small molecule that gets recognized and binds specifically to a receptor is called ligand or linking molecule.

Linkage: When two genes are present very close to each other on a single chromosome and inherited together in subsequent generations, it is called linkage.

Linkage disequilibrium: The presence of two genes in a population at a higher frequency than the expected frequency of the individual gene is called linkage disequilibrium.

LPS (lipopolysaccharide): These are constituent of Gram negative bacterial strain that activates B cells.

Lymphokines: These are the product lymphocytes also called interleukins and take part in signaling between two cells to trigger immune response.

Lymphokines activated killer cells (LAKs): These are cells generated due to IL-2 or other cytokine signaling.

Ly antigens: These are murine T-cells surface marker now assigned as CD system.

Lysosome: It is a cellular vesicle containing enzymes, adhesion molecules, and toxic molecules.

Lytic pathway: It is a C5-C9 sub-unit mediated complement pathway responsible for cell lysis.

MALT (mucosa-associated lymphoid tissue): It is a lymphoid tissue associated with gastrointestinal tract and other mucosa including bronchial tree.

MAP kinases (mitogen activated protein kinases): It is an enzyme cascade that forms a signaling pathway to activate transcription factors (TFs).

Marginal zone: A zone that separates spleenic lymphoid white pulp from the red pulp.

Mast cells: These are cells that contain granules full of inflammatory mediators which are present near blood vessels of tissues.

MCPs (macrophage chemotactic proteins): These are chemokine proteins present in macrophage.

Memory cells: These are long lived differentiated lymphocytes that store memory of earlier encountered antigens to get differentiated into effector cells after second strike of similar antigens.

MHC (major histocompatibility complex): These are proteins responsible for the rejection of grafts in any recipient individuals taking part in signaling between lymphocytes and antigen expressing cells.

Microglia: Macrophages residing in the central nervous system, brain, and spinal cord.

MHC restriction: Immune reactions restricted only to the MHC haplotypes.

MIF (migration inhibition factor): These are chemokine proteins blocking migration of macrophage.

MIRL: It is membrane inhibitor of reactive lysis or CD59 that works in the same way as RF.

Mitogen: A protein that encourages lymphocytes to undergo cell division.

MLR/MLC (mixed lymphocyte reaction/mixed lymphocyte culture): T-lymphocyte recognition reaction.

Mononuclear phagocyte system: It is a system constituting blood monocytes and tissue residing macrophages having a longer life span.

Myeloid cells: These are bone marrow derived phagocytes consisting of monocytes, neutrophils, and eosinophils.

Myeloma: It is a bone invading B-cell lineage lymphoma.

N-regions: These are part of recombined antigen receptor genes which are normally absent in germ-line DNA.

Neoplasm: It is a synonym of cancerous tissue.

Neutrophils: These are polymorphonuclear granulocytes that are stained by neutral dyes.

NF-κB: Nuclear factor-κB is a transcription factor also called the master-switch of immune system that activates signal in many leukocytes.

NK (natural killer) cells: Those lymphocytes that are able to destroy virally infected cells and tumour cells are called as NK cells.

Nude mice: These are transgenic mice that lack hair producing transcription factors.

Opsonization: It is a process in which opsonin (C3b and Ig) facilitates phagocytosis of antigens.

PAF (platelet activating factor): These are proteins released from basophils that activate platelets to get aggregated.

PALS (periarteriolar lymphatic sheath): It is a white pulp sheath of the spleen formed due to accumulation of lymphoid tissues.

Paracrine: A process in which cytokines produced by one cell acts on distinct cell.

Passenger cells: These are leukocytes of donor graft tissues that sensitize graft recipient tissues.

Patch test: It is a test used to detect hypersensitivity type-IV for antigens present on skin, as a patch.

Pathogen: Disease causing organisms are called pathogen.

PC (phosphoryl choline): Theses are bio-molecules situated on microbial surface membrane and behave as hapten.

Perforin: It is a C9 complement protein which is homologous to granules associated molecules of cytotoxic cells to create pore in the target cells.

Peyer's patches: These are aggregated lymphoid cells which are present in the gut of birds to form secondary lymphoid organs.

PFC (plaque forming cells): These are antibody producing cells that form plaque in vitro by complement-mediated lysis of erythrocytes.

PHA (phytohaemagglutin): It is a mitogen for T lymphocytes.

Phagocytosis: A process by which phagocytic cells engulf solid external molecules to form a vesicle called phagosome in the cytoplasm.

Phenotype: The observable characteristics of an individual.

Plasma cell: These are differentiated B lymphocytes that are able to produce antibodies.

Polymorphs: These are polymorphonuclear cells, that is basophils, neutrophils, and eosinophils.

Prick test: A type of test used to detect type-I hypersensitivity by introducing antigen into the skin.

Primary lymphoid tissues: These are initial maturation site of antigen, that is fetal lever, bone marrow, and thymus in vertebrates while bursa of Fabricius in birds.

Primary response: An immune response generated after the first encounter of the body with antigen.

Protaglandins: Metabolite of arachidonic acid produced by cyclooxygenase pathway by monocytes and macrophages that down regulate expression of class-II MHC on B cells.

Proteasomes: These are multi-functional protease complex which degrade peptides in an ATP-dependent manner for presentation on APC cell surface.

Protein A and Protein G: Proteins found on the cell membrane of *Staphylococcus aureus* (protein-A) and *Streptococcus sp.* (protein-G) respectively, and bind to the Fc region of the antibody with high affinity. Genetic engineering is used to generate hybrid protein A-G by combining best features of both the proteins.

Pseudogenes: Genes that do not encode any protein product but have homologous structure to other genes, e.g., some of class-I MHC gene, segment of Vλ, Jλ, and Cλ genes.

Radioimmunoassay (RIA): A technique used to measure the concentration of antigens or antibodies based on their interaction ability by radio labeling.

RAG-1 and RAG-2: It is recombination activating genes. Both the genes are required for recombination of V, D, and J gene segments for the generation of functional B-cells antigen receptor genes called BCR.

Reactive oxygen/Nitrogen intermediates (ROIs/RNIs): Phagocytic cells produced bactericidal metabolites include hydrogen peroxide, hypohalites, and nitric oxide.

Receptor: A cell surface molecules having specificity to bind extra-cellular molecules.

Recombination: It is a process by which genetic information is rearranged during meiotic division. Genes encoding antibody molecules (BCRs) and T-cell antigen receptors (TCRs) are formed by recombination of DNA.

Relative risk: A number which expresses how much more likely (>1) or less likely (<1) an individual is prone to develop a particular disease, if he or she possess a particular genotype, e.g. HLA allele. It is calculated by dividing the patient's genetic population with the genetic of general population.

Respiratory burst: It is the increase in the amount of oxidative metabolites in phagocytic cells after engulfing opsonized particles.

SCID (severe combined immunodeficiency): It is a genetic disease which leads to major immunological deficiencies due to absence of both B cells and T cells.

Secondary response: It is the immune response after second time subsequent encounter with a particular antigen. This response is an enhanced response as compared to primary response.

Secretory component: A polypeptide produced by cells of some secretory epithelia which is involved in transporting polymeric IgA secreted across the cell membrane and protecting it from digestion in the gastrointestinal tract.

Selectins: These are adhesion molecules namely P-selectin (CD62P), E-selectin (CD62E), and L-selectin (CD62L), which are involved in slowing the movement of leukocytes during their transition through venules.

Self-MHC restriction: It is a mechanism in which both T cells (CD4 and CD8) are able to recognize antigen only when presented on cell surface by self-MHC molecules.

Serotonin: It is a type-I hypersensitivity degranulation product used to increase vascular permeability and contraction of smooth muscles.

SLE (Systemic Lupus Erythematosus): A non-organ specific autoimmune disease in humans which normally involves anti-nuclear antibodies.

Spleen: A secondary lymphoid organ situated in the peritoneal cavity, present next to the stomach.

Somatic mutation: B-cell maturation process that affects those regions of genes that determine antibody specificity.

STATs: A group of proteins that constitute components of signal transducer and activation of transcription factors following activation by kinases.

SCF (stem cell factor): Also known as steel factor, it is a cytokine required for the earliest stages of leukocyte development in the bone marrow.

Super-antigens: Antigens that stimulate different antigen specificity T-cell clones by TCR V-genes.

Suppressor T Cell: T-cell populations that reduce the immune responses of other T cells or B cells.

Synergism: A phenomenon in which cooperative interaction of two agents produces a greater effect in comparison to the effects produced by the individual agents.

Syngeneic: After repeated inbreeding, the animal strain produced possess identical pair of autosomal traits.

T cells: These are lymphocytes that differentiate in the thymus. The sub-groups of T-lymphocytes are: cytotoxic T cells (TC) and T-helper cells (Th, Th1 and Th2).

T15: These are idiotype of anti-phosphorylcholine antibodies, named after the TEPC15 myeloma prototype sequence.

TAP transporters: Protein molecules produced by TAP gene used to transport peptide either from cytoplasm to rough endopolasmic reticulum or between intracellular compartments inside the cell. Blocking TAP-gene expression blocks the assembly of class-I MHC molecules.

TCR (T-cell receptor): T cells express receptors, either γδ chain TCR or αβ chain TCR, get associated with CD3 molecular complex.

T-dependent antigens: These require both T cells and B cells to trigger immune response.

T-independent antigens: Antigens that directly stimulate B cells for antibody production.

Thoracic duct: It is the largest lymphatic vessel drain into left vein near the heart.

Thromboxanes: These are metabolite of arachidonic acid produced by cyclooxygenase pathway by monocytes and macrophages that cause inflammation, platelet aggregation, and blood vessels constriction.

Thymus: It is the primary lymphoid organ lying in the thoracic cavity above the heart.

Tissue typing: It is a procedure that identifies blood group antigens and MHC antigens.

TNF (tumour necrosis factor): Cytokines released from activated macrophages that shows structural similarity with the lymphotoxin produced by T cells and cause cell death.

Tolerance: Immunological unresponsiveness is called tolerance.

Toll receptors: Cell surface receptor evolved evolutionarily in the earliest stage that takes part in inflammatory signal transduction.

TGFs (transforming growth factors): These are immunosuppressive fibroblast growth promoting cytokines.

Transgenic animal: An animal in which there has been a deliberate modification of the genome and is able to express one or more new genes in particular tissues or organs.

V-domains: These are N-terminal part of both heavy and light chain of an antibody and α, β, γ, and δ chains of TCR.

Vaccination: Immunization against infectious diseases. The term is derived from Vaccinia virus which is used for immunization against smallpox.

Vasoactive amines: Degranulation products of basophils and mast cells acting on mucous glands, platelets, small blood vessels, eosinophils, sensory nerves, and smooth muscle of the nearby vasculature.

Very late antigens (VLA-1 to VLA-6): These are cell adhesion molecules on pro-B-cells, that is, set of integrins which share a common b1 chain (CD29).

Western blotting: A technique used to detect protein based on their antibody specificity.

White pulp: The component of spleen consisting of primary follicle, marginal zone, and periarteriolar lymphoid sheaths (PALS) which is populated mainly by T lymphocytes.

Xenogeneic: It refers to interspecies antigenic differences.

Index

About the Author

Dr Arvind Kumar is Associate Professor at Banaras Hindu University. He has a PhD in Biotechnology and is recipient of NIH-project fellow at Baylor College of Medicine, Visiting Scientist at University of Alabama Birmingham and Research Associate at University of Pittsburgh USA. He has published research paper in various discipline of molecular biology and immunology, and reviewed many books. At present he is teaching immunology and cell biology at School of Biotechnology, Faculty of Science, BHU apart from pursuing research work in the area of immunology specifically focusing on the broader topic such as immune-modulatory roles of cytokines/chemokines and there regulation by various transcription factors in Balb/c cancer mice model in response to MNU, Urethane, and selected medicinal plants.